Contexts of the Dark Side of Communication

LIFESPAN COMMUNICATION

Children, Families, and Aging

Thomas J. Socha
GENERAL EDITOR

Vol. 10

The Lifespan Communication series
is part of the Peter Lang Media and Communication list.
Every volume is peer reviewed and meets
the highest quality standards for content and production.

PETER LANG
New York • Bern • Frankfurt • Berlin
Brussels • Vienna • Oxford • Warsaw

Contexts of the Dark Side of Communication

Edited by Eletra S. Gilchrist-Petty
and Shawn D. Long

PETER LANG
New York • Bern • Frankfurt • Berlin
Brussels • Vienna • Oxford • Warsaw

Library of Congress Cataloging-in-Publication Data

Names: Gilchrist-Petty, Eletra S., editor. | Long, Shawn, editor.
Title: Contexts of the dark side of communication / edited by Eletra S. Gilchrist-Petty, Shawn D. Long.
Description: New York: Peter Lang, 2016.
Series: Lifespan communication: children, families, and aging; Vol. 10
ISSN 2166-6466 (print) | ISSN 2166-6474 (online)
Includes bibliographical references and index.
Identifiers: LCCN 2015040631 (print) | LCCN 2016006603 (ebook)
ISBN 978-1-4331-2750-2 (hardcover: alk. paper)
ISBN 978-1-4331-2749-6 (paperback: alk. paper) | ISBN 978-1-4539-1801-2 (e-book)
Subjects: LCSH: Interpersonal communication. | Communication in organizations. Cyberbullying. | Gossip in the workplace.
Classification: LCC HM1166 .C696 2016 (print) | LCC HM1166 (ebook)
DDC 302—dc23
LC record available at http://lccn.loc.gov/2015040631

Bibliographic information published by **Die Deutsche Nationalbibliothek**.
Die Deutsche Nationalbibliothek lists this publication in the "Deutsche Nationalbibliografie"; detailed bibliographic data are available on the Internet at http://dnb.d-nb.de/.

The paper in this book meets the guidelines for permanence and durability of the Committee on Production Guidelines for Book Longevity of the Council of Library Resources.

29 Broadway, 18th floor, New York, NY 10006
www.peterlang.com

Printed in the United States of America

Contents

Context 2: *The Dark Side of Organizational Communication*

Context 3: *The Dark Side of Health Communication*

Context 4: *The Dark Side of Computer-Mediated Communication*

Foreword: Welcome to the Dark Side

Mark P. Orbe

In many cultures, a dichotomous distinction exists between light(ness) and dark(ness). Lightness is associated with optimism, affirmation, goodness, purity, softness, gentleness, delicacy, joy, happiness, carelessness, and anything viewed as positive. Darkness, in contrast, is situated with evil, anger, fear, hatred, greed, violence, brutality, cynicism, shame, fear, denial, manipulation, deceit, exploitation, abuse, and anything viewed as negative. Historically, "dark topics" have been regarded as taboo within personal, familial, *and* intellectual conversations.

"Taboo" (or tabu) is a Polynesian word that refers to a general ban on a particular object; alternatively, a taboo is "marked off"—implying that certain things are unsafe for casual contact or conversation (Evans, Avery, & Pederson, 2000). Taboo topics, then, refer to a strong social prohibition against words, actions, and discussions that are considered offensive and/or undesirable. A taboo can exist on an action (something that you should not do), on discussion (things that you do but do not discuss), or anything associated with certain actions (including thinking about it, creating a label for it, etc.). Topics generally exist on a continuum from those that are acceptable for discussion to those that are strictly taboo. Taboo topics can vary depending on context (personal, relational, cultural, and/or societal), but the commonalty is that all involve subjects that are perceived to be painful, embarrassing, and/or humiliating to self and/or others. Within the field of communication, taboo topics largely are associated with the "dark side" of communication (Anderson, Kunkel, & Dennis, 2011).

Distinctions between "light" and "dark" topics can be seen within the field of communication. Specifically, scholarship historically has most often focused on "light" topics, with little if any attention to topics deemed as "dark" or taboo. In the past couple of decades, "the dark side of communication"—a body of literature that engages topics traditionally regarded off limits and consequently largely unexamined through scholarly endeavors—has emerged and makes significant contributions to our field. This edited volume makes a significant contribution to this growing body of scholarship. I encourage readers to embrace the opportunity to engage the taboo topics covered. Larger society cautions us against "walking on the dark side" due to the unpredictable dangerous association with such activity. Yet, I would argue that addressing these topics is the primary means to understanding the power that they have over us. Ignoring "dark topics" allows them to remain in the shadows, where they continue to impact our relationships without scrutiny.

As you read through this book's chapters, three important points must be acknowledged. First, distinctions between the "light side" and "dark side" of communication topics are a socio-political construction. Marking certain topics as off-limits (or taboo) is not natural, inherent, or innate; instead, it reflects human activity filled with bias, subjectivity, and intent. Traditional thinking situates certain topics as either "dark" or "light"; however, all communication topics are best understood as having both positive and negative attributes. Second, unless we embrace the responsibility to study subject areas traditionally marked as taboo, the story of human communication will remain incomplete. All relationships are communicative and include experiences that are uncomfortable, humiliating, and/or embarrassing. Engaging these topics does not give them more power, they actually work to reduce the power they have in our lives. Third, and finally, explorations into the "dark side" generally, and the "dark side" of communication more specifically, are critically important. By acknowledging, accepting, and embracing chances to explore these topics we create safe opportunities to release some of the pressures associated with these communicative behaviors. As such, engaging the "dark side of communication" ultimately can work to empower all relational partners—transforming "safe spaces" to "brave spaces" (Arao & Clemens, 2013).

This edited volume features an eclectic collection of studies spanning a variety of communication topics, including those situated within interpersonal, organizational, health, computer-mediated, and blended contexts. As you consider engaging each chapter individually and the entire volume collectively, I encourage you to "take a walk on the dark side" and embrace the opportunities to gain invaluable insight on an array of topics typically marked as off-limits. You may be amazed at what insights can be gleaned through such an exciting exploration.

References

Anderson, M., Kunkel, A., & Dennis, M. R. (2011). "Let's (not) talk about that:" Bridging the past sexual experiences taboo to build healthy relationships. *Journal of Sex Research, 48*(4), 381–391.

Arao, B., & Clemens, K. (2013). From safe spaces to brave spaces: A new way to frame dialogue around diversity and social justice. In L. M. Landreman (Ed.), *The art of effective facilitation: Reflections from social justice educators* (pp. 135–150). London, UK: Stylus Publishing.

Evans, R. W., Avery, P. G., & Pederson, P. V. (2000). Taboo topics: Cultural restraint on teaching social issues. *The Clearing House, 73*(5), 295–303.

Acknowledgments

Collectively, we would like to thank each of the contributing authors who eagerly embraced this challenge with us. Without your research expertise, anecdotal experiences, time, and dedication, this volume would not have been possible. We also extend heartfelt appreciation to Peter Lang Publishing Inc., who made this scholastic work a reality. We especially thank Mary Savigar who believed in this project even in its infancy. Finally, Dr. Tom Socha, the series editor, we thank you for recognizing that the dark side of communication is intimately linked to our holistic understanding of lifespan communication.

Eletra S. Gilchrist-Petty

To my immediate family, Norris, London, Mom, Dad, Taylor, and Fredrick, your love and support are priceless, and each day you inspire me to reach beyond my greatest potential. Landri and Logan, you are forever in my heart. I would like to thank my Charger family at The University of Alabama, in Huntsville, for recognizing the importance of liberal arts education and scholarship. To my research assistants, Heather Patrick Beard and Moeshia Williams, I thank you for exhibiting diligence and keen editing skills. To my co-editor, Dr. Shawn D. Long, thank you for embarking on this journey with me. Above all, I thank my heavenly father for blessing me with the strength, intuition, and ability to complete another book project—truly, my cup runneth over.

Shawn D. Long

Many people have contributed to the conceptualization, execution, and support of this book. Special thanks to my co-editor, Dr. Eletra S. Gilchrist-Petty, for your excellent, insightful, and diligent work on this project. It was an honor to work with you. I would like to thank all of the dedicated authors of this book. You are committed thought-leaders in dark-side scholarship across a number of contexts. It was my pleasure to work with you to bring this book to life. Thanks to my research assistant, Alex Kello, for his assistance in the early stages of this project. The support of the College of Liberal Arts and Sciences, Department of Communication Studies, and the Organizational Science Program at the University of North Carolina at Charlotte has been invaluable in completing this project. Finally, and most importantly, I would like to thank my family and friends—near and far—for their long-lasting and unwavering support of my personal and professional endeavors.

Introduction: A Contextual Perspective of the Dark Side of Communication

Eletra S. Gilchrist-Petty

Scholars and practitioners have long understood the importance of communication in human interaction. However, aside from the few select popular topics of conflict, transgressions, and relationship dissolution, traditional interpersonal and/or human communication pedagogy and theories have focused more on the positive, or at least the more neutral, aspects of communication. During this time, social scientists largely presented communication as a kind of "panacea"—a tool that when wielded effectively could provide redress for problematic situations (Spitzberg & Cupach, 2007, p. 14). We now know that this historical perception is idyllic and fails to fully unravel the inextricable complexities that co-exist with bright optimisms. In other words, traditional communication pedagogy neglected to tell the complete story, and instead advanced an "ideology of the pursuit of goodness" (Spitzberg & Cupach, 2007, p. 7).

This one-sided representation of communication and human experiences persisted in academic texts and most curricula until the mid-1990s when Cupach and Spitzberg ignited a paradigm shift by arguing that not all communication is positive; in contrast, they posited that relationships can be sources of extreme problematic interactions. Cupach's and Spitzberg's groundbreaking text, *The Dark Side of Interpersonal Communication* (1994), offered the first collection of chapters that focused exclusively on the not-so-sunny side of communication and included necessary conversations relative to equivocation, relationship uncertainty, unrequited love, and even abuse. This initial dark side volume served as a forerunner for examining the pages left unturned by sanguine progenitors and presented a more complete view of the multiple layers of human engagement. Spitzberg and Cupach continued these pertinent discussions with their subsequent volumes *The Dark Side of Close Relationships* (1998a) and the second edition of *The Dark Side of Interpersonal Communication* (2007). The authors used these pioneer volumes to establish the basic framework for conceptualizing the dark side of communication.

What Is the Dark Side of Communication?

Previous researchers have taken a relatively general approach when conceptualizing the dark side. Though Spitzberg and Cupach are the perceived originators of the dark side concept, they "have to date restrained from formally defining the dark side" (2007, p. 4). Instead the authors have cast a wide net and distinguished seven "sins" that describe the dark side: (1) the three d's—dysfunctional, distressing, and destructive aspects of human interaction; (2) deviance, betrayal, transgression, and violation; (3) exploitation of the innocent; (4) the unfulfilled, underestimated, and unappreciated endeavors of life; (5) physically unattractive, the ugly, the distasteful, and repulsive; (6) dehumanization; and (7) paradoxical, dialectical, and mystifying aspects of human action (Spitzberg & Cupach, 1998b, p. xi–xxii). Olson, Baiocchi-Wagner, Wison-Kratzer, and Symonds (2012) synthesized Cupach and Spitzberg's original work and defined dark communication as "verbal and/or nonverbal messages that are deemed harmful, morally suspect, and/or socially unacceptable" (p. 11). These varied and wide-reaching approaches open the door for a slew of behaviors to be studied under the multi-encompassing "dark side" umbrella. In this volume, we follow suit and choose not to limit proverbial dark side behaviors to finite margins that determine what is good and evil. It is our position that the dark side represents behavior that upon first consideration does not readily appear ideal, optimal, or overwhelmingly positive, yet it is part of human existence and, therefore, warrants scholarly consideration.

In conceptualizing the dark side of communication, it is also important to note that it contains both *functional* and *normative* dimensions. Spitzberg and Cupach (2007) addressed these dimensions with their quadric representation depicting the various boundaries of the dark side of communication. Briefly put, the quadric representation contends much of what we consider "dark" can be characterized as either (1) functionally productive and normatively destructive, (2) functionally destructive and normatively productive, (3) functionally destructive and normatively destructive, or (4) functionally productive and normatively productive.[1] As alleged by Spitzberg and Cupach, there are few communication activities that are purely and clearly evil from both a functional (actual) and normative (perceptual) stance. Instead, residing at the heart of dark side investigations is the idea that many behaviors previously blanketed as "dark," "dysfunctional," or even "evil," are in fact "functionally ambivalent" (Spitzberg & Cupach, 2007). Certain behaviors traditionally perceived as socially inappropriate or dark can produce positive and maybe even desirable outcomes. For example, when gossip is done in a playful manner it induces laughter, sociability, entertainment, group identity, and comfort for the interactants, but gossip as mockery is reflective of maliciousness, envy, jealousy, and competitiveness (Ferreira, 2014). This bifurcated view of gossip speaks to its ambiguous nature and suggests that what is considered "dark" is situationally dependent, dialectical, and in "the eye of the beholder." With this volume, we seek to continue challenging the ubiquitous practice of positioning "light" and "dark" as opposing binaries and, instead, further elucidate the coexistence of "light" and "dark" behaviors that define the dark side as a multidimensional, situationally driven, paradox influenced by contextually driven functional and normative dimensions.

Contexts of the Dark Side of Communication

All credit is rightfully due to Cupach and Spitzberg for insisting that the complete story of human communication be told because "dark" and "light" behaviors exist simultaneously. Since Cupach and Spitzberg's initial work in this area, many scholars have deemed it worthy to extend the dark side scope (e.g., Fox & Spector, 2005; Goodwin & Cramer, 2002; Greenberg, 2010; Griffin & O'Leary-Kelly, 2004; Harden Fritz, & Omdahl, 2006; Hearn & Parkin, 2001; Kirkpatrick, Duck, & Foley, 2006; Kowalski, 1997; Kowalski, 2001; Leather, Brady, Lawrence, Beale, & Cox, 1999; Lutgen-Sandvik, & Sypher, 2009; Miller, 2004; Olson et al., 2012; Randall, 2001; Segrin, 2001). Each of the aforementioned works, however, have a more narrow and unilateral scope, in that they focus solely on dark side issues within singular contexts of the workplace, organizations, or general interpersonal relationships. This volume transcends this traditional one-dimensional perspective and presents a wide span of communication topics relating to five specific communication contexts (i.e., Interpersonal, Organizational, Health, Computer-Mediated, and Blended).

We duly note that the five contexts addressed in this volume are not the sole areas of inquiry within the communication discipline; however, it would be an impossible feat for any volume to serve as an ultimate text covering all domains of human communication scholarship. Therefore, the five contexts presented are related to our collective areas of training, pedagogy, and scholarly pursuit. Additionally, these context-driven areas serve as categories for organizing and illuminating communication-in-action vis-à-vis the multiple ways human engagement is experienced in our day-to-day interactions. We would like to stress that this volume is not conceptualized as simply a book on *negative* communication, because sometimes difficult conversations and human interactions are needed, possibly remain unresolved, and exist within a consensus-building fashion or dialectic. Our goal is simply to present a volume that serves as a vanguard text by offering a more complete view of how dark side behaviors permeate human communication across various domains.

In addition to presenting a premier volume that addresses the dark side of human communication from a contextual perspective, the chapters presented in this collection merge theory with practical application. Specifically, the volume addresses a myriad of theoretical positions, such as Black Feminist Epistemology, Matching Hypothesis, Structuration Theory, and Expectancy Violations, just to name a few. The intense scrutiny of previous scholarship ensures that the chapters are theoretically grounded to capture the essence of academic excellence. Moreover, the volume relates to the reader on a humanistic level by presenting authentic lived experiences. Chapter contributors write candidly and unapologetically about how they and various populations under investigation mitigate a wealth of dark side behaviors spanning sexualization, cyberstalking, bereavement, and various illnesses. Toward this end, the volume offers a number of diverse perspectives and lenses through which students, academics, and the general population can enhance their understanding of how dark side behaviors are experienced and communicated.

Contextual Content

This volume is comprised of five contexts pertinent to our understanding of the dark side of human communication. The first context, The Dark Side of Interpersonal Communication, explores various dark side elements germane to relational interactions. Striley initiates this section by discussing ostracism in adolescents' daily lives. Through an electronic journaling study, Striley found that though ostracism can bring much pain and despair, the experience of rejection can transform and awaken adolescents to deeper insights about the social world, fostering resiliency and imagination. The second chapter by Basinger and Wehrman investigated how bereaved individuals experience grief as a relational process. The authors interviewed students who lost either a parent or sibling and found that the participants generally regarded supportive communication negatively. Sexualization is the focal point of the volume's third chapter. Here, Stephenson argues that perceptions of attractiveness and the sexualization of those deemed physically "attractive" can become a communication dark side with real consequences. Gilchrist-Petty penned the fourth chapter in the interpersonal context and used Expectancy Violation Theory as a framework to better understand the hurtful events that confront dating and married couples. Via quantitative content analysis, Gilchrist-Petty concludes couples collectively experience a mix of relational transgressions that may or may not be experienced equally among dating and married persons, and as indicative of Expectancy Violation Theory, both dating and married couples have relational expectations that are perceived as hurtful when violated. The last chapter in the interpersonal context, authored by Jordan Jackson, examines verbal aggression in the classroom by analyzing the comment portions of student evaluations of teaching. Findings lend further support to previous research that alleges instructors who are women and/or racial minorities receive more negative evaluations than White male faculty.

The volume's second context examines The Dark Side of Organizational Communication. In this context, the authors address a myriad of challenging situations that affect workplace environments. Theiss and Webb open the section with their study of workplace bullying. Thematic analysis of interview data collected from academic managers revealed that managers use an assortment of collaborative and dominant strategies in their interactions with perpetrators and victims of workplace bullying. Next, Richardson and Gravely explore retaliation against whistle-blowers. Based upon interview data collected from whistleblowers, the authors explore how retaliation against whistleblowers is expressed, reasons retaliation is used against whistleblowers, and the outcomes of retaliation for whistleblowers, retaliators, and the organization. Chapter eight features a study by Kartch and Valde who address destructive gossip as an aggressive form of communication from a resource-control theoretical perspective. Using participant observer ethnographic data, the authors chronicle a real-life gossip situation within an organization and address how organizational members use gossip as a means of resource control. Results indicate gossip was prevalent and used as relationally and socially aggressive tactics to manipulate workplace relationships, build cliques, and exclude others. Padgett, Gupton, and Snider interrogate workplace diversity and discrimination in chapter nine. The authors critique cases that are emblematic of workforces that have struggled with issues of diversity and inclusion, and they advance a dialogic approach designed to not only confront workplace discrimination, but create and sustain an inclusive organizational environment. Long, Woznyj, Coleman, Makkawy, and Spivey collaborated on chapter ten and examine microaggressive communication in organizational settings. Specifically, the authors critique a series of case studies that illuminate organizational microaggressions, considering implications for diversity and inclusion. Murrary and Brown, authors of the final chapter in the organizational communication context, scrutinize hazing as a tool of destructive organizational identification and loyalty. Survey data from 287 respondents were used to draw conclusions about organizational members' perceptions of hazing.

The Dark Side of Health Communication is the focus of this volume's third context, which concerns how dark side communication can impact the construction, dissemination, and application of information applicable to one's health. Thompson, Brown-Burton, and Jackson launch the section with a qualitative study examining lack of familial social support among African Americans diagnosed with prostate cancer. Their findings confirm that African American men experience physical and psychological challenges because of prostate cancer and perceive neither emotional nor tangible support from family members, and some feel that the well-intentioned efforts of loved ones undermine their confidence to cope with the disease. Chapter thirteen features Hernandez's qualitative thematic analysis of Mexican-American women's birth experiences within the U.S. health care system. Data collected from semi-structured interviews revealed patients had low levels of medical understanding, power struggles with health care providers, and anxieties about the birth and postpartum care processes. Chapter fourteen by Gatison uses case study as a research methodology to analyze communication strategies of the National Breast Cancer Coalition. Gatison stresses the need for diversity and inclusiveness, especially in relation to Black women breast cancer victims. The final chapter in the health communication context examines the supermax detention cell as one of the key architectural components of contemporary counterinsurgency warfare. Vicaro, author of the chapter, argues that new high-tech prison cells used for both military and civilian detention are able to produce conditions tantamount to torture without requiring overtly violent physical contact with detainees. Vicaro further critiques health implications of long-term solitary confinement.

Context number four, The Dark Side of Computer-Mediated Communication (CMC), considers dark behaviors that occur through either synchronous or asynchronous electronic systems. Degroot and Carmack commence this discussion with their investigation of deception during online dating. Through case study methodology, they describe the disenfranchised grief that ensues from catfishing. In chapter seventeen, Mumpower and Bassick examine organizational rumors' impact on relationships via traditional and CMC channels. The authors surveyed 165 sorority and fraternity members and found that rumors were most frequently spread via face-to-face interactions, followed by text messaging, social networking sites, and other media. Chapter eighteen features the scholarship of Kulovitz and Mabry, who tested cyberbullying group behavior in the online video game *Left 4 Dead 2* by focusing on game outcome, group cohesion, and leader influence. Through participant observations, Kulovitz and Mabry found the presence of both cyberbullying and prosocial behavior within the same gaming sessions. Davies composed chapter nineteen and used qualitative measures to examine the lived experiences of technological stalking victims. Results indicate that relationships marred with physical, psychological, and emotional abuses often serve as precursors to technological stalking. Johnson, Bostwick, and Anderson round out our discussion of the dark side of CMC with their inquiry of how technology negatively affects interpersonal relationships. Data acquired through a series of focus groups revealed six main challenges CMC use presents to human engagement.

The final portion of the volume, The Dark Side of Blended-Communication Contexts, speaks to the multifaceted nature of communication and features chapters on dark side behaviors that infiltrate two or more areas of human communication. For example, chapter twenty-one by Eckstein marries the interpersonal and CMC disciplines by exploring technologically mediated abuse in settings of intimate partner violence. Eckstein used a mixed-method design to study the experiences of 495 self-identified victims of intimate partner violence and found supportive evidence for a future Technologically Mediated Abuse Scale. Findings also revealed the varied ways victims of intimate partner violence experience technologically mediated abuse. Fox and Anderegg also present a blended interpersonal and CMC study in chapter twenty-two that investigated the dark side of social networking sites for romantic couples. Their chapter highlights various conflicts couples experience related to uses of social networking sites, including negative relational maintenance, partner monitoring, romantic jealousy, and relationship dissolution. Interpersonal and organizational communication are intertwined in Dixon and Liberman's study, which probed the challenges of work/life balance as perceived by working adults representing different family structures. Through the tenets of Structuration Theory, three themes emerged that describe the dark structures working adults encounter when balancing work and non-work obligations. Barrett authored chapter twenty-four, blending the CMC and organizational contexts. Specifically, Barrett interrogated email overload in the workplace and concluded from qualitative interview data that individual personalities, face sensitivities, technological dependencies, and workplace preferences contribute to employees' perceptions of email overload. Chapter twenty-five—this volume's final chapter—is authored by Hoffman and DeGroot, who meld together the interpersonal and CMC contexts with an examination of various Facebook jealousy triggers that affect romantic relationships. Analysis of survey data revealed both rival-based and partner-based jealousy triggers.

Holistically, the various contexts and corresponding chapters featured in this volume serve to further enlighten our understanding of the dark side of human communication. In bringing forth this scholarly collection, the editors desire simply to expand our knowledge base, initiate thought-provoking conversations, and inspire future studies that advance the limitless inquisitions of contextual dark side research.

Notes

1 See Spitzberg and Cupach (2007) p. 4–8 for a detailed description of the four dimensions.

References

Cupach, W. R., & Spitzberg, B. H. (Eds.). (1994). *The dark side of interpersonal communication.* Hillsdale, NJ: Lawrence Erlbaum.

Ferreira, A. A. (2014). Gossip as indirect mockery in friendly conversation: The social functions of "sharing a laugh" at third parties. *Discourse Studies, 16*(5), 607–628. doi:10.1177/1461445614538564

Fox, S., & Spector, P. E. (Eds.). (2005). *Counterproductive work behavior: Investigations of actors and targets.* Washington, DC: American Psychological Association.

Goodwin, R., & Cramer, D. (Eds.). (2002). *Inappropriate relationships: The unconventional, the disapproved, and the forbidden.* Mahwah, NJ: LEA.

Greenberg, J. (2010). *Insidious workplace behavior.* New York, NY: Routledge.

Griffin, R. W., & O'Leary-Kelly, A. M. (Eds.). (2004). *The dark side of organizational behavior.* San Francisco, CA: Jossey-Bass.

Harden Fritz, J. M., & Omdahl, B. L. (Eds.). (2006). *Problematic relationships in the workplace*. New York, NY: Peter Lang.

Hearn, J., & Parkin, W. (2001). *Gender, sexuality and violence in organizations*. London, UK: Sage.

Kirkpatrick, D. C., Duck, S., & Foley, M. K. (Eds.). (2006). *Relating difficulty: The processes of constructing and managing difficult interaction*. Mahwah, NJ: LEA.

Kowalski, R. M. (Ed.). (1997). *Aversive interpersonal behaviors*. New York, NY: Plenum.

Kowalski, R. M. (Ed.). (2001). *Behaving badly: Aversive behaviors in interpersonal relationships*. Washington, DC: American Psychological Association.

Leather, P., Brady, P., Lawrence, C., Beale, D., & Cox, T. (1999). *Work-related violence: Assessment and intervention*. New York, NY: Routledge.

Lutgen-Sandvik, P., & Sypher, B. D. (2009). *Destructive organizational communication: Processes, consequences, and constructive ways of organizing*. New York, NY: Routledge.

Miller, A. G. (Ed.). (2004). *The social psychology of good and evil*. New York, NY: Guilford.

Olson, L. N., Baiocchi-Wagner, E., Wison-Kratzer, J. M. W., & Symonds, S. E. (2012). *The dark side of family communication*. Cambridge, UK: Polity.

Randall, P. (2001). *Bullying in adulthood: Assessing the bullies and their victims*. New York, NY: Taylor and Francis.

Segrin, C. (2001). *Interpersonal processes in psychological problems*. New York, NY: Guilford Press.

Spitzberg, B. H., & Cupac v h, W. R. (Eds.). (2007). *The dark side of interpersonal communication*. (2nd ed). Mahwah, NJ: Lawrence Erlbaum.

Spitzberg, B. H., & Cupach, W. R. (Eds.). (1998). *The dark side of close relationships*. Mahwah, NJ: Lawrence Erlbaum.

Spitzberg, B. H., & Cupach, W. R. (1998b). Introduction: Dusk, detritus, and delusion—A prolegomenon to the dark side of close relationships. In B. H. Spitzberg & W. R. Cupach (Eds.), *The dark side of close relationships* (pp. xi–xxii). Mahwah, NJ: Lawrence Erlbaum.

Context One

The Dark Side of Interpersonal Communication

ONE

Unlimited: Ostracism's Potential to Awaken Us to Possibility and Mystery

Katie M. Striley

I grew up in a school that said if I can't succeed in the system that's laid out for me, then my life will be unremarkable.... I was taught to believe in limitations, that I must color inside the lines, that I must connect the dots in numerical order. And that's fine—if all you want is a picture of an octopus. But if you want a picture of an octopus that wears a human for a backpack so it can walk around on land and protest seafood restaurants, you're gonna have to go about things a little differently.... We live in a constantly changing world, and in that world systems break because they are rigid and unbending. If we spend our lives trying to adjust to something broken, we break ourselves in the process.

—Shane Koyczan (2014)

Humans are limited only by our imaginations; anything in the world could be otherwise if we can envision it. However, many of us exist in limiting social systems. Often, we face social consequences for rebelling against limitations social actors place on us. For example, adolescents quickly learn they must look and act certain ways, or else they will experience peer rejection. Ostracism and social rejection are tools of limitation. Adolescents are bombarded with limiting messages: *You can't. You shouldn't. You mustn't. Or, you will face exile.* However, some do rebel against these limitations and envision alternate possibilities for interaction.

This chapter discusses ostracism's ability to awaken adolescents to what coordinated management of meaning theory (CMM; Pearce, 1989) calls *mystery.* Mystery is an awareness that our understanding of social worlds is finite and limited. Mystery tells us the current construction of a social world is not the only possible construction. In this chapter, I discuss research about adolescent ostracism and ask if ostracism can ever foster positive outcomes in adolescents' lives. I describe an electronic journaling study I conducted with intellectually gifted adolescents, and utilize CMM in my analysis of adolescents' journals.

The Ostracized Adolescent

Humans have the potential for great acts of compassion or hateful acts of cruelty. Adolescent peer groups, in particular, manifest cruelty through social rejection and exclusion (Sunwolf, 2008). Through acts of ostracism, we exile and marginalize others. Ostracism is ignoring others in order to exclude them from meaningful social participation; it is the experience "of feeling invisible" (Williams, 2001, p. 2). Ostracism is a communicative attempt to create a world *without* another person or group.

Few life events are more painful than feeling invisible to those from whom we seek acceptance. Ostracism is particularly upsetting to adolescents because the human need for belonging is strongest when we are young (Kerr & Levine, 2008; Rawlins, 1992; Sunwolf, 2008). Adolescents spend much of their time seeking social acceptance through identification with peer groups (Sunwolf, 2008), and need to feel accepted by peers in order to live well-adjusted lives (Sullivan, Farrell, & Kliewer, 2006). Group acceptance helps adolescents develop a sense of identity, self-worth, and validation (Onoda et al., 2010; Rawlins, 1992; Stokholm, 2009). Ostracism symbolizes a lack of group acceptance that leaves young people feeling alone and alienated.

Exclusion yields severe consequences. Academically, ostracized adolescents can experience poor academic performance (Wentzel & Caldwell, 1997), underachievement, and disengagement from classroom participation and school (Buhs, Ladd, & Herald, 2006). Psychologically, ostracized youth might exhibit increased thoughts about suicide (Hawker & Boulton, 2001), depression (DeWall, Gilman, Sharif, Carboni, & Rice, 2012), anger (Pharo, Gross, Richardson, & Hayne, 2011), and aggression (Chow, Tiedens, & Govan, 2008), compared to their included counterparts. Physically, ostracized adolescents tend to experience more illness than do included peers (Buhs & Ladd, 2001; Sunwolf, 2008). Socially, ostracized youth might stop caring about peer relationships (Leary, 2001) and withdraw from social participation (Wood, Cowan, & Baker, 2002). Social effects of ostracism can negatively affect adolescents' relationships even into adulthood (Buhs & Ladd, 2001).

Evidence suggests ostracism can create chain reactions of anti-social behavior. Social rejection can turn youth to drugs, crime (MacDonald, 2006), or violence (Leary, Kowalski, Smith, & Philips, 2003; Matthews, 1996; Wesselman, Nairne, & Williams, 2012). Excluded youth might be more likely to engage in delinquent behavior, such as stealing, property damage, truancy, and alcohol or cigarette use (Sullivan et al., 2006). An analysis of 15 United States school shootings from 1995 to 2001 demonstrated that 12 out of these 15 shooters experienced chronic peer ostracism during adolescence (Leary et al., 2003). One individual's ostracism can affect society writ large.

Is Ostracism Always Negative?

Prevailing ideas about ostracism offer a pessimistic view, whereby scholars typically discuss negative consequences of ostracism. For instance, Williams (2001) suggested experiences of ostracism harm four basic human needs. First, ostracism jeopardizes our sense of belonging because others communicate that we are unwanted group members. Second, ostracism threatens our self-esteem by creating negative self-evaluations that prompt questioning our worth as relational partners. Third, ostracism takes away our sense of control because others take control of our social world. Finally, ostracism threatens our need for meaningful existence by making us feel worthless.

Additionally, short- and long-term effects of ostracism also paint a grim picture. Williams (2001) described reactions to ostracism in temporal stages. Immediately after ostracism, we feel intense pain (Williams, 2001; Williams & Gerber, 2005). The short-term effects of ostracism drive us to replenish lost needs (Williams, 2001; Williams & Gerber, 2005; Wirth & Williams, 2009; Zadro & Williams, 2006), perhaps by forging new friendships, or attempting more participation in the group that ostracized us (Williams & Sommer, 1997). Finally, long-term ostracism might make us feel hopeless, despairing, or that we belong nowhere. This view of ostracism leaves little room for a positive side of long-term ostracism.

Scholars utilizing Williams's theory of ostracism tend to view ostracism as a painful event with lasting negative repercussions. However, *is ostracism always negative?* Could ostracism have potentially positive effects on individuals? For example, perhaps ostracism can help adolescents develop their identities. For instance, a child ostracized from the "popular" crowd might realize she does not share the group's values and seek participation within another group with whom there is more in common. Additionally, perhaps ostracism can be freeing. Peer acceptance often requires giving up freedom by following certain social norms. An excluded individual need not worry about following social norms with which she disagrees or conforming to peers to avoid rejection because the individual is already rejected. Perhaps, in rejection lies the freedom to act unencumbered by others' judgments.

Journaling about Ostracism

In order to explore ostracism in adolescents' daily lives, I asked 45 intellectually gifted adolescents to participate in an electronic journaling study. Intelligent Quotient tests are currently the only widely accepted tool to identify giftedness (Vaivre-Douret, 2011); the National Association of Gifted Children (2008) defines individuals scoring within the top 10% of the population as "gifted." Gifted adolescents quickly acquire, retain, conceptualize, synthesize, and apply new information (McCollins, 2011). However, they also tend to feel lonely and isolated from their peers (Cassady & Cross, 2006; Niehart, 2002). Williams and Gerber (2005) reported, "gifted children often complain that their worst obstacle is dealing with being ostracized by other children in their classroom" (p. 364). Some gifted adolescents feel that their advanced intellect sets them apart from peers, making them easy targets for ostracism.

I recruited participants by contacting parents with the help of the national organization Supporting Emotional Needs of the Gifted (SENG), the Belin-Blank Center at the University of Iowa, and the Davidson Institute. Forty-five intellectually gifted adolescents participated in the study; 22 were girls and 23 were boys. Ages ranged from 10 to18; 25 middle school and 20 high school students participated. Participants came from 18 different U.S. States and four different countries (United States, Colombia, Canada, and Germany). Most participants self-identified as Caucasian (39); two individuals identified as Hispanic, one as Native American, one as African American, and one as Asian. One participant declined disclosing racial identity.

I conducted one-hour telephone interviews with participants' mothers, and then I interviewed participants. After preliminary interviews, I asked participants to keep an electronic journal of their ostracism experiences for a minimum of one month (some journaled for up to five months). Electronic journaling offered a modern twist to the diary method (Zimmerman

& Wieder, 1977) of data collection. This method captures intimate descriptions of naturally occurring daily experiences (Lämsä, Rönkä, Poikonen, & Malinen 2012; Nicholl, 2010) and is ideal for exploring social rejection among adolescents (Sanstrom & Cillessen, 2003). Diary studies have the potential to reduce problems with participant recall (Flook & Fuligni, 2008; Nicholl, 2010; Suveg, Payne, Thomassin, & Jacob, 2010), elicit more detailed descriptions than interviews (Palmero, Valenzuela, & Stork, 2004), and provide more complete views of children's emotional states (Suveg et al., 2010).

I instructed participants to write about four experiences: (a) anytime they felt ignored, excluded, or left out during the journaling period; (b) anytime they ignored, excluded, or left someone else out during the journaling period; (c) anytime they observed someone else being ignored, excluded, or left out during the journaling period; and (d) if they did not have the first three experiences on a given day, then they could write about a memorable past experience of feeling ignored, excluded, or left out. Participants averaged three to four weekly journal entries. I checked in with participants via telephone, text messaging, or email (based on their preference) every week they journaled. Finally, I conducted a one- to two-hour telephone interview when participants finished journaling.

Despite scholars' tendencies to suggest predominantly negative ostracism responses, my work with intellectually gifted adolescents suggests many of them transformed ostracism into something positive. Adolescents are not always condemned to lives of loneliness when exiled by peers. Instead, they may continue living meaningful lives. Ultimately, participants exhibited numerous positive responses to chronic ostracism, including an ability to re-imagine their social worlds through the recognition of CMM's mystery.

Coordinated Management of Meaning

CMM is a social constructionist theory about how humans interact (called coordination), make sense of the world (called coherence), and recognize their creative potential (called mystery) (Pearce, 1989). According to CMM, everyday talk *creates* our social environments (Pearce, 2005). When I say a world is created, I do not mean it is imaginary or fake. Instead, I mean that communication produces and reproduces patterns of interaction that mold social worlds (and we are molded by our social worlds). Collectively, we determine what it means to attend school, to celebrate holidays, to have a funeral, and how to act appropriately in these settings. Humans are simultaneously physical and symbolic entities. The same mechanical, earthly processes confine humans and all other matter in the universe, but only *we* live lives of moral significance; we are never *simply* biological events.

Coordination and Coherence

According to CMM, we create social worlds through processes of coordination and coherence. Coordination is matching our actions with others. For instance, when someone introduces her/himself, often she/he stretches a hand towards you. Most likely, you also extend your hand and the two of you engage in a handshake. You are coordinating your actions together by recreating a common North American pattern of behavior. In the case of ostracism, several adolescents might coordinate their actions to exclude another adolescent from an activity. Therefore, their actions create an exclusive reality where one individual is unwelcomed.

As humans coordinate actions together, we must also make sense of our actions. Therefore, coherence is our ability to make sense of actions in our social world together. In the example of a handshake, you and your new acquaintance likely both realize that a handshake is an accepted way of greeting another person. The understanding you have associated with a handshake is coherence. Coherence is the sense we make individually and collectively. In the case of ostracism, the excluded adolescent might make sense of her rejection by believing she is "too weird" to socialize with the excluding group, something is wrong with her, or that she behaved in some way to merit ostracism from the group. Thus, she might determine that she exists in a world where one must look, act, or dress a certain way to find acceptance.

Mystery

Patterns of interaction and sensemaking become so ingrained in us that we often forget that we can act differently or create an alternative to our realities. For Pearce (1989), language's continuity enmeshes us in our particular reality, impels us to make sense of the world in predetermined ways, and coerces us to recreate habitual interpersonal interactions. According to Pearce (1989), communication coerces us into accepting that our symbolic worlds are real. Pearce terms this suspension of disbelief and acceptance of our symbols *as* reality, enmeshment. When we exist in states of deep enmeshment, we are blinded to alternative possibilities because enmeshment makes us overlook the fabrication of social worlds. For instance, an adolescent experiencing chronic ostracism might give in to despair and depression, as suggested by Williams (2001), and become stuck in patterns of exclusion (Cacioppo & Hawkley, 2005; Rosen, Milch, & Harris, 2009) if he or she fails to realize that other worlds are possible.

However, CMM's concept of mystery offers hope that we might escape negative social worlds and create new realities in their place. Mystery reminds us that the worlds we live in are only *some* of the worlds that could have been or might yet be. Mystery suggests that lines drawn by communication are arbitrary constructions (Pearce, 1989; Pearce & Pearce, 2004) and nothing inherent in realities requires us to understand them in any particular way (Gergen, 1999). For Pearce (1989), the natural human condition is to live deeply enmeshed in social worlds so that we might easily achieve coordination and coherence. However, multiple interpretations of events are possible. We are anything but fixed in place. Mystery is a recognition that every story *must* leave something out. *Everything* we think we know could be otherwise (Pearce & Pearce, 2004). Mystery, therefore, is the nexus of emancipation from particular social worlds. Mystery offers an awareness of our role in creating social worlds and our ability to direct the process of creation.

In the following analysis, I demonstrate that exclusion has a transformative potential that can awaken us to mystery. Ostracism and social rejection do not always break us. Sometimes, as the scars heal, we become *better* than we ever could have been without the pain of exclusion. Incredible beauty can grow from the sorrow of once shattered lives. In the current study, chronic ostracism sometimes allowed participants to recognize the constructed nature of social reality. In other words, exclusion provided a window into mystery. When some participants experienced the trauma of exile from their peers' realities, they encountered a rupture in coordination and coherence as they scrambled to repair their social worlds. This rupture provided a momentary awareness of mystery that liberated some participants from their negative social realities. They became unfixed in their world and recognized the potential to build better social worlds in the ashes of old.

Life after Exile

Humans experience both joys and sorrows in life. Sometimes we find ourselves living in repressive, negative worlds. Sometimes we face exile from social groups, and we feel like we will never find acceptance again. However, there is life after exile. Many of my participants did not succumb to despair when excluded from peers' social worlds. Some participants found a bright side to sorrow. For example, Jonas, a 13-year-old 8th grader, reflected on the healing after ostracism. He wrote about his and his classmates' powerful reaction to Shane Koykzan's Ted Talk performance of his poem, *To This Day*. Jonas wrote,

> Some classmates cried. I didn't cry, but I felt bad for the guy in the video and I felt bad for my friends that cried. We all related to past experiences where we were bullied, excluded, etc. I felt true empathy for my friends that day, as their true emotions about bullying and exclusion came out that day—as well as my own—but I am now happy that we shared our experiences. I now truly know that I am not the only one who has ever experienced it, but also I know that there are people, like my friends in that classroom that day, that care for me.

Jonas and his friends bonded through their realization of shared painful experiences. Ultimately, Jonas saw beauty in their shared sorrow and their potential to heal each other. Jonas wrote that Koyczan's words reminded him that, "healing is possible, and we have to be each other's lights." For Jonas, Koyczan's words helped him to realize he could turn the memory of exclusion into something positive.

Many participants' experiences of ostracism transformed them in different ways. Their loneliness fostered mindful reflection on their lives and, often, ignited change. Now, I will share participants' insights about overcoming ostracism in their lives. All participants have pseudonyms to protect their identities. I have provided participant quotes exactly as they appear in journal entries; I opted not to change grammatical errors or typos in participant responses. Some participants experiencing chronic ostracism were able to develop deeper understandings about reality because of their exclusion. Other participants were able to alter their perceptions of and desires about the world to find freedom from others' judgments. Still others found freedom from following unquestioned social norms. Some participants were even able to catch glimpses of other possible realities as their worlds fractured and ruptured before their eyes.

Deeper Understandings about Humanity

For some gifted youth in my study, chronic ostracism allowed them to see the world more clearly, as they developed profound understandings of the human condition. Umberto, a 17-year-old male living in Colombia, best exemplified the cultivation of a deeper understanding. Umberto described his greater perception of people because of his rejection. He wrote,

> At first, social awkwardness was my life. I was rejected all the time. This caused me to turn my intellectual attention to people, I had to perceive them better. Now, I can communicate much better than most people. I can easily capture even the slightest gestures, and make sense of it all in a matter of seconds…[understanding] body language and gesticulation is something that I VERY RARELY fail at. Now, You could call me a social genius.

Umberto's rejection compelled him to understand humans better; now, Umberto uses his advanced perceptions to "read" people. He further explained that he has figured out several ways to make people happy. For instance, he always remembers "characteristics and likes of others to show that you pay attention."

In order to escape his life of rejection, Umberto developed a survival tactic by consciously choosing to study human behavior. Anzaldua (2007) argued that rejected individuals develop perception shifts as coping mechanisms, and that "this shift in perception deepens the way we see concrete objects and people" (p. 61). Umberto often discussed his deep love for humanity and his desire to read people as a way to make them happy. As someone who could bring happiness to others, he eventually found acceptance after years of rejection.

Freedom from Others' Judgments

After experiencing chronic ostracism, several participants decided they would change their reactions to others' judgments. For example, Tesla, a 16-year-old 12th grader, explained that he used to care what his peers thought, "but now, I no longer concern myself with their judgments. I am free to just be myself and not confined by my thoughts about their thoughts about me." Tesla stopped desiring peer acceptance, and thus, no longer viewed exclusion negatively. He said in an interview, "I no longer want to be liked at school because most people are painfully, stereotypically, normal. My life has been consumed by the desire to learn. I don't need anyone for that." Tesla redefined social acceptance as negative and valued learning over peer interaction. Therefore, the likelihood that his peers would reject him no longer bothered Tesla. Although his view of his peers is somewhat negative, he feels free to pursue what he loves, unhampered by the thoughts of others.

Several participants recognized inclusion in some groups meant giving up the freedom to forge their own identities. For instance, Ginny, a 16-year-old 12th grader, used to want acceptance from the "popular" crowd. After years of ostracism and exclusion from the popular girls, she recognized that she no longer sought their acceptance because she found acceptance with a group of peers who let her be herself. She wrote, "my friends in band don't ask me to dress like them (we all dress differently) or act like them (we're all weirdos and act completely randomly). So why would I want to join a group that expects me to change for them?" Similarly, Marie, a 16-year-old 11th grader who experienced the pain of social rejection so strongly that she switched schools, realized she was also tired of caring what others thought about her. She wrote,

> My view now is that not everyone has to like me. In fact, I don't want judgmental people to like me. I'm sick of living by their judgments and rules. With this new mentality, I can now just be me.

Both Ginny and Marie realized acceptance sometimes came with the price of conformity. Their experiences of rejection accentuated their freedom to just be themselves.

Several participants' experiences of ostracism fostered a realization that they were free to be non-conformists. Poppy, a 14-year-old 9th grader, said in an interview, "people who are always included are followers. They always follow other people and don't think or act for themselves. People who are excluded think differently, don't go with all the trends, but go their own way and are independent." Several participants exemplified the freedom expressed by Poppy and performed their nonconformity. Jennifer, a 13-year-old 7th grader, wrote,

> I'm tired of caring what others think, so I don't. they never accepted me anyway. So now, I'll have my hair how I want it, and I will wear my clothes how I want, and if you have a problem with how I look, it sucks to be you, doesn't it.

Exclusion gave Jennifer the strength and desire to reject conformity, and live free of the constraints others often feel. Katniss, a 15-year-old 10th grader, wrote, "Go ahead and call me crazy, I will consider it a compliment. I go my own way and do what I want. It might seem crazy to you, but it is the only sane response I see to an insane world." Like Jennifer, after years of facing social rejection, Katniss now felt free to act unconstrained by others' judgments. Umberto wrote, "I accept myself as somebody often deemed 'weird'. I am different. I am not normal. I can live without being normal." Many participants celebrated their "weirdness" and reveled in newfound realizations of freedom that sometimes accompanied exclusion.

Awakening to Mystery

Ostracism sometimes allowed participants to recognize the constructed nature of social reality by awakening them to mystery and possibility. Enmeshment within our social worlds allows mindless engagement in coherence and coordination with others. We tend not to question our social worlds when things are going well. However, when profoundly negative experiences occur, it can jar us into questioning the world. For example, Umberto described how happiness could ensnare us and sadness could free us from old perceptions. When we are sad from loneliness,

> We have nothing to lose, but happiness gives us everything to lose. So, when we are happy, we are not free. Sadness makes you introspective. To figure out what is wrong you look inward. Sadness carries depth, urges you to fix yourself. So sadness and loneliness make us free to act and think how we want.

For Umberto, exclusion and the accompanying sadness, allowed him to both see mystery and recognize his own freedom to act without the fear of losing anything. Sadness allowed Umberto to be free and re-imagine his social world.

When some participants found themselves in a world where peers actively denied them, they began to question their reality. For example, Amelia, a 17-year-old 11th grader living in Germany, said in an interview,

> Ostracism and exclusion made me realize that reality, at least the reality of high school and middle school is a game. We made up the game and everyday we create the game over and over. People didn't want me to be part of the game, so they ignored me. I just decided not to play their game anymore and create my own.

For Amelia, the discomfort of her reality fostered mindfulness about her social world; the recognition of life's mystery induced her mindfulness. She experienced a rupture in reality when she realized others were coordinating reality without her; thus, she had to repair her social world. This rupture in reality is a moment when we might become aware of mystery. For example, Phoenix, a 17-year-old 12th grader, said in an interview, "being gifted makes you very non-conforming. You are so used to rejection and so smart that you realize life is not 'yes' or 'no.' There is so much more to life than most people will ever see." Sharing a similar sentiment, Umberto wrote,

> I was alone for so long–or I should say lonely—that I began to intimately study and think about reality. I felt like I was getting ever-closer to another reality that hides behind reality itself...This is what I call the 'different echo.' Most people are blind to this, they just live their lives asleep. I've been spending my life waking up.

Phoenix, Amelia, and Umberto sought comfort in mystery by realizing there is more to life than the reality at their school.

The rupture into mystery liberates us, as the snares of language become visible. Through the experience of a highly negative social reality, some participants comforted themselves by recognizing the myriad other realities possible. For example, Harper, a 14-year-old 10th grader, reminded herself that other possibilities existed. She wrote,

> We chose the reality we live in. I could have chosen to be sad all day when people didn't like me in middle school, or I could chose to make a better world. I want to make my world and other people's world better...There is no REASON that the popular kids are on the top. Really, if we all rebelled we could create a much more fair school environment where everyone was treated equally. I think no one thinks its possible so they don't try, BUT IT IS POSSIBLE! I love the quote that says "If you can imagine it, you can achieve it; if you can dream it, you can become it." I think not enough people realize that.

With the realization of mystery, Harper became unfixed in the world. She was able to recognize possibilities that remained clouded to others. According to Anzaldua (2007), persecuted people develop the capacity to "see in surface phenomena the meaning of deeper realities, to see the deep structure below the surface...[making us] excruciatingly alive to the world" (p. 60). For Anzaldua, "it's a kind of survival tactic that people, caught between worlds, unknowingly cultivate" (p. 61). Many participants embraced mystery because it was freeing.

Implications for Seeing a Brighter Side to Sorrow

The major implication of analysis presented in this chapter is that if we are able to survive ostracism, the experience of rejection can transform us and awaken deeper insights about our social worlds. Ostracism is not always negative; it can also have positive effects. For some participants, ostracism empowered them to see the world not as it is, but as it could be. Therefore, ostracism fostered resiliency, imagination, and freedom. Findings suggest individuals are more versatile than some ostracism scholars might report. Williams (2001) suggested ostracism usually results in pain and despair. However, most of my participants were well-adjusted and happy, despite years of chronic ostracism. Future studies should explore the positive side of ostracism. Why do some individuals become stronger after ostracism, while others fall into despair? How can we harness the positive potential of ostracism? Understanding positive effects of ostracism might help children and adolescents cope more effectively with exclusion. Maybe parents and educators can learn to cultivate ostracism's transformative potential in victims of exclusion. Ostracism's apparent ability to foster mystery suggests future research should explore the connections between rejection and imagination. Does rejection create an opportunity to recognize our creative potential and forge new paths of interaction?

Exclusion communicates our existence is unwanted. However, even when ostracized we find ways to be heard, to define ourselves, and to become something *more*. The human spirit is resilient. I hope this chapter is a first step in understanding how to navigate better ways to cope with exclusion. We do not have to passively accept rejection from a social world. We can reclaim our agency when others attempt to steal it. Exclusion is a very human experience from which we can learn a great deal. We do not need to become lost in the impulse to "get better" because, sometimes, living with the pain of ostracism makes us better.

References

Anzaldúa, G. (2007). *Borderlands: The new mestiza: la Frontera* (3rd ed.). San Francisco, CA: Aunt Lute Books.

Buhs, E., & Ladd, G. W. (2001). Peer rejection as an antecedent of young children's school adjustment: An examination of mediating processes. *Developmental Psychology, 37*(4), 550–560.

Buhs, E. S., Ladd, G. W., & Herald, S. L. (2006). Peer exclusion and victimization: Processes that mediate the relation between peer group rejection and children's classroom engagement and achievement? *Journal of Educational Psychology, 98*(1), 1–13.

Cacioppo, J. T., & Hawkley, L. C. (2005). People thinking about people: The vicious cycle of being a social outcast in one's own mind. In K. D. Williams, J. P. Forgas, & W. von Hippel (Eds.), *The social outcast* (pp. 91–108). New York, NY: Psychology Press.

Cassady, J. C., & Cross, T. L. (2006). A factorial representation of suicidal ideation among academically gifted adolescents. *Journal for the Education of the Gifted, 29*(3), 290–304.

Chow, R. M., Tiedens, L. Z., & Govan, C. (2008). Excluded emotions: The role of anger in antisocial responses to ostracism. *Journal of Experimental Social Psychology, 44*(3), 896–903.

DeWall, C. N., Gilman, R., Sharif, V., Carboni, I., & Rice, K. G. (2012). Left out, sluggardly, and blue: Low self-control mediates the relationship between ostracism and depression. *Personality and Individual Differences, 53*(7), 832–837.

Flook, L., & Fuligni, A. J. (2008). Family and school spillover in adolescents' daily lives. *Child Development, 79*(3), 776–787.

Gergen, K. J. (1999). *An invitation to social construction.* Thousand Oaks, CA: Sage.

Hawker, D. S. J., & Boulton, M. J. (2001). Subtypes of peer harassment and their correlates: A social dominance perspective. In J. Juvonene & S. Graham (Eds.), *Peer harassment in school: The plight of the vulnerable and victimized* (pp. 378–397). New York, NY: Guilford Press.

Kerr, N. L., & Levine, J. M. (2008). The detection of social exclusion: Evolution and beyond. *Group Dynamics: Theory, Research, and Practice, 12*(1), 39–52.

Koyczan, S. [TEDxYouth]. (2014, January 6). Blueprint for a breakthrough: Shane Koyczan at TEDxYouth@SanDiego 2013. Retrieved from https://www.youtube.com/watch?v=aV805a2XJgA

Lämsä, T., Rönkä, A., Poikonen, P., & Malinen, K. (2012). The child diary as a research tool. *Early Child Development and Care, 182*(3–4), 469–486.

Leary, M. R. (2001). Towards a conceptualization of interpersonal rejection. In M. R. Leary (Ed.), *Interpersonal rejection* (pp. 4–20). New York, NY: Oxford University Press.

Leary, M. R., Kowalski, R. M., Smith, L., & Phillips, S. (2003). Teasing, rejection, and violence: Case studies of the school shootings. *Aggressive Behavior, 29*(3), 202–214. doi: 10.1002/ab.10061

MacDonald, R. (2006). Social exclusions, youth transitions and criminal careers: Five critical reflections on "risk." *The Australian and New Zealand Journal of Criminology, 39*(3), 371–383.

Matthews, M. W. (1996). Addressing issues of peer rejection in child-centered classrooms. *Early Childhood Education Journal, 24*(2), 93–97.

McCollin, M. J. (2011). The history of giftedness and talent development. *Advances in Special Education, 21*, 289–313.

National Association for Gifted Children. (2008). *Redefining giftedness for a new century: Shifting the paradigm.* Accessed from http://www.nagc.org/index.aspx?id=6404

Neihart, M. (2002). Gifted children and depression. In M. Neihart, S. M. Reis, N. M. Robinson, & S. M. Moon (Eds.). *The social and emotional development of gifted children: What do we know?* (pp. 93–101). Waco, TX: Prufrock Press.

Nicholl, H. (2010). Diaries as a method of data collection in research. *Pediatric Nursing, 22*(7), 16–20.

Onoda, K., Yasumasa, O., Nakashima, K., Nittono, H., Yoshimura, S., Yamawaki, S., Yamaguchi, S., & Ura, M. (2010). Does low self-esteem enhance social pain? The relationship between trait self-esteem and anterior cingulated cortex activation induced by ostracism. *Social Cognitive Affect Neuroscience, 5*(4), 385–391.

Palermo, T. M., Valenzuela, D., & Stork, P. P. (2004). A randomized trial of electronic versus paper pain diaries in children: Impact on compliance, accuracy, and acceptability. *Pain, 107*(3), 213–219.

Pearce, W. B. (1989). *Communication and the human condition.* Carbondale, IL: Southern Illinois University Press.
Pearce, B. (2005). The coordinated management of meaning. In W. B. Gudykunst (Ed.), *Theorizing about intercultural communication* (pp. 35–54). Thousand Oaks, CA: Sage.
Pearce, W. B., & Pearce, K. A. (2004). Taking a communication perspective on dialogue. In R. Anderson, L. A. Baxter, & K. N. Cissna (Eds.), *Dialogue: Theorizing difference in communication studies* (pp. 39–56). Thousand Oaks, CA: Sage.
Pharo, H., Gross, J., Richardson, R., & Hayne, H. (2011). Age-related changes in the effect of ostracism. *Social Influence, 6*(1), 22–38.
Rawlins, W. K. (1992). *Friendship matters: Communication, dialectics, and the life course.* New Brunswick, NJ: Aldine Transaction.
Rosen, P. L., Milich, R., & Harris, M. J. (2009). Why's everybody always picking on me? Social cognition, emotion regulation, and chronic peer victimization in children. In M. J. Harris (Ed.), *Bullying, rejection, & peer victimization: A social cognitive neuroscience perspective* (pp. 79–100). New York, NY: Springer Publishing Company.
Sandstrom, M. J., & Cillessen, A. H. (2003). Sociometric status and children's peer experiences: Use of the daily diary method. *Merrill-Palmer Quarterly, 49*(4), 427–452.
Silverstein, S. (1974). *Where the sidewalk ends.* New York, NY: HarperCollins.
Stokholm, A. (2009). Forming identities in residential care for children: Manoeuvring between social work and peer groups. *Childhood, 16*(4), 553–570.
Sullivan, T. N., Farrell, A. D., & Kliewer, W. (2006). Peer victimization in early adolescence: Association between physical and relational victimization and drug use, aggression, and delinquent behaviors among urban middle school students. *Development and Psychology, 18*(1), 119–137.
Sunwolf, S. (2008). *Peer groups: Expanding our study of small group communication.* Thousand Oaks, CA: Sage.
Suveg, C., Payne, M., Thomassin, K., & Jacob, M. L. (2010). Electronic diaries: A feasible method of assessing emotional experiences in youth? *Journal of Psychopathology Behavior Assessment, 32*(1), 57–67. doi: 10.1007/s10862-009-9162-0
Vaivre-Douret, L. (2011). Developmental and cognitive characteristics of "high-level potentiality" (highly gifted) children. *International Journal of Pediatrics, 2011*, 1–14. doi:10.1155/2011/420297
Wentzel, K. R., & Caldwell, K. (1997). Friendships, peer acceptance, and group membership: Relations to academic achievement in middle school. *Child Development, 68*(6), 1198–1209.
Wesselmann, E. D., Nairne, J. S., & Williams, K. D. (2012). An evolutionary social psychological approach to studying the effects of ostracism. *Journal of Social, Evolutionary, and Cultural Psychology, 6*(3), 309–328.
Williams, K. D., (2001). *Ostracism: The power of silence.* New York, NY: The Guilford Press.
Williams, K. D., & Gerber, J. (2005). Ostracism: The making of the ignored and excluded mind. *Interaction Studies, 6*(3), 359–374.
Williams, K. D., & Sommer, K. L. (1997). Social ostracism by coworkers: Does rejection lead to loafing or compensation? *Personality and Social Psychology Bulletin, 23*(7), 693–706.
Wirth, J. H., & Williams, K. D. (2009). "They don't like our kind": Consequences of being ostracized while possessing a group membership. *Group Processes & Intergroup Relations, 12*(1), 111–127.
Wood, J. J., Cowan, P. A., & Baker, B. L. (2002). Behavior problems and peer rejection in preschool boys and girls. *Journal of Genetic Psychology, 163*(1), 72–89.
Zadro, L., & Williams, K. D. (2006). How do you teach the power of ostracism? Evaluating the train ride demonstration. *Social Influence, 1*(1), 81–104.
Zimmerman, D. H., & Wieder, D. L. (1977). The diary: Diary-interview method. *Urban Life, 5*(4), 479–498.

TWO

"They Don't Get It, and I Don't Want to Try to Explain It to Them": Perceptions of Support Messages for Individuals Bereaved by the Death of a Parent or Sibling

Erin Basinger and Erin Wehrman

The death of a loved one can be a debilitating and stressful event with effects lasting for years after the loss (Stroebe, 2010). Grief, or the emotional reaction to bereavement, can create negative experiences including anxiety and depression (Stroebe, 2001). Although grief can be an individual process, many researchers have begun to view grief relationally (e.g., Rosenblatt, 1988; Stroebe, 2001). In this study, we examine grief as it occurs in relationships by focusing on supportive communication between bereaved individuals and their social networks. Our goal is to understand why individuals evaluate supportive communication in the context of grief as more or less helpful. In the following sections, we review research on bereavement and support, and we describe a study designed to answer research questions about support messages in the context of grief.

In recent years, scholars have characterized grief as an interpersonal process, arguing that grief occurs within relationships (Hayslip & Page, 2013; Stroebe, 2001). One way people grieve in the context of their relationships is through supportive communication. Many bereaved individuals find that support helps them adapt to the loss (Hayslip & Page, 2013) and lowers stress levels (Kaunonen, Tarkka, Paunonen, & Laippala, 1999). Supportive communication comes in various forms, and Toller (2011) found that bereaved parents appraised several types of social support as helpful, including tangible aid, emotional support, and nurturant support, which involves talking to similar others about the loss. Similarly, Kaunonen et al. (1999) found that when individuals share a common loss (e.g., both parties have lost a child), support is particularly helpful. Together, these studies indicate that under certain conditions, support is beneficial.

Unfortunately, social support is not always advantageous. The bereaved can be negatively affected by misguided support messages. Wilsey and Shear (2007) found that grieving individuals felt distressed when messages from others were rude or aggressive. Moreover, Toller (2011) reported that parents who lost a child perceived messages that attempted to control their grief (e.g., telling the parents what to do or how to feel) as frustrating and unhelpful. In another study, messages providing information about why the person died (e.g., offering religious messages like, "It was in God's hands") or telling the person how to grieve (e.g., focus on the positives) were particularly devastating (Breen & O'Connor, 2011). Although scholars have discovered that not all supportive communication benefits grieving individuals, it is vital to elucidate what leads people to appraise messages as helpful or not.

Support messages clearly play an important role in the interpersonal process of grief. Unfortunately, the literature lacks insight into how bereaved individuals manage various forms of support (both positive and negative), so understanding precisely what affects individuals' perceptions of support is important. In the next section, we review theoretical models that emphasize the interpretation of supportive communication.

Theories of Social Support

Scholars in many fields have provided insight into why social support can be helpful. The buffering hypothesis, an early framework for understanding support, stated that social support shields individuals from the negative effects of stress (Cohen & Wills, 1985). Although it sounds straightforward, the relationship between stress and support is actually quite complex. Cohen and Syme (1985), for instance, recognized that the direct effects of support are influenced by characteristics of the support provider, the type of support offered, the problem the distressed person is facing, and other factors. Later, Cohen (1992) qualified the buffering hypothesis with the matching hypothesis, asserting that support is only effective when the type of support offered is compatible with the needs of the distressed person.

Whereas the matching hypothesis focuses on selecting the most effective *type* of support, communication scholars have concentrated on elucidating which support *messages* are effective. Burleson (1982), for instance, focused on the person-centeredness of support messages, suggesting that individuals who take another person's perspective into account produce higher-quality messages. Typically, researchers recognize three levels of verbal person-centeredness in support messages. When messages are characterized by the lowest level of person-centeredness, individuals deny, challenge, or ignore the support receiver's feelings, and moderately person-centered messages acknowledge the distressed person's perspective but also include some attempt to distract the person or "explain away" his or her feelings (Burleson, 1982, p. 1581). The most comforting messages are highly person-centered and recognize others' feelings while including a genuine attempt to help manage their distress from their own perspective (Burleson, 1982). Burleson's contribution suggests that certain messages have a greater impact than others for providing adequate support.

Recently, Burleson and colleagues (e.g., Bodie, Burleson, & Jones, 2012; Burleson, 2009a, 2009b) proposed the dual-process model of supportive communication, which suggests that many factors influence how distressed individuals respond to support. In particular, message characteristics, the support provider, the context, and the support recipient impact the evaluation of support (Burleson, 2009b). A primary assumption of the dual-process approach is that the content of the message is potentially the most important part of the supportive exchange; however, if recipients are not able or motivated to process the message, they might pay attention to other cues. Therefore, people who are able or motivated to process message content will be most influenced by the quality of the message. Empirical work using the dual-process model has sought to (a) elucidate the qualities that influence how people process message content and (b) clarify how other cues in the support interaction influence the impact of messages. Although this work has shed light on some of the processes involved in message evaluation, there are still a number of characteristics of the support interaction left unexplored.

Goldsmith's (2001; Goldsmith & Fitch, 1997) normative/rhetorical approach also focuses on the evaluation of messages as effective or appropriate, and both perspectives recognize the myriad influences on supportive interactions, including the situational context. However, the normative approach is unique in many ways. Fundamentally, it suggests that individuals' communication is a response to their management of potentially conflicting identity, relational, and task goals (Goldsmith, 2001; Goldsmith & Fitch, 1997). As such, messages can be judged by the degree to which they are successful at meeting communication goals. Goldsmith (2001) proposed that scholars should focus on understanding why some communication is judged as better than other communication; therefore, our goal in this study is to clarify the reasons that some support messages are deemed more helpful than others in the context of grief. In particular, we examine the support messages received by individuals who have experienced the death of a parent or sibling and the conditions that influence evaluations of helpfulness of those messages. We offer the following research questions:

> RQ_1: How do individuals bereaved by the death of a parent or sibling manage supportive interactions?
>
> RQ_2: How do individuals in this context evaluate support messages as effective or ineffective?

Method

We conducted semi-structured, in-depth interviews with individuals who had lost either a parent or a sibling. Then, we used interpretive analyses to gain insight into individuals' perceptions of support from those in their social networks.

Participants

Participants in the study were 21 college students at a large Midwestern university. Most were female (n = 15, 71.4%), and the mean age was 21.24 years (range 19–28). Thirteen participants were European American/White (61.9%), six were African American/Black (28.6%), one was Asian American (4.8%), and one was Pakistani (4.8%). Students discussed their experiences following the death of a parent (n = 11 fathers, 52.4%; n = 4 mothers, 19.0%) or a sibling (n = 4 brothers, 19.0%; n = 2 sisters, 9.5%). Family members died from a variety of causes, but most were health related (n = 17, 81.0%), including cancer, pulmonary embolism, heart-related issues, and general health complications. Other causes of death included accidents (n = 2) and murder (n = 2).

Procedures

After obtaining Institutional Review Board approval, the first author recruited students enrolled in basic communication courses in exchange for a small amount of extra credit. Eligible students had to (a) be 18 years old or older and (b) have had a parent or sibling die while the student was alive. Individual interviews were held in an interaction lab where participants were invited to complete informed consent and a demographic questionnaire. After clarifying the study procedures, the researcher began audio-recording. Data for this project on social support were taken from a larger data set about family bereavement. Interviews began with general questions about the deceased family member and the events surrounding his or her death. The rest of the interview was structured in two sections: one focused on conversations with family members, and the other focused on conversations with people outside the family.

Analysis

After each interview, the authors created memos documenting notable moments, and each audio recording was transcribed verbatim. We used grounded theory methods to analyze the data (Corbin & Strauss, 2008), using constant comparison to allow themes to emerge from the data (Glaser & Strauss, 1967). We read through the transcripts and identified incidents of supportive communication, defining social support as any interaction between participants and their social networks in which the main focus of the conversation was a topic related to the loss. We compiled each incident of social support from individual transcripts into a single document, creating 140 pages of double-spaced text. Each author line-by-line coded sections of the transcript, developing lists of social support codes. We met to discuss our individual lists and created a codebook of the most prevalent codes. Finally, each author and an undergraduate research assistant engaged in focused coding on the entire data set. Our goal was to gain an in-depth understanding of each code and to learn about connections among the codes. We used constant comparison during this step, allowing each code to develop in distinction while noticing new and emergent themes (Corbin & Strauss, 2008). This process led to the discovery of another prevalent experience, which was added to the codebook. After coding with the revised coding scheme, no new insights emerged from the data, suggesting we reached a high degree of theoretical saturation (Strauss & Corbin, 1998).

Results

Findings revealed that bereaved individuals' perceptions of their interactions with others about the death of their parent or sibling are intricate. They avoided talking about the death, anticipating that the interaction would be difficult. When conversations did occur, participants described them as negative. In the following paragraphs, we answer each research question in turn, offering participant quotes illustrative of each experience. All participants' identities have been protected through the use of pseudonyms.

Managing Supportive Interactions (RQ_1)

Participants described several ways they actively manage their interactions with others. Primarily, their management strategies were characterized by avoidance. Our data analysis revealed two types of avoidance: *avoiding the topic* and *avoiding sympathy.* Both types were rooted in the emotions participants felt during interactions about the deceased family member. More than anything, individuals said that both they and the person providing support felt uncomfortable and awkward whenever the deceased family member came up in conversation. Joy described her interactions with others about her father's death: "But now when people ask like, 'Oh what happened to your dad?' 'Oh, he died.' And people will be like, 'Oh I'm so sorry.' They don't know how to react." Similarly, Megan perceived that others became uncomfortable because they did not know what to say or do when she disclosed that her father had died: "I say, 'Well he's dead,' and then it brings up an awkward...they don't know how to handle that." Discomfort governed participants' interactions with their social network members.

The bereaved perceived that the awkwardness stemmed from the fact that others did not know the "right" thing to do or say when the deceased came up in conversation. As a result, they used cultural scripts like "I'm so sorry," attempting to be supportive. Chandra noted, "It's just like they don't know what to say, so they just say what feels appropriate." These messages made participants feel pitied rather than supported, which made interactions feel forced. Given the discomfort they felt during interactions, most participants made an effort to avoid conversations about their loved one. By avoiding the topic and avoiding sympathy, participants controlled their interactions in a way that allayed their feelings of discomfort about discussing their loved one. Each type of avoidance is described below.

Avoiding the topic. With few exceptions, participants avoided supportive interactions. Their avoidance took different forms. Some people avoided the topic by not bringing up the death at all. Tanya described her approach to coping with her father's death saying, "I don't want to think about the bad things. I'm an avoider if anything, so I would rather not talk about it." Similarly, Joy's family as a whole took an avoidant approach to their grief over the loss of her father: "We just kind of don't like to bring him up." Often, participants like Tanya and Joy would steer the conversation away from their family member if it was headed in that direction.

Other forms of avoidance were less clear cut than total avoidance. Some individuals described leaving out details or being vague when they talked about their family member, like Ellen who described how she navigates conversations about her deceased father: "It's something that I wouldn't go into details." Most of the time, participants chose to only talk about positive memories or qualities of the person, which allowed them to engage in a conversation about their loved one while maintaining some control of the interaction. Finally, some individuals managed interactions about their family member by offering reassurance that they were okay in order to stop the other person from continuing to ask questions. Similarly, some would avert their conversational partner from the topic by moving quickly onto other subjects. Alex, who lost his brother, would avoid talking about the death with the following message:

> I'd say I have two brothers and a sister. And then if they say how old, I would say, "Well the oldest brother passed away." But then I would say, "But oh, that was two years ago. So don't worry about it" if they said sorry or something. Then I would quickly go on and say I have another brother and he's however old, and then I have a sister and she's 10.

Tactics like Alex's allowed participants to acknowledge their loved ones without offering much information about them.

Participants' reasons for avoiding talking about the deceased person varied. Many individuals expressed that they do not talk about their loved one because it is "none of their business," as Chandra mentioned. When others asked questions or brought up the death, participants felt that their privacy had been invaded, so they changed the subject or spoke in vague language. Others avoided the topic because they felt that although their relationship with the deceased was an important part of their identity, it was not the part of themselves they wanted to show first or most often. Tanya wanted others to know about her father, for example, but did not want it to be her most obvious quality: "I don't want it to be on my t-shirt. But I don't want it to not be seen at all." Hannah echoed Tanya's sentiment in describing her thoughts about disclosing her brother's death: "I'm not ashamed to admit it because he's my brother; whether he's dead or alive, he's still my brother." Finally, people described how their avoidance changed throughout the grief process. Later in the experience, it seemed the person naturally came up less often. In addition, as participants came to terms with their grief, they realized they only had to discuss their loved one when and with whom they wanted. Megan's advice to others who are bereaved was to own their grief experience: "If it makes you uncomfortable to talk about it then don't." Our participants exercised control of their interactions by avoiding the topic when and in the ways they wanted.

Avoiding sympathy. A second type of avoidance was avoiding sympathy. Over and over again, participants reflected on how much they hated it when others felt sorry for them. One common support message individuals heard was "I'm sorry for your loss." When others provided such messages, participants' primary response was to feel pitied. Makayla was adamant that she did not want others to think less of her because of her mother's death, saying, "I don't want everybody to sympathize for me or think that I'm at a disadvantage because I grew up without my mom." Hannah described how she perceived others who expressed sympathy for her loss: "I hate, 'Oh I'm so sorry!' and they just look at you with puppy eyes." The bereaved individuals in our study opposed the idea that others felt sorry for them following their loss, so one of their goals was to avoid conversations that induced sympathy.

They described several reasons for their adverse reactions to others' attempts at support. First, they discussed how pity made them feel like others thought of them as weak. Chandra explained, "I don't want to feel weak…I want to project this image of being strong, like I'm dealing. I've dealt with this." To prevent others from thinking of them as weak, some participants would say things like, "I'm fine; you don't have to pity me" or "You don't have to say you're sorry." When Ellen disclosed her father's death, she would tell others not to be sad for her: "I'll be like, 'Stop, it's okay. You don't need to feel uncomfortable. You don't need to feel sad for me.'" A second reason individuals avoided sympathy is because they did not want to bring down their mood. Receiving support often reminded participants of the sadness they felt surrounding the death. Their preference was to keep conversations about the person light. Jessica enjoyed talking about her brother, but only if the conversations were focused on positives: "I don't like to talk about how he died. I like to talk about, you know, his life…If it's just a casual conversation, I do like to bring him up in funny memories and stuff like that." Finally, some people said receiving messages like "I'm sorry" from others was unproductive. It did not change their sadness or grief. Chandra repeated several times about her mother's death that "It is what it is," so there was no sense in other people saying they were sorry. Tanya's sarcastically mentioned that "I hate it when people say, 'Oh I'm so sorry for your loss.' I'm like thanks. Your apology does so much for me, you know."

Evaluating Support Messages (RQ_2)

Our goal in this study was to learn how bereaved individuals evaluate support messages, including investigating the cues they used in perceiving which messages were helpful. However, our analyses revealed that for students bereaved by a parent or sibling's death, the message itself was less important than the person who delivered it. In particular, participants evaluated all support messages as ineffective with two exceptions: (a) those delivered by other bereaved individuals and (b) those who had very close relationships with the bereaved.

Recognizing differences. Most people provided comfort in ways participants interpreted as ineffective (e.g., saying "I'm sorry" or asking probing questions about the death). When participants reflected on why they perceived support as unhelpful, they noted that most people simply did not understand their perspective. That is, they recognized fundamental differences between themselves and people who had not lost someone close to them. Sylvia said about the people in her life, "I feel like they don't get it and I don't want to try to explain it to them." Support from those who did not understand made participants feel angry or annoyed. Even among the bereaved, participants were sensitive to the fact that there were differences in their experiences. Ellen reflected on the uniqueness of each person's bereavement when she said, "I feel like a lot of people might try, but my situation is a hundred percent different than theirs, so if they try to find similarities, they can't." Since most other people did not have similar experiences to our participants, they perceived that others could not offer adequate support.

One reason participants felt unable to relate to other people was because they experienced a shift in perspective as part of their grief. Many things that seemed important before their loss no longer mattered to them. Friends who had not lost a parent or sibling could not relate to this change, and it annoyed and frustrated participants when their friends stressed out about things that were "unimportant." Hannah noted how she felt different from her peers after her younger brother died: "I feel that I'm a little bit of a more older [*sic*] soul compared to some people my age because I do have a little different perspective on things." Katie, whose brother also died, said she had fundamentally changed in how she related to others: "I feel like I've lost my empathy for people a little bit. Because when someone complains about trivial stuff, I'm like, 'Come on.' I can't." Losing a family member is a unique experience that caused our participants to rank the importance of things in their lives differently than their friends.

Because of their perceived differences, participants evaluated their conversations with others as not beneficial. Supportive conversations were not perceived as cathartic or helpful. Instead, they were painful and uncomfortable. Makayla noted how conversations with others were exhausting and unproductive: "After a certain amount of time, I just get tired of talking to people about it because I feel like it doesn't get me anywhere." In the same way, Daniel stated, "I don't talk to people a lot, besides my family…except for my really close friends because it doesn't really benefit me that much. I don't really care what you know. I don't need to talk about it that much." The differences the bereaved felt between themselves and other people in their lives made them want to avoid supportive interactions.

Evaluating messages from similar and close others. Although most support messages were appraised as unhelpful, all of our participants recognized that having support could be valuable if it came from the correct people. Megan captured this notion when she said, "Loss is a team sport." However, participants chose strategically who was on their team. In particular, family members and friends whom they perceived had similar experiences were allowed to provide support. In addition, even if they had not experienced a death, some friends or family who were exceptionally close could also talk or ask about the grief experience.

Similar others included family members who had experienced the same death as the participants, as well as friends who had also lost a parent or sibling. Ellen described how rare it is to truly relate to others: "Nobody else understands. It's like a club. You're not part of it until you're part of it." Hannah's family found comfort in connecting with other families who had lost a member: "In a weird sense it's a comfort in knowing that you're not alone. However, in the same aspect it's like you wish that these people never had to understand the pain." As a result, their support messages were perceived as effective because they came from a place of understanding, rather than pity. The similarity among family members or others who were bereaved created a framework through which supportive messages could be perceived as effective.

Others who could provide effective support messages were close friends or romantic partners. Although these individuals often could not relate to the bereaved person, their closeness to the participant permitted them to talk about the deceased family member. Participants often mentioned these people as exceptions, like Scott, who said, "To certain people I'm glad to tell. But then other times it's just, I don't know. It's just not the same with people. It depends on how close you are to people I guess." However, even though close friends were sometimes able to provide support, they were still separate from similar others. For example, Jessica said her best friend was so close to the family that her best friend also felt like she had lost a brother. However, there were still qualitative differences in their grief experiences. Jessica reflected: "She felt definitely similar feelings that I did, but at the same time knew that it was about my healing and not necessarily hers as well. She knew to just be there for me and not make it about her if I just wanted someone to just be there for me." Close friends, therefore, still maintained some degree of separation from the participant. In sum, participants (a) avoided interactions about their deceased loved one and (b) placed a great deal of emphasis on the sender of support messages, evaluating messages from most people as unhelpful.

Discussion

Bereavement is a unique context for support, and those who experience it go through an intricate process of coping. Generally, social network members called on cultural scripts when they provided support to the bereaved. When participants in our study reflected on supportive interactions, they suggested that, overall, they avoided talking about their deceased family member with others. Conversations about their loved one felt uncomfortable and awkward, and they wanted to avoid feeling pitied. Moreover, the actual messages participants received were less consequential to them than the person providing the support. Everyone was unqualified to provide quality support except for (a) those who had similar experiences and (b) those who were exceptionally close to the bereaved.

Theoretical Implications

Our findings add to bereavement literature by qualifying research that nominates grief as a relational process (e.g., Rosenblatt, 1988; Stroebe, 2001). Scholars have noted that grief occurs within family and social systems and that social support is important to those who are grieving (Hayslip & Page, 2013). Indeed, our participants appraised some interactions with others as valuable; however, not everyone was permitted to participate in the grief process. In other words, the results of our study suggest that grief is conditionally relational, as only some people can make helpful contributions to the process. Extant research on grief has left some questions unanswered about what, exactly, makes some support effective and other support frustrating (e.g., Breen & O'Connor, 2011; Toller, 2011; Wilsey & Shear, 2007). Our analyses respond to this dearth in research by suggesting that when people do not actually understand an experience like grief, their attempts at support are appraised as "socially inept" (Dyregrov, 2003, p. 31). Moreover, many supportive interactions were face-threatening for the bereaved, who felt pitied rather than supported by most people in their networks. Perhaps this explains why our participants preferred to avoid interactions about their loved one, to avoid potential face threats coming from talking to those who could not relate (Toller, 2011). Overall, our results contextualize other grief research by specifying some of the conditions affecting perceptions of support.

Our results also add to social support research by adding depth of understanding to supportive communication in the context of bereavement. Both the dual process model of social support (Bodie et al., 2012; Burleson, 2009a, 2009b) and the normative approach (Goldsmith, 2001; Goldsmith & Fitch, 1997) focus on evaluations of supportive messages as effective or ineffective. Similarly, our goal in this study was to elucidate the circumstances under which individuals bereaved by a family member's death identified support as helpful. One of the most enlightening findings in our study is that the sender of the message is of more importance than the message itself. The dual process model of social support suggests this might be because our participants were unmotivated or unable to process others' supportive communication, and instead, they relied on other cues to interpret the message. In this case, they developed the following heuristic for how to evaluate messages: if the person is similar or close, the message was helpful. Otherwise, it was not. Many of our participants' interactions about the deceased were with casual friends or others who did not know the participant well, which means more times than not, their supportive interactions were unsatisfying. Although some dual process model scholars have investigated support in the context of bereavement (e.g., Burleson et al., 2009; Burleson et al., 2011), our in-depth analyses suggest that the role of the message sender cannot be overestimated, a constraint not accounted for in extant research.

Some of our findings also raise questions about the consideration of cultural scripts in previous support research. For instance, Bodie et al. (2012) suggested their participants evaluated some support (even if it was not highly person-centered) as being better than none. Our participants would likely disagree with this notion. Cultural norms dictate that when a social network member faces a loss, the most appropriate response is something similar to "I'm sorry for your loss." Although our participants recognized that this is what their friends were "supposed" to say, it evoked far more negative emotions than positive ones, and in fact, they explicitly noted they would rather people said nothing at all. To date, nearly all of the studies of the dual process model have utilized hypothetical support messages and/or stress scenarios. Certainly, this work has given support scholars a more precise understanding of how individuals evaluate support; however, this method has also precluded the kind of response our participants offered, highlighting the need for methodological triangulation.

Practical Utility

Our investigation both focuses on a dark context (i.e., bereavement) and highlights a dark set of findings: most supportive communication following the loss of a family member is not beneficial. Perhaps the most striking point to note here is that the deeply engrained cultural scripts most people use when offering support evoke viscerally negative reactions in the message recipient. However, it might also be damaging to ignore the pain of others following a death in their family. As such, we call for future research to investigate the properties of supportive conversations after the death of a loved one that make the interaction satisfying for bereaved individuals and their families. When we asked our own participants for a good alternative to the cultural script, they generally said that acknowledging how hard the loss was without providing sympathy might be appropriate, but overall, they wanted the conversation to be factual rather than emotional, particularly if the person was not a close friend or a family member. Individuals should be sensitive to the possibility that cultural scripts, though widely used, might not be the best option in a given conversation. This finding might be used to produce a body of work dedicated to understanding how to construct support for the bereaved.

Limitations and Future Directions

Though our study makes some important contributions, it is also constrained by some limitations. The homogeneity of our sample makes the application of our results to other contexts or populations difficult. All of our participants were emerging adults enrolled in college. Although some of them lost loved ones when they were young, they likely have different insight into the support process than those who are vastly younger or older than them. Future research should investigate how individuals in other stages of the life cycle interpret supportive interactions. Second, research on the dual process model of social support suggests there are differences in how males and females interpret supportive communication (e.g., Burleson et al., 2009; Burleson et al., 2011). Although we did not notice any qualitative gender differences in our study, we had far more female participants than male participants. As such, some important differences might have been obscured. Similarly, more participants in our study had lost a parent than a sibling. As researchers continue to investigate individuals' perceptions of support, they should be sensitive to within-group differences that might exist among the bereaved that we did not detect in our study.

Conclusion

This study examined how individuals who lost a parent or sibling managed supportive interactions and evaluated support messages. Participants managed interactions primarily by avoiding the topic of the death and by avoiding sympathy. In addition, their evaluations of messages were based on the sender's similarities to and relationship with the participant; however, most messages were perceived as unhelpful because they came from individuals who did not understand the grief experience. These findings have both theoretical and practical implications for understanding what makes supportive communication beneficial in the context of bereavement.

References

Bodie, G. D., Burleson, B. R., & Jones, S. M. (2012). Explaining the relationships among supportive message quality, evaluations, and outcomes: A dual-process approach. *Communication Monographs, 79,* 1–22. doi:10.1080/03637751.2011.646491

Breen, L. J., & O'Connor, M. (2011). Family and social networks after bereavement: Experiences of support, change and isolation. *Journal of Family Therapy, 33,* 98–120. doi:10.1111/j.1467-6427.2010.00495.x

Burleson, B. R. (1982). The development of comforting communication skills in childhood and adolescence. *Child Development, 53,* 1578–1588. doi:10.2307/1130086

Burleson, B. R. (2009a). Explaining recipient responses to supportive messages: Development and tests of a dual-process theory. In S. W. Smith & S. R. Wilson (Eds.), *New directions in interpersonal communication* (pp. 179–199). Thousand Oaks, CA: Sage.

Burleson, B. R. (2009b). Understanding the outcomes of supportive communication: A dual-process approach. *Journal of Social and Personal Relationships, 26,* 21–38. doi:10.1177/0265407509105519

Burleson, B. R., Hanasono, L. K., Bodie, G. D., Holmstrom, A. J., McCullough, J. D., Rack, J. J., & Rosier, J. G. (2011). Are gender differences in responses to supportive communication a matter of ability, motivation, or both? Reading patterns of situation effects through the lens of a dual-process theory. *Communication Quarterly, 59,* 37–60. doi:10.1080/01463373.2011.541324

Burleson, B. R., Hanasono, L. K., Bodie, G. D., Holmstrom, A. J., Rack, J. J., Rosier, J. G., & McCullough, J. D. (2009). Explaining gender differences in responses to supportive messages: Two tests of a dual-process approach. *Sex Roles, 61,* 265–280. doi:10.1007/s11199-009-9623-7

Cohen, S. (1992). Stress, social support, and disorder. In H. O. F. Veiel & U. Baumann (Eds.), *The meaning and measurement of social support* (pp. 109–124). New York, NY: Hemisphere Press.

Cohen, S. C., & Syme, S. L. (1985). Issues in the study and application of social support. In S. Cohen & S. L. Syme (Eds.), *Social support and health* (pp. 3–22). San Francisco, CA: Academic Press.

Cohen, S. C., & Wills, T. A. (1985). Stress, social support, and the buffering hypothesis. *Psychological Bulletin, 98,* 310–357. doi:10.1037//0033-2909.98.2.310

Corbin, J., & Strauss, A. (2008). *Basics of qualitative research* (3rd ed.). Los Angeles, CA: Sage.

Dyregrov, K. (2003). Micro-sociological analysis of social support following traumatic bereavement: Unhelpful and avoidant responses from the community. *Omega: Journal of Death & Dying, 46,* 23–44. doi:10.2190/T3NM-VFBK-68R0-UJ60

Glaser, B. G., & Strauss, A. L. (1967). *The discovery of grounded theory: Strategies for qualitative research.* Chicago, IL: Aldine.

Goldsmith, D. J. (2001). A normative approach to the study of uncertainty and communication. *Journal of Communication, 51,* 514–533. doi:10.1093/joc/51.3.514

Goldsmith, D. J., & Fitch, K. (1997). The normative context of advice as social support. *Human Communication Research, 23,* 454–476. doi:10.1111/j.1468-2958.1997.tb00406

Hayslip, B., & Page, K. S. (2013). Family characteristics and dynamics: A systems approach to grief. *Family Science, 4,* 50–58. doi:10.1080/19424620.2013.819679

Kaunonen, M., Tarkka, M., Paunonen, M., & Laippala, P. (1999). Grief and social support after the death of a spouse. *Journal of Advanced Nursing, 30,* 1304–1311. doi:10.1046/j.1365-2648.1999.01220

Rosenblatt, P. C. (1988). Grief: The social context of private feelings. *Journal of Social Issues, 44,* 67–78. doi:10.1111/j.1540-4560.1988.tb02077

Strauss, A., & Corbin, J. (1998). *Basics of qualitative research: Techniques and procedures for developing grounded theory* (2nd ed.). Thousand Oaks, CA: Sage.

Stroebe, M. S. (2001). Bereavement research and theory: Retrospective and prospective. *American Behavioral Scientist, 44,* 854–865. doi:10.1177/00027640121956430

Stroebe, M. S. (2010). Bereavement in family context: Coping with the loss of a loved one. *Family Science, 1,* 144–151. doi:10.1080/19424620.2010.576081

Toller, P. (2011). Bereaved parents' experiences of supportive and unsupportive communication. *Southern Communication Journal, 76,* 17–34. doi:10.1177/00027640121956430

Wilsey, S. A., & Shear, M. (2007). Descriptions of social support in treatment narratives of complicated grievers. *Death Studies, 31,* 801–819. doi:10.1080/07481180701537261

THREE

On Sexualization and Attraction…A Communication "Dark" Side

D. L. Stephenson

Sexualization is defined as the construction of an identity intended to be sexually arousing and stimulating (American Psychological Association Task Force, 2007; Brooks, 2006; Cabrera, 2007; Cross, 2004; Crowley, 2006; Durham, 2008; Egan & Hawkes, 2007; Egan & Hawkes, 2008; Haug et al; 1999; Levin, 2005; MacRae & Sears, 2007; McPherson, 2006; McRobbie, 1991; Robinson, 2005; Rush & La Nauze, 2006; Spigel, 1993). Often focusing on commercialization, researchers critically examine female sexuality deployed in films, advertising, music videos, and other media (Garner, 2012). Generally, however, "[s]exualization is a process caught up in issues of gender, power, everyday experiences, practices, and identities" (Garner, 2012, p. 325). The sexualization of women and girls in and out of the media is viewed as normal in a culture that places considerable emphasis on female attractiveness and beauty (Haug et al., 1999). Few believe that routine sexualization is harassing, harmful, or even problematic. In fact, there is considerable controversy around what actually constitutes sexual harassment. Sexual harassment is legally defined as repeated, unwanted, behavior of a sexual variety or sexually explicit verbal and nonverbal communication (Eisenberg & Goodall, 2004). It is also defined as unwanted and unwelcome sexual advances or overtures, characterized by promises of career advancement or other professional help in exchange for sexual favors, and it can include direct and/or indirect threats intended to coerce a subordinate or vulnerable person into a sexual relationship (Eisenberg & Goodall, 2004).

Sexualization, unlike sexual harassment, does not include threats and/or quid pro quo inducements for sex or sexual favors, although sexualization is an aspect of sexual harassment. Sexualization and its corollary, objectification, involve the imposition of an identity, schema, and persona—sexual or otherwise—onto another regardless of whether that identity, schema, or persona are accurate or accepted by the person being sexualized (Calogero, Davis, & Thompson, 2005; Fredrickson & Roberts, 1997; Szymnski & Henning, 2007). In objectification, the individual or subject is treated as and turned into a "thing" (Nussbaum, 1999, p. 218). Human beings—and in this discussion women—are reduced to their sexual essence and stripped of their own sense of sexual identity (Vaes, Paladino, & Puvia, 2011). And like all objects, sexualized individuals must adapt to having their actions, words, and motives viewed through a framework over which they have no control and that renders them primarily as sexual beings (Haug et al., 1999). However, when both males and females are sexualized and objectified—attributing sexual intent to individuals who do not intend to indicate sexual interest, desire, or availability—it appears normal and natural. There are many examples of routine sexualization, but perhaps one of the most obvious occurred when a dentist decided his dental assistant's attractiveness was grounds for her dismissal.

Fired: The Myth of Uncontrollable Male Sexual Desire

In 2013, in Fort Dodge, Iowa, a married, male dentist fired his married, female, dental assistant, Melissa Nelson, 33, because he feared his sexual attraction to her would lead to an extra-marital affair (Foley, 2013). The dentist, according to court transcripts, decided to terminate his employee of 10 years after his wife became aware of text messages the dentist, James Knight, 54, sent to his assistant expressing his physical attraction (Strauss, 2013). Nelson brought a lawsuit against Knight alleging discrimination. The male Iowa District Court judge dismissed the case before trial, and the all-male Iowa Supreme Court upheld the Iowa District Court judge's decision. Even when the case was reconsidered, the Iowa Supreme Court ruled that Nelson's firing was not discriminatory. According to *Associated Press* reporter Ryan Foley (2013),

> Coming to the same conclusion as it did in December, the all-male court found that bosses can fire employees they see as threats to their marriages, even if the subordinates have not engaged in flirtatious or other inappropriate behavior. The court said such firings do not count as illegal sex discrimination because they are motivated by feelings, not gender. (para. 2)

The dominant discourse on feelings and emotions in the U.S. ignores that human beings are always acting according to *social* conventions even when they experience and respond to their feelings and emotions (Jaggar, 1997). According to Jaggar (1997) "…'feeling' is often used colloquially as a synonym for emotion, even though the more central meaning of 'feeling' is physiological sensation" (p. 389). Feelings can be defined as physiological sensations, as Jaggar makes clear, and are "likely to be interpreted as various emotions, depending on the context of [the] experience" (p. 389). The feeling of attraction, for example, can lead one person to experience the emotion of love, while leading another to experience the emotion of shame or guilt. Nonetheless, emotions are often ignored or discounted as important and integral to the construction of meaning (Jaggar, 1997). Because emotions and feelings have been marginalized "[w]ithin the western philosophical tradition, emotions usually have been considered as potentially or actually subversive of knowledge" (p. 385).

If Jaggar (1997) is correct in her assessment, then it can be argued that feelings of sexual desire and the emotions it inspires (i.e., sexual attraction and desire) are viewed by the judges in Nelson's lawsuit (and those who agree with their ruling) as outside of the realm of reason and as adequate reasons to terminate an employee. While men identified as White, male, straight, educated, and materially privileged are stereotyped as rational and devoid of and estranged from their emotions, their sexuality, paradoxically, is often viewed as wild and outside of reason (Holloway, 1984; Paglia, 1992; 1994). And because male sexual desire is viewed as "out of control," the dentist's decision to fire his assistant appears to the bench a reasonable response to a threat to the Knights' marriage. The judges' ruling that Nelson was fired because of the feelings she evoked in Knight ignores the fact that the combination of Nelson's gender and his subjective evaluation of her physical attractiveness engendered Knight's sexual feelings. Had Nelson been male or a woman he found physically unappealing, Knight would have more than likely felt no sexual interest. Further, the judges ignored Nelson's stated lack of sexual feelings for Knight, suggesting that the dentist' feelings and emotions were more important or relevant than the emotions of the *sexualized object* of his desire. Arguing that emotions inform knowledge, Jaggar (1997) states,

> Like everything else that is human, emotions in part are socially constructed; like all social constructs, they are historical products, bearing the marks of the society that constructed them. Within the very language of emotion, in our basic definitions and explanations of what it is to feel pride or embarrassment, resentment or contempt, cultural norms and expectations are embedded. (p. 395)

The judges' ruling reflects the ways in which their knowledge is informed by their *feelings* about what they assume will stem naturally from Knight's (or any man's) sexual desires. However, sexual desire is a feeling like all others—to be negotiated, indulged, or ignored. How each individual behaves or acts, despite his or her sexual feelings and the emotions those feelings elicit, is a *choice*. Nonetheless, in the cultural discourse on male and female heterosexuality, male sexuality, and sexual desire are viewed as incapable of restraint, dangerous, and aggressive (Holloway, 1984; Paglia, 1992). Yet, this is not the case in the actual social-sexual experience of most men and women. There are many instances when aroused men control their sexual urges. The mind ponders, assesses, and chooses—no matter how "instinctive" a behavior or action may seem, no matter how much an impulse may vie for expression. We've all heard the old adage that "men and women can't be just friends" because of the inevitable sexual tensions and feelings that will eventually break loose and undermine the platonic relationship (Hansen, 1985; Monsour, 1996; Swain, 1992). The reality is that many opposite-sex friendships are not based on sexual attraction and desire (Kuttler, LaGreca, & Prinstein, 1999; Messman, Canary, & Hause, 2000; Rubin, 1985). The rhetorical chestnut "men and women can't be friends" reveals the wide-spread sexualization of male-female heterosexual interaction. The common wisdom holds that men and women, particularly those deemed attractive, are destined to find each other desirable, and, as a result, will "fall in love" or, at the least, engage in some form of sexual activity (Monsour, 1996). Regardless of whether men and women find each other sexually attractive or want to act on their sexual attractions and feelings should they exist, the cultural assumption is that they will.

In reality, the Iowa dentist could have ignored his sexual feelings or channeled his sexual feelings into his marriage. Instead, he *chose* to pursue a woman who did not reciprocate his feelings, and when his wife learned of his actions, he *chose* to fire his assistant rather than risk further angering his wife by keeping his dental assistant in his employ. Though routine and normal, the sexualization of "attractive" women, across race lines, poses a serious social (and sometimes economic) danger to virtually all women in a society that privileges men's sexual feelings. During one interview, a female reporter asked Nelson if she ever flirted, acted provocatively, or dressed in a way that would have encouraged Knight. Nelson responded that she never flirted and always wore scrubs to work. What's interesting about this line of questioning is the reporter's attempts to determine whether Nelson is blameless in attracting Knight's sexual interest. Nelson must demonstrate that she is chaste, pure, and innocent and not a "loose" woman (LeMoncheck, 1997). Nelson must be framed as the kind of woman with whom the public may sympathize. A married, monogamous, educated, middle-class, White, mother, who wears clothing that conceals her neither too heavy nor too thin body, can be taken seriously and believed. For LeMoncheck (1997) naming and defining are instances of patriarchal power that limit women's sexuality and its presentation and representation:

> Women's heterosexual subordination by men is a subordination of *identity*. In a patriarchal society, women are defined in terms of our heterosexuality and reproductivity in order to serve the needs and maintain the privileges of men. Therefore, women's sexuality under patriarchy must be very carefully circumscribed lest it gain an independent credibility and power of its own. Men's ideal of women is that we be sexual only *in a certain way*. (p. 56)

For Nelson and all women deemed attractive, their sexuality must always manifest itself in ways that are "serviceable" (LeMoncheck, 1997). Firestone (1970) argues, "In a male-run society that defines women as an inferior and parasitical class, a woman who does not achieve male approval in some form is doomed" (p. 131). Women must also gain the approval of other women, who also judge—and often quite harshly. The jurists' rulings and the reporter's questions re-inscribe the wide-spread belief that women's sexuality must be deployed only in sanctioned relationships and that women are responsible for men's actions and reactions (*Meritor Savings Bank v. Vinson,* 1986; Ranney, 1998). Had Nelson dressed "provocatively" or "flirted" with the dentist or acted in any way culturally determined as inappropriate, then her firing would appear valid to those who subscribe to the idea that women's behavior has the power to prompt men's actions. What's important to consider here is not whether Nelson acted "inappropriately" with the dentist. What's important is how her actions will be interpreted by the public, in the agenda-setting news and social media, and by the judges hearing her case.

However, the belief that women determine men's behavior constitutes a scapegoat that divests individuals of responsibility (Szasz, 1996). According to Szasz (1996), "In every age and in every society, people have generated and society has legitimated certain excuses for evading responsibility, by attributing it to causes or agents outside the self" (p. 36). Historically, women in Western societies have been viewed as "polluted" or as actual "pollution" (Szasz, 1974, p. 106). If this is true, it stands to reason that women's beauty, while both coveted and celebrated, is also dangerous. The paradox of female beauty is that it also has the power to *pollute* men's ability to reason and act rationally. In literature and history, female characters and figures have earned their notoriety, depicted as having used their beauty to attract men and cloud male judgment. In what is considered the best of Western literature, the fall of all humankind from God's grace (Eve), devastating wars that destroyed a society (Helen), and personal betrayal or epic proportions (Delilah), all have been attributed to women whose beauty beguiled, bewitched, and betrayed men.

If the dental assistant's stated lack of sexual interest in the dentist and her assertion that she had no desire to pursue a sexual relationship with her employer are accepted at face value, a *mutual* sexual relationship between the two is clearly impossible, making untenable the grounds on which Nelson's firing rests. It is irrational to believe that a man's sexual interest in and desire for a woman—absent the woman's *mutual* sexual interest—could lead to a sexual relationship. Had the dentist proceeded to have sex with his demonstratively unwilling assistant, he would be a *rapist* subject to legal and social sanctions. Absent an actual attack of a sexual nature, there would be no sexual contact between the two. However, sexualization and objectification as routine, normative cultural practices reject or ignore that most human beings make *intellectual* decisions about their sexual behavior regularly—if for no other reason than to stay out of trouble. Because attraction weighs so heavily in the choosing of sexual partners and the maintenance of relationships, mutual sexual attraction is necessary for a sexual relationship to develop. However, the dental assistant's firing hinges on the patriarchal myth that when women refuse men's sexual advances, women are simply playing hard to get or acting coy. Women are, once again, perceived as in control or acting in ways that threaten existing patriarchal controls deemed necessary. According to Firestone (1970), men are encouraged "...to look at women as only things whose resistance to entrance must be overcome" (p. 140).

One of the requirements of a sexual or romantic relationship is that it be mutual and understood as such by all of the parties involved (Hendrick & Hendrick, 1992; Miller & Steinberg, 1975). Sexual behavior may appear to be dominated by instinctive urges, but men and women make choices and decisions regarding their sexual behavior. Despite what are often strong, immediate sexual feelings and emotions, human beings are social beings who *think* about their attraction(s). Despite the dominant ideology in the United States that normalizes the idea that heterosexual sexual behavior is primitive and instinctive, heterosexual men and women are not barn animals. Men and women are social beings with the ability to delay gratification, consider consequences, and act based on deliberation, calculation, and reason. As social actors influenced by cultural norms and expectations regarding how to respond to sexual attraction, reason acts in tandem with powerful sexual impulses. Moreover, two of the hallmarks of being human are each individual's ability to curtail his or her impulses and to select *specific* sexual and romantic partners. It should go without saying, then, that all straight or heterosexual men are not attracted to all women and all straight or heterosexual women are not attracted to all men.

Nelson's firing was, to her and her lawyer, an egregious example of sex and gender discrimination (Foley, 2013, para 5). However, because Knight did not threaten the assistant or accost her and because the assistant did not complain about her boss's alleged "inappropriate" texts and comments, Nelson's firing does not fall within the legal parameters of sexual harassment. One could argue that the dental assistant's firing, because it was not deemed illegal, was not discriminatory. However, simply because the dental assistant's firing did not fall within the narrow legal definition of sexual harassment, it does not mean that her firing was not unethical, sexist, and an example of how attractiveness and routine sexualization can be dangerous.

Attractiveness as Social Currency

In the United States, as well in just about every other country in the world, attractiveness in women is considered a positive attribute and all women and girls—across race and ethnic lines—are strongly encouraged to make being "attractive" a goal (Haug et al., 1999). As Haug et al. (1999) point out,

> Since women enter the market place as commodities, they are under pressure to make themselves externally presentable—to use attractive packaging to bump up their market price, or to make themselves saleable in the first place. Implicit in the image of the carefully groomed woman is an assumption that the exchange value of women is open to manipulation, that it may be added to a given use value as an external appendage. (p. 131)

Attractive women are held up as examples to other women and frequently awarded and rewarded with male sexual and romantic attention—the primary cultural goal of heterosexual female attractiveness. However "attractive" women are often accused, frequently by other women, of using their physical appeal as a means to gain advantages or curry favor (Strelan & Hargreaves, 2005). From a cultural perspective, women whom men find attractive should not be offended when their comely traits are noted, but they also must take full responsibility for the attention and even the mistreatment they receive because of them (Wood, 2005). As a consequence, not only is it extremely difficult for women who have been sexually harassed or raped and/or sexually assaulted to seek redress, it is virtually impossible for women who experience objectification and sexualization to complain. In a world in which rape and sexual harassment are real and dangerous crimes but are still viewed as somehow the fault of women, sexualization seems like a non-issue. The key to understanding sexualization as a salient social issue lies, at least in part, in realizing that despite feminist influences on social life, women are still viewed in the United States (and in other seemingly "enlightened" societies) as sexualized objects that must expect and accept unwanted sexual attention as a condition of being perceived as attractive. The sexualization of women in patriarchies maintains male sexual power over women and dictates how both men and women will respond to women (LeMoncheck, 1997).

Sexualization *is* not only about attraction; it also about a sense of *entitlement* to act on that attraction. In sexualization, the feelings and beliefs a man has toward his object of desire are considered natural, normal, appropriate, and reasonable—and to be expected. In short, sexualization *is* about the routine imposition of sexual attention onto "attractive" women. Heterosexual sexualization *is* about how women are made to believe that their physical appearance is something they must be ashamed of, apologize for, or alter in order to deflect unwanted sexual attention. In sexualization, women's styles of dress, grooming, and overall physical appearance, and physical desirability are *always* held up as the rationale for why they are treated in the ways they are. In sexualization, women's demeanor, personality, and interpersonal styles are routinely interpreted in ways that make them a seemingly legitimate target (e.g., "She looks yummy!"). As a result, women are often viewed as being responsible for their own sexualization (Ranney, 1998).

Sexualization often manifests itself as the verbal expression of desire, sexual/romantic feelings, attraction, or liking. Sexualization is often verbalized with offhand or matter-of-fact comments asserting that a woman looks good, attractive, hot, beautiful, and/or pretty. On the surface these are compliments meant to flatter women. In fact, most women never complain when they receive these compliments, being socialized to believe they should be gracious.

Sexualization also manifests in knowing looks, gazing, or staring (Henley, 1977). This kind of looking should not be confused with the kind frequently detailed in sexual harassment literature (Eisenberg & Goodall, 2004). Staring at a woman's breasts conveys sexual desire. Facial expressions that reveal emotions and feelings can be overtly harassing (Eisenberg & Goodall, 2004). Commonly, sexualization outside the workplace, particularly when men and women are socializing is often explained away and dismissed as harmless flirting.

Not all hugs and kisses indicate a desire for sex and sexual contact. However, the person doing the hugging and kissing may be perceived negatively by a person who does not want or like to be greeted this way. Few people, if any, ever ask if they may hug or touch another before doing so. It is a cultural norm for people to simply throw their arms around those they are greeting or plant kisses on cheeks during a farewell—and many of us do this. Touching in this manner assumes that the one being touched welcomes the touch as much as the one doing the touching.

Women who voice their desire to restrict and/or avoid touch are often perceived as cold and unfriendly or even psychologically damaged. A woman's body is simply never off limits (LeMoncheck, 1997). In fact, it is virtually impossible to refuse hugs and kisses. Unwanted touching operates as a milder, socially acceptable entitlement to other people's bodies. Men's (or women's) presumption of the right to hug and kiss others, without asking and without considering how their behavior is interpreted by the person subject to it, reinforces the social myth that hugs and "pecks" are nothing more than "friendly" gestures that demonstrate warmth, good will, and kindness. However, intimate touching, viewed critically, also may be displays of dominance and control (Henley, 1977).

Suggesting that men simply lose control and fail to correctly interpret signals of disinterest or discomfort is less reasonable than the alternative: some men simply *choose* to ignore disinterest or discomfort. Leaning over and brushing against a woman, caressing a woman's arm or face, or touching her hair. These gestures may appear innocuous because contact is made with a smile and a disarming comment (e.g., "you look like you could use a hug") in order to convey that the touch is non-sexual and filial and *desired.*

Sexualization may also include the use of pet names and endearments. Calling a colleague, student, or coworker *honey, hon, baby, babe, sweetie, dear* and *sweetheart* are examples of verbal sexual harassment when the one being addressed objects and the one doing the naming persists (Eisenberg & Goodall, 2004). In routine sexualization, the person using these terms often does so to create a more intimate connection and/or show affection. Through their body language, some men attempt to convey to others (especially other males) their sexual interest. Sexualization is indicated through both "proxemics" and "haptics" or what can be referred to as the *posture* (Hall, 1959; Trenholm & Jensen, 2011). The man who sits next to a woman rather than across from her may attempt this kind of posturing because he finds her attractive (Trenholm & Jensen, 2011). Standing in a woman's personal space rather than at a social distance while speaking, placing an arm or hand on a woman's chair, or leaning into her so that their legs, arms, or heads touch are some common examples of how the use of space and touch demonstrate attraction, and by extension, sexualize a relationship (Reynolds, 2010).

It has been noted that some men will mistake a woman's friendliness for sexual interest (Abbey, 1982). What is important is that the man's decision to act on his attraction is *not* treated as the central problem; the woman's anger, frustration, and resentment are. Anger is an "outlaw emotion" (Jaggar, 1997, p. 396). The sexualized object who complains may be viewed as an imprudent woman who places herself in "bad situations" (MacKinnon, 1987). If the sexualized object had not gone to that office, had declined dinner, refused a ride, or rejected a male friend's invitation to his home, then the attempted touching, kissing, and/or hugging never would have occurred. The assumption of women's culpability in their own sexualization suggests that certain spaces men and women enter are sexualized as much as the men and women who enter them. When a man and woman enter a restaurant, there is often an assumption that they are a couple. Cars and taxis are also sexualized when a man and woman enter them together. When opposite sex colleagues eat lunch together, it is not uncommon for coworkers to gossip about their presumed sexual relationship. In fact, sexualization is so normative, that the "couple" may even agree not to eat lunch in certain restaurants lest it be assumed they are romantically involved. Because these spaces are so fraught, men will often assume that agreeing to eat in a restaurant or meet in a bar or lounge means—regardless of the stated purpose for the meeting—that the woman is sexually or romantically interested.

According to Jaggar (1997),

> People who experience conventionally unacceptable, or what I call 'outlaw,' emotions often are subordinated individuals who pay a disproportionately high price for maintaining the status quo. The social situation of such people makes them unable to experience the conventionally prescribed emotions: for instance, people of color are more likely to experience anger rather than amusement when a racist joke is recounted, and women subjected to male sexual banter are less likely to be flattered than uncomfortable or even afraid. (p. 396)

Samantha Brick, a British, freelance journalist received criticism from readers when she attributed the positive attention she received from men to her beauty (Brick, 2012). It is common knowledge that women who adhere to cultural standards of attractiveness generate considerable attention from men. However, there is a cultural taboo about appearing to revel in that attention. Indeed, readers derided both Brick's physical attractiveness and character. (Brick, 2012; Curry, 2012; Freeman, 2012; Watson & Reilly, 2012; West, 2012). Though beauty is celebrated, women who acknowledge their own beauty are not (Dragon, 2010). To be sexualized and resist is to be viewed as a difficult woman; to be sexualized and enjoy it is to be viewed as a narcissist.

References

American Psychological Association Task Force. (2007). *Report of the APA task force on the sexualization of girls.* American Psychological Association. Retrieved from http://www.apa.org/pi/wpo/sexualization.html

Abbey, A. (1982). Sex differences in attributes for friendly behavior: Do males misperceive females' friendliness? *Journal of Personality and Social Psychology, (42)*5, 830–838.

Brick, S. (2012). The I'm so beautiful backlash: In yesterday's Mail, Samantha Brick claimed other women loathe her for being too attractive. It provoked a worldwide internet storm. Here she says: This bile just proves I'm right. *Daily Mail.* Retrieved from http://www.dailymail.co.uk/femail/article-2124782/Samantha-Brick-says-backlash-bile-yesterdays-Daily-Mail-proves-shes-right.html

Brooks, R. (2006). No escaping sexualization of young girls: With JonBenet back in the headlines, it's hard for a parent to avoid paranoia. *Los Angeles Times.* Retrieved from http://wwwlatimes.com

Cabrera, Y. (2007, November 15). Too sexy too soon? *The Orange County Register.* Retrieved from http://www.ocregister.com/column/kilbourne-says-parents

Calogero, R. M., Davis, W. N., & Thompson, J. K. (2005). The role of self-objectification in the experience of women with eating disorders. *Sex Roles, 52*(1–2), 43–50.

Cross, G. (2004). *The cute and the cool: Wondrous innocence and modern American children's culture.* Oxford, UK: Oxford University Press.

Crowley, M. (2006). No strings attached sex: Teen girls are buying into the sleaze we're selling. *Reader's Digest* online. Retrieved from http://www.rd.com/print

Curry, C. (2012). "Too pretty" columnist Samantha Brick ridiculed. ABC News. Retrieved from http://abcnews.go.com/US/columnist-samantha-brick-ridiculed-complaining-women-hate-beauty/story?id=16062169#.T3s4i79ST80

Dragon, G. (2010). The curse of being a beautiful woman. *Hall of the Black Dragon: Online Magazine for the Modern Gentleman.* Retrieved from http://halloftheblackdragon.com/reel/the-curse-of-being-a-beautiful-woman
Durham, M. G. (2008). *The Lolita effect: The media sexualization of girls and what we can do about it.* Woodstock, NY: Overlook Press.
Egan, R. D., & Hawkes, G. (2007). Producing the prurient through the pedagogy of purity: Childhood sexuality and the social purity movement. *Journal of Historical Sociology, 20*(4), 443–461.
Egan, R. D., & Hawkes, G. (2008). Girls, sexuality and the strange carnalities of advertisements: Deconstructing the discourse of corporate paedophilia. *Australian Feminist Studies, 23*(57), 307–322.
Eisenberg, E. M., & Goodall, H. L. Jr. (2004). *Organizational communication: Balancing creativity and constraint* (4th ed.). Boston, MA: Bedford/St. Martin's.
Firestone, S. (1970). *The dialectic of sex: The case for feminist revolution.* New York, NY: William Morrow.
Foley, R. (2013). All-male Iowa Supreme Court rules firing of woman for being too attractive was legal. *AP.* Retrieved from http://www.huffingtonpost.com/2013/07/12/iowa-supreme-court-attractive-woman-firing_n_3586861.html.
Fredrickson, B. L., & Roberts, T. A. (1997). Objectification theory: Towards understanding women's lived experiences and mental health risks. *Psychology of Women Quarterly, 21,* 173–206.
Freeman, H. (2012). The *Mail* simply threw Samantha Brick to the wolves. *The Guardian.* Retrieved from. http://www.theguardian.com/commentisfree/2012/apr/04/samantha-brick-thrown-to-wolves
Garner, M. (2012). The missing link: The sexualization of culture and men. *Gender and Education, 24*(3), 325–331.
Hall, E. T. (1959). *The silent language.* New York, NY: Anchor.
Hansen, G. L. (1985). Dating jealousy among college students. *Sex Roles, 12*(7–8), 713–721.
Haug, F., Andresen, S., Bünz-Elfferding, A., Hauser, K., Lang, U., Laudan, M., Lüdemann, M., Meir, U., Nemitz, B., Niehoff, E., Prinz, R., Räthzel, N., Scheu, M., & Thomas, C., (1999). *Female sexualization: A collective work of memory.* E. Carter (Trans.). New York, NY: Verso.
Hendrick, S. S., & Hendrick, C. (1992). *Romantic love.* Thousand Oaks, CA: Sage.
Henley, N. M. (1977). *Body politics: Power, sex, and nonverbal communication.* Englewood Cliffs, NJ: Prentice-Hall.
Holloway, W. (1984). Gender difference and the production of subjectivity. In J. U. Henriques, C. Venn, & V. Walkerdine (Eds.), *Changing the subject* (pp. 223–261). London: Methuen.
Jaggar, A. M. (1997). Love and knowledge: Emotion in feminist epistemology. In D. T. Myers (Ed.), *Feminist social thought: A reader* (pp. 384–405). New York, NY: Routledge.
LeMoncheck, L. (1997). *Loose women, lecherous men: A feminist philosophy of sex.* New York, NY: Oxford University Press.
Levin, D. (2005). So sexy so soon: The sexualization of childhood. In S. Olfman (Ed.), *Childhood lost: How American culture is failing our kids* (pp. 137–154). New York, NY: Praeger.
MacKinnon, C. A. (1987). *Feminism unmodified: Discourses on life and law.* Cambridge, MA: Harvard University Press.
MacRae, F., & Sears, F. (2007, February 20). The little girls sexualized at the age of five. *Mail Online.* Retrieved from http://www.dailymail.co.uk/news/article-437343/The-little-girls-sexualised-age-five.html
McPherson, K. (2006). Is childhood becoming oversexed? *Cycnet Features.* Retrieved from http://www.Cyc-net.org/features/ft-oversexed.html
McRobbie, A. (1991). *Feminism in youth culture.* London: Macmillan.
Meritor Savings Bank v. Vinson, 477 U.S. 57 (1986).
Messman, S. J., Canary, D. J., & Hause, K. S. (2000). Motives to remain platonic, equity, and the use of maintenance strategies in opposite-sex friendships. *Journal of Social and Personal Relationships, 17*(1), 67–94.
Miller, G. R., & Steinberg, M. (1975). *Between people: A new analysis of interpersonal communication.* Chicago, IL: Science Research Associates.
Monsour, M. (1996). Communication and cross-sex friendships across the life-cycle: A review of the literature. In B. Burleson (Ed.), *Communication yearbook 20* (pp. 375–414). Thousand Oaks, CA: Sage.
Nussbaum, M.C. (1999). *Sex and social justice.* New York, NY: Oxford University Press.
Paglia, C. (1992). *Sex, art, and American culture.* New York, NY: Vintage.
Paglia, C. (1994). *Vamps and tramps: New essays.* New York, NY: Vintage.
Ranney, F. J. (1998). *Posner on legal texts: Law, literature, (economics), and "welcome harassment." College Literature, 25*(1), 163–183.
Reynolds, G. (2010). What are the physical signs a man is attracted to a woman? Body language clues to look for. Sooper Articles. Retrieved from http://sooperarticle.com/relationship-articles/dating-flirting
Robinson, K. (2005). Childhood and sexuality: Adult constructions and silenced children. In J. Mason & T. Fattore (Eds.), *Children taken seriously: In theory, policy and practice* (pp. 66–79). London: Jessica Kingsley
Rubin, L. (1985). *Just friends.* New York, NY: Harper & Row.

Rush, E., & La Nauze, A. (2006). Corporate paedophilia: Sexualisation of children in Australia. Canberra: The Australia Institute.

Spigel, L. (1993). Seducing the innocent: Childhood and television in postwar America. In W. W. Solomon & R. W. McChesney (Eds.), *Ruthless criticism: New perspectives in U.S. communication history* (pp. 132–152). Minneapolis, MN: University of Minnesota Press.

Strauss, E. M. (2013). Iowa woman fired for being attractive looks back and moves on. ABC.com. Retrieved from http:// http://abcnews.go.com/Business/iowa-woman-fired-attractive-back-moves/story?id=19851803

Strelan, P., & Hargreaves, D. (2005). Women who objectify other women: The vicious circle of objectification? *Sex Roles, 52*(9–10), 707–712.

Swain, S. O. (1992). Men's friendships with women: Intimacy, sexual boundaries, and the informant role. In P. M. Nardi (Ed.), *Men's frienships: Vol 2. Research on men and masculinities* (pp. 153–172). Newbury Park, CA: Sage.

Szasz, T. (1974). *Ceremonial chemistry: The ritual persecution of drugs, addicts, and pushers.* Garden City, NY: Anchor.

Szasz, T. (1996). *The meaning of mind: Language, morality, and neuroscience.* Westport, CT: Praeger.

Trenholm, S., & Jensen, A. (2011). *Interpersonal communication* (7th.ed). New York, NY: Oxford University Press.

Vaes, J., Paladino, P., & Puvia E. (2011). Are sexualized women complete human beings? Why men and women dehumanize sexually objectified women. *European Journal of Social Psychology, 41*(6), 774–785.

Watson, L., & Reilly, J. (2012). Samantha Brick the spoof sensation: *Daily Mail* writer has gone viral after her controversial "why do women hate me for being beautiful?" article. *Daily Mail.* Retrieved from http://www.dailymail.co.uk/femail/article-2125138/Samantha-Brick-Daily-Mail-writer-goes-viral-controversial-Im-beautiful article.html?ito=feeds-newsxml

West, L. (2012). Yes, Samantha Brick is obnoxious, but the *Daily Mail* is trolling us all. *Jezebel.* Retrieved from http://jezebel.com/5898848/yes-samantha-brick-is-obnoxious-but-the-daily-mail-is-trolling-us-all

Wood, J. T. (2005). *Gendered lives: Communication, gender, & culture* (6th ed.). Belmont, CA: Wadsworth.

FOUR

Didn't Expect You to Hurt Me This Way: A Typology of Hurtful Events in Dating and Marital Relationships

Eletra S. Gilchrist-Petty

Being in a committed romantic relationship can be either one of the most rewarding or disappointingly hurtful human experiences. As in any intimate relationship, expectations for romantic partners develop over time as people observe the other party's behaviors in different situations and begin to anticipate how they will act in the future (Cohen, 2010). People assume that their romantic partners will treat them well, and when partners' actions deviate from this expectation their actions are often unexpected and hurtful (Leary, Springer, Negel, Ansell, & Evans, 1998). Such deviations are referred to as relationship transgressions/hurtful events,[1] and may include anything from exhibiting a lack of sensitivity to committing sexual infidelity (Metts & Cupach, 2007).

According to Expectancy Violation Theory (Burgoon, 1993), relationship transgressions are hurtful because they violate the rules and expectations to which romantic partners hold one another accountable. Such violations typically produce some level of relationship uncertainty or dissonance in the injured party. To cope with this uncertainty, a hurt partner undergoes a sequence of interpreting the event and evaluating the relationship (Afifi & Metts, 1998). Different factors may contribute to how couples experience this sequence, as well as to the transgressions' ultimate impact on the relationship. A major factor impacting how partners perceive relationship transgressions is the relationship's status. For example, previous research indicates that more committed partners, such as those who are married, may be more inclined to reconcile after a transgression (Finkel, Rusbult, Kumashiro, & Hannon, 2002; Wong & Fellows, 2005). Other differences between married and dating couples that may affect how couples experience and respond to transgressions include evidence that dating couples report higher satisfaction in their relationship (Sabatelli, 1988) and, in some conditions, exhibit healthier forgiveness strategies (Sheldon, Gilchrist-Petty, & Lessley, 2014), while married couples report higher overall relationship quality (Brown & Bulanda, 2008) and exhibit more tendencies to facilitate disclosure of emotions, sustain personal development, and help partners with task demands (Roisman, Clausell, Holland, Fortuna, & Elieff, 2008). Research continues in its quest to better understand the hurtful events that confront dating and married couples, and this chapter aims to add to this growing line of dark side research. Thus, through a discussion of how relationship partners classify hurtful events from an Expectancy Violation theoretical perspective, this chapter examines the similarities and differences between how dating and married couples experience relationship transgressions.

Classifications of Relationship Transgressions/Hurtful Events

Romantic relationship transgressions may be defined as any situation where one partner violates the implicit or explicit "rules" for how to behave in the relationship (Metts & Cupach, 2007). Several researchers have explored the various types and rates of transgressions in romantic relationships. For example, through content analysis, Cameron, Ross, and Holmes (2002) found the most common type of transgression in college students' dating relationships to be broken promises, followed by overreacting to the victim's behavior, inconsiderate behavior, violating the victim's desired level of intimacy, neglecting the victim, threat of infidelity, actual infidelity, verbal aggression toward the victim, unwarranted disagreement, and violent behavior toward the victim. Subsequent to Cameron et al., Feeney (2004) proposed the following five categories of hurtful events in romantic relationships: active disassociation (rejection), passive disassociation (being ignored), criticism, infidelity, and deception. Building off Feeney's work, Metts and Cupach (2007) suggested nine categories of romantic relationship transgressions, including inappropriate interaction, lack of sensitivity, extra relational involvement, relational threat confounded with deception, disregard for primary relationship, abrupt termination, broken promises/rule violations, deception, and abuse.

Some researchers have distinguished dating and married couple types when identifying categories of hurtful events. For example, Roloff, Soule, and Carey (2001) asked undergraduate students to recall transgressions committed against them by dating partners. The most frequently reported transgressions among the dating couples were romantic infidelity, lack of openness/honesty, inconsiderate/insensitive behavior, dominating behavior, and considering the relationship to be of low priority. Harvey (2004), on the other hand, surveyed long-term married couples and found that the most often reported transgression was mismanagement of money, followed by parenting disagreements, health-related traumas, alcohol and drug dependence, disagreements about in-laws, verbal aggression, personality differences, infidelity, anger, deception, neglect, unilateral decisions, and public embarrassment.

The various typologies offered by previous research suggest that couples experience a variety of hurtful events. While the typologies may differ to some extent, all of them reference events that violate relationship expectations, as noted by Expectancy Violation Theory (Burgoon, 1993). The following section explores differences in how married and dating couples experience deception and infidelity, which are two major types of transgressions that, according to Bachman and Guerrero (2006b), are the most commonly mentioned among relationship partners.

Deception

Trust is a critical part of romantic relationships, as it encourages openness and pro-social relationship maintenance strategies (Kalbfleisch, 2001). Despite this, around 90% of people can recall lying to their romantic partner about an important matter (Knox, Schacht, Holt, & Turner, 1993), often with positive intentions. For example, Boon and McLeod (2001) found that a majority of participants in their study thought it better to mislead a partner than to tell the truth in order to protect their partner's feelings. Despite these positive intentions, partners who uncover a deceptive act may experience increased uncertainty about their relationship. According to Planalp and Honeycutt (1985), this increased uncertainty can strongly affect beliefs about the honesty of a romantic partner, which may lead to negative emotional responses, communication patterns of avoidance, and consequences for relationships, such as termination.

Lying to a romantic partner is a common form of deception. Research by Metts (1989) investigated lying in dating and married couples as ranging from overt to covert misrepresentations of information. For dating couples, overt lies were found to be more common, and avoiding relational trauma was significantly more often endorsed as a reason for lying. This is likely because partners in these developing relationships are still attempting to establish relational parameters, and are actively trying to get to know one another. Requests for information make it more difficult to conceal facts about oneself, making falsification seem a better option than avoiding a question or disclosing something that might harm a relationship in its early stages. As for married couples, covert lies (omissions) were found to be more common, likely because married couples already know each other very well, making falsification of information more difficult. Married couples also significantly more often endorsed protecting a partner's face/self-esteem as their reason for lying. Regardless of the reasons previous research has identified for lying, deception is likely to reduce trust within a relationship (Boon & McLeod, 2001). Trust is especially reduced when a relationship partner commits infidelity, which is explored in the next section.

Infidelity

Sexual exclusivity is typically an inherent condition of any committed romantic relationship. Because staying in line with this type of expectation is so crucial to the continuance of the relationship, infidelity may be one of the most uncertainty-producing transgressions that a couple can experience. As evidence suggests that relational satisfaction is a significant predictor of whether one commits relational infidelity (Treas & Giesen, 2000), one might expect infidelity to be less of a problem amongst dating couples, who are more likely to be going through a "honeymoon" stage and to express satisfaction in their relationship (McAnulty & Brineman, 2007; Sabatelli, 1988). Nevertheless, research indicates that sexual unfaithfulness is more common for unmarried couples (Allen & Baucom, 2006; Treas & Giesen, 2000). In a survey of 618 respondents, around 40% reported at least one instance of sexual infidelity in a current or previous serious dating relationship (Wiederman & Hurd, 1999) while in a separate survey of 1,717 married respondents, 15.5% reported sexual infidelity (Treas & Giesen, 2000). Suggested reasons for this divide include lower levels of formal commitment in dating couples and that cheating on one's spouse is typically considered a greater offense than cheating on a dating partner (McAnulty & Brineman, 2007).

In a study comparing sexual unfaithfulness in married and dating couples, Allen and Baucom (2006) asked both married and dating individuals about their reasons for engaging in extradyadic (outside of the relationship) affairs. Married individuals were more likely than dating individuals to express feeling neglected and lonely in the primary relationship, love for the extradyadic partner, and needing a self-esteem boost as their reasons for sexual unfaithfulness. Though married and dating partners expressed similar levels of remorse after an affair, dating individuals reported significantly lower levels of concern regarding hurting their relationship partner, being judged negatively by others, or feeling like the extradyadic involvement was inconsistent with their typical values and behaviors. Allen and Baucom concluded that married individuals may require higher levels of multiple motives to betray their spouses. Thus, while dating infidelity may be more common, it seems that marital infidelity may result in more severe emotional consequences and uncertainty.

Although deception and infidelity represent two salient types of transgressions that are commonly cited among relationship partners (Bachman & Guerrero, 2006b), they are by no means the only ones perceived as hurtful or detrimental to romantic relationships. The types of transgressions identified by previous research have ranged from mild cases of being inconsiderate of the relationship to more severe forms of violence (Cameron et al., 2002). Extant research has clearly established hurtful events as a dark side of romantic relationships, but the formulation of a typology of hurtful events for both dating and marital relationships is warranted to enhance our growing body of knowledge in this area. Toward this end, the following questions are posed:

> RQ_1: What hurtful events do couples experience?
>
> RQ_2: To what extent are dating and married couples similar in the types of hurtful events they experience?
>
> RQ_3: To what extent are dating and married couples different in the types of hurtful events they experience?

Methods

Participants and Procedures

Study participants included a sample of 174 adults who were at least 19 years old and were either married or in a dating relationship. The participants were recruited via their affiliation with a medium-size research university in the southeastern United States. Of the 174 participants, 94 were married and 80 were dating.[2] Of those who were married, 76.6% were females, and 23.4% were males, whereas 75% of the dating respondents were females and 25% were males. The average age for the married participants was 38.90 (*SD* = 14.2) and 24.39 (*SD* = 6.3) for the dating. On average, the marriages have lasted for 12 years (*SD* = 10.58), and the romantic dating relationships have lasted for two years (*SD* = 1.82). In terms of race, the participants self-identified as 80% Caucasian, 15% African-American, 2% Hispanics, 1% Asian-American, and 2% other.

Participants were recruited via convenient and network sampling and were either surveyed during the regular meetings of a variety of communication and psychology classes, or they completed the survey at their convenience via SurveyMonkey.com. Participants first completed a consent form, followed by a demographic questionnaire in which they listed their age, sex, and relationship status. They were then instructed to describe (in as much detail as possible) one salient relational transgression that caused conflict either in their current or in a past relationship.

Data Analysis

With the data collected from the married and dating couples' written accounts of relationship transgressions, the goal was to categorize occurrences of hurtful events. Thus, quantitative content analysis was the appropriate data analysis procedure. According to Neuendorf (2002), *content analysis* is a "summarizing, quantitative analysis of messages that relies on the scientific method (including attention to objectivity-intersubjectivity, a priori design, reliability, validity, generalizability, replicability, and hypothesis testing) and is not limited as to the types of variables that may be measured or the context in which the messages are created or presented" (p. 10). With this in mind, content analysis is a prime method for quantitatively summarizing message content.

These participants' descriptions were coded independently by two researchers, as recommended by Frey, Botan, and Creps (2000) for enhancing the study's validity when performing content analysis. Researchers followed Bachman and Guerrero's (2006a) and Frey et al.'s (2000) coding procedure and initially unitized the written statements into *thematic units*, which is a deductive process that involves reading through written responses and identifying recurrent patterns. The thematic unitization resulted in the following 11 category classifications of hurtful events: (a) partner made a negative evaluation of value, worth, or quality, (b) partner violated a confidence, (c) partner was deceptive, (d) partner made unilateral decision, (e) partner spent money we did not have, (f) partner dated or flirted with someone else, (g) partner made an unfair accusation, (h) partner changed or forgot an important plan or occasion, (i) partner threatened physical harm, (j) partner abused alcohol or drugs, and (k) partner broke off the relationship (see Appendix A for a description of each hurtful event). Each written statement was then coded into one of the 11 categories of hurtful events. (Note: To be considered a legitimate category, the theme had to be present at least twice in either the dating or married couple's written responses, see Appendix B). There was high intercoder reliability (Cohen's kappa = .87).

Results

The primary objective of this study was to determine hurtful relationship events experienced by married and dating individuals. From the data analysis 11 total categories of hurtful events emerged. The top hurtful event experienced was *partner made a negative evaluation of value, worth, or quality* (married, n = 20; dating, n = 25). This theme reflected any type of verbal or nonverbal communication that is perceived as degrading to the relationship, disrespectful to the nature of the relationship, or reduces a partner's self-worth. For example, one male participant said, "She didn't think I was a good enough Christian."

Partner violated a confidence (married, n = 10; dating, n = 12) was the second most dominant hurtful message to emerge from the data and concerns revealing personal information about the person or relationship without prior approval, which can create mistrust within the relationship. A male respondent said he was hurt in his relationship when his female partner "was talking about our private business with others."

The third prevailing theme was *partner was deceptive* (married, n = 8; dating, n = 10), which encompasses dishonest behavior, such as lying, omissions, manipulation, or withholding relevant information. A sample statement reflective of this theme included: "My boyfriend lied about where he was going one night. He said he was going home. The next day when I saw him I looked through his phone when he left the room and saw texts between him and his friend about meeting up at a bar for drinks and to watch the game."

Partner made unilateral decision (married, n = 10; dating, n = 3) ranked fourth among the emerging themes and occurred when a relationship partner made decisions without communicating with or consulting the other partner. "They made a purchase without consulting me," one respondent said in reference to this theme.

Fifth, the data revealed another hurtful event occurred when the *partner spent money we did not have or used money in a selfish manner* (married, n = 10; dating, n = 1). This theme highlighted financial irresponsibility related to shared funds. For instance, a female participant said, "My partner gave his mother money to pay her rent when we needed the money."

Partner dated or flirted with someone else (married, n = 2; dating, n = 7) was the sixth-highest-ranking theme and referenced behavior that exhibits infidelity, perceived infidelity, or improper interaction with a person outside of the relationship. As recalled by one participant, "I had evidence that my partner had arranged a date on a dating website."

Concerning the seventh emerging theme, *partner made an unfair accusation* (married = 3, dating = 5), the participants cited words or actions that placed false blame on a relationship partner. A sample statement included, "She accused me of something when it wasn't true."

Eight, study participants said their *partner changed or forgot an important plan or occasion* (married = 2; dating = 4). This involved a lack of integrity or commitment related to a shared plan or event. "He forgot my birthday," a female respondent said reflective of this theme.

Partner threatened physical harm (married = 4; dating = 1), was the ninth category of occurrence. Here, the partner's words or actions conveyed a likelihood of physical harm or aggression. According to one female participant, "My husband yelled at me in the car and slammed his hand on the dashboard."

Rounding out the top ten was *partner abused alcohol or drugs* (married = 1; dating = 3). This theme referenced harmful behavior influenced by alcohol or drugs. One participant expressed hurt when her partner was "drinking all the time" and "he got arrested for drunk driving."

The last theme to emerge was *partner broke off the relationship* (married = 1; dating = 2), and it included behavior or communication from a partner that ultimately terminated the relationship. One participant recalled, "We were arguing in a very dramatic way over something I can't remember. In his anger he threw up the word 'divorce'..."

Discussion

Since the dawn of time, individuals have practiced coupling and engaged in romantic relationships. Love, support, trust, companionship, financial stability, intimacy and many other perks legitimize the value of romantic relationships. However, in spite of the positive attributes of romantic relationships, the dark side is also quite evident. Specifically, researchers have long understood that romantic relationships are riddled with hurtful events (e.g., Cameron et al., 2002; Feeney, 2004; Metts & Cupach, 2007), and this study's findings indeed support past research. Given the prevalence of relationship transgressions, a prime goal of this study was to distinguish the hurtful events experienced by dating and married couples. The first research question queried the types of hurtful events couples collectively report. Data from this study revealed couples identified 11 distinct hurtful events. The sheer quantity of typological hurtful events found in this study (i.e., 11) give further credence to Expectancy Violation Theory, which alleges individuals have certain expectations in relationships and when those expectations are violated hurt ensues (Burgoon, 1993). The typology of hurtful events that emerged in the study also supports previous research findings that suggest transgressions in romantic relationships are varied and multidimensional in scope (Cameron et al., 2002; Feeney, 2004; Metts & Cupach, 2007). The findings from this study are also meaningful because they indicate both similarities and differences in the types of hurtful events confronting dating and married couples.

The second research question probed the extent that dating and married couples are similar in the types of hurtful events they experience. The data indicated 7 of the 11 categories of hurtful events (i.e., approximately 64%) were experienced quite similarly among dating and married couples: violated a confidence, deception, made an unfair accusation, changed or forgot an important plan or occasion, threatened physical harm, abused alcohol or drugs, and broke off the relationship. Findings relative to RQ_2 suggest there is a mix of relationship transgressions that transcends relationship type, adding further credibility to previous scholars who found certain hurtful events impact both dating and marital couples (e.g., Metts, 1989).

While there are many similarities in the types of hurtful events confronted by dating and married couples, the data also revealed some striking differences in how the couple types experienced 4 of the 11 categories (i.e., approximately 36%): made a negative evaluation of worth, value, or quality; made a unilateral decision; spent money we did not have or used money in a selfish manner; and dated or flirted with someone else. Per these findings, in comparison to married couples, dating couples are more prone to experience comments that degrade or disrespect the relationship or the relational partner. Since Aristotle first introduced the term rhetoric (i.e., communication), we have known about the power of language. Our words have a remarkable ability to either build up or tear down an individual's character or self-esteem. The findings from this study suggest that dating couples engage in more destructive rhetoric than marital couples. We can only speculate the many reasons why demeaning language is seen more in dating couples: perhaps the message sender is not serious about the relationship; maybe the hurtful comment was meant more as a joke; or perchance the negative statement was designed to test the relationship. While understanding the sender's motive is beyond the scope of this study, it is clear that dating partners are more bothered by demeaning messages than married couples.

According to the data, dating couples are also more likely than married couples to be hurt by their partners flirting or dating someone else. This finding is very consistent with previous research findings, including Cameron et al. (2002) who found that infidelity or the threat of infidelity were common transgressions in college students' dating relationships. Likewise, this finding meshes with former research by Allen and Baucom (2006) and Treas and Giesen (2000) that indicate unfaithfulness is more common for unmarried couples. Perhaps flirting and infidelity are perceived as more hurtful among dating couples because, as posited by

McAnulty and Brineman (2007), there is a lack of formal commitment in dating couples. This lack of formal commitment could also lead dating partners to be more insecure in their relationship and overly sensitive to any type of flirtatious behavior or communication that could potentially trigger infidelity.

Compared to dating couples, married couples are more likely to be hurt by their partner making unilateral decisions. This finding supports previous research by Harvey (2004), who found married couples readily reported unilateral decision-making as a major transgression. This finding is not surprising given that there is generally a presumption of oneness, interdependency, and togetherness among most traditionally married couples (Fitzpatrick, 1988). So, if most married persons have a natural presumption that they make major decisions together, it is very plausible that they would be hurt by their spouse making a decision without consulting them. Dating couples, in contrast, may be less likely to view their partner making a unilateral decision as hurtful because they have not entered into a contractual agreement stating they are a legally bound couple. Instead, many dating couples have separate residencies, bank accounts, and responsibilities; therefore, their actions have little impact on their partner's life.

The results also indicated that married couples perceive the misuse of money as a hurtful event more often than dating couples. This result is plausible considering that multiple studies have found that financial issues are among the most common sources of married couples' disagreements (Betcher, & Macauley, 1990; Chatzky, 2007; Harvey, 2004; Papp, Cummings, & Goeke-Morey, 2009). It is customary for many married couples to share residencies, children, and expenses; thus, there is often a direct or implied expectation that finances will be combined and shared, or at least spent with the welfare of both parties in mind. Therefore, when a spouse misuses money it is seen as disrespectful and an egregious violation of marital norms. On the other hand, dating couples may disagree minimally about money because they less often share the same financial pool, joint expenses, or are equally dependent on each other for financial security.

Conclusions

Given that the sample was rather homogeneous with most participants being White, female, American, and students, the present research findings cannot be generalized to all populations. Additionally, the study relied on self-reports, which can skew data objectivity. For example, participants might not be totally honest in their answers, or they may not accurately recall an experienced transgression. To rectify these limitations, future research should aim for a larger, more diversified sample. It may also be prudent for additional studies to survey both relationship partners to ensure data accuracy, as well as query motives prompting the transgressions. Future research may add breadth and depth to this study's findings by exploring how relational partners respond to hurtful events by either exercising forgiveness and preserving the relationship or transitioning following the relationship's dissolution.

In spite of the obvious limitations, this study is significant in that it advances our understanding of relational transgressions as a salient dark side issue confronting both married and dating couples. From the respondents' self-reports, 11 categories of hurtful events emerged suggesting couples experience a mix of relational transgressions. The typology of hurtful events is also indicative of Expectancy Violation Theory, which contends that individuals have relational expectations that are perceived as hurtful when violated (Burgoon, 1993). Furthermore, this study has clearly illustrated that some transgressions are experienced almost equally among dating and married persons (i.e., violating confidence, deception, unfair accusations, changing or forgetting an important plan or occasion, threatening physical harm, abusing alcohol or drugs, and breaking off the relationship). However, the data also unveiled some differences regarding the hurtful events experienced by the two groups. Transgressions that are more specific to married than dating couples include making unilateral decisions and misusing money. On the other hand, compared to married couples, dating partners experience more hurtful events relative to making negative evaluations of worth, value, or quality and dating or flirting with someone else. Taken together, findings from this study suggest dating and married individuals must be prepared to potentially mitigate a variety of hurtful events—many of which are experienced quite similarly by the two groups, while some are vastly different.

Notes

1 This chapter uses relationship transgressions and hurtful events interchangeably.
2 The sample included both heterosexual and same-sex relationship partners.

References

Afifi, W. A., & Metts, S. (1998). Characteristics and consequences of expectation violations in close relationships. *Journal of Social and Personal Relationships, 15*(3), 365–392.

Allen, E. S., & Baucom, D. H. (2006). Dating, marital, and hypothetical extradyadic involvements: How do they compare? *The Journal of Sex Research, 43*, 307–317.

Bachman, G. F., & Guerrero, L. K. (2006a). Forgiveness, apology, and communicative responses to hurtful events. *Communication Reports, 19*, 45–56.

Bachman, G. F, & Guerrero, L. K. (2006b). Relational quality and communicative responses following hurtful events in dating relationships: An expectancy violations analysis. *Journal of Social and Personal Relationships, 23*(6), 943–963.

Betcher, W., *&* Macauley, R. (1990). *The seven basic quarrels of marriage: Recognize, defuse, negotiate, and resolve your conflicts*. New York, NY: Villard Books.

Boon, S. D., & McLeod, B. A. (2001). Deception in romantic relationships: Subjective estimates of success at deceiving and attitudes toward deception. *Journal of Social and Personal Relationships, 18*(4), 463–476. doi:10.1177/0265407501184002

Brown, S. L., & Bulanda, J. R. (2008). Relationship violence in young adulthood: A comparison of daters, cohabitors, and marrieds. *Social Science Research, 37*(1), 73–87.

Burgoon, J. K. (1993). Interpersonal expectations, expectancy violations, and emotional communication. *Journal of Language and Social Psychology, 12*(1–2), 30–48.

Cameron, J. J., Ross, M., & Holmes, J. G. (2002). Loving the one you hurt: Positive effects of recounting a transgression against an intimate partner. *Journal of Experimental Social Psychology, 38*(3), 307–314.

Chatzky, J. (2007). 5 smart tips to manage money with your honey. *Retrieved from* http://www.msnbc.msn.com/id/20108870/from/ET.

Cohen, E. L. (2010). Expectancy violations in relationships with friends and media figures. *Communication Research Reports, 27*(2), 97–111.

Feeney, J. A. (2004). Hurt feelings in couple relationships: Towards integrative models of the negative effects of hurtful events. *Journal of Social and Personal Relationships, 21*(4), 487–508.

Finkel, E. J., Rusbult, C. E., Kumashiro, M., & Hannon, P. A. (2002). Dealing with betrayal in close relationships: Does commitment promote forgiveness? *Journal of Personality and Social Psychology, 82*(6), 956–974. doi:10.1037/0022-3514.82.6.956

Fitzpatrick, M. A. (1988). *Between husbands and wives: Communication in marriage.* Newbury Park, CA: Sage.

Frey, L., Botan, C., & Kreps, G. (2000). *Investigating communication: An introduction to research methods* (2nd ed.). Boston, MA: Allyn & Bacon.

Harvey, J. (2004). *Trauma and recovery strategies across the lifespan of long-term married couples.* Phoenix, AZ: Arizona State University West Press.

Kalbfleisch, P. J. (2001). Deceptive message intent and relational quality. *Journal of Language & Social Psychology, 20*(1/2), 214–230.

Knox, D., Schacht, C., Holt, J., & Turner, J. (1993). Sexual lies among university students. *College Student Journal, 27*, 269–272.

Leary, M. R., Springer, C., Negel, L., Ansell, E., & Evans, K. (1998). The causes, phenomenology, and consequences of hurt feelings. *Journal of Personality and Social Psychology, 74*, 1225–1237.

McAnulty, R. D., & Brineman, J. M. (2007). Infidelity in dating relationships. *Annual Review of Sex Research, 18*(1), 94–114.

Metts, S. (1989). An exploratory investigation of deception in close relationships. *Journal of Social and Personal Relationships, 6*(2), 159–179. doi:10.1177/026540758900600202

Metts, S., & Cupach, W. R. (2007). Responses to relational transgressions: Hurt, anger, and sometimes forgiveness. In B. H. Spitzberg & W. R. Cupach (Eds.), *The dark side of interpersonal communication* (2nd ed; pp. 243–274). New York, NY: Routledge.

Neuendorf, K. A. (2002). *The content analysis guidebook*. Thousand Oaks, CA: Sage.

Papp, L. M., Cummings, E. M., & Goeke-Morey, M. C. (2009). For richer, for poorer: Money as a topic of marital conflict in the home. *Family Relations, 58*, 91–103.

Planalp, S., & Honeycutt, J. M. (1985). Events that increase uncertainty in personal relationships. *Human Communication Research, 11*, 593–604.

Roisman, G. I., Clausell, E., Holland, A., Fortuna, K., & Elieff, C. (2008). Adult romantic relationships as contexts of human development: A multimethod comparison of same-sex couples with opposite-sex dating, engaged, and married dyads. *Developmental Psychology, 44*(1), 91–101.

Roloff, M. E., Soule, K. P., & Carey, C. M. (2001). Reasons for remaining in a relationships and responses to relational transactions. *Journal of Social and Personal Relationships, 18*, 362–385.

Sabatelli, R. M. (1988). Exploring relationship satisfaction: A social exchange perspective on the interdependence between theory, research, and practice. *Family Relations, 37*, 217–222.

Sheldon, P., Gilchrist-Petty, E., & Lessley, J. A. (2014). You did what? The relationship between forgiveness tendency, communication of forgiveness, and relationship satisfaction in married and dating couples. *Communication Reports, 27*(2), 78–90. doi:10.1080/08934215.2014.902486

Treas, J., & Giesen, D. (2000). Sexual infidelity among married and cohabiting Americans. *Journal of Marriage and Family, 62*(1), 48–60.

Wiederman, M. W., & Hurd, C. (1999). Extradyadic involvement during dating. *Journal of Social and Personal Relationships, 16*(2), 265–274.

Wong, N., & Fellows, K. (2005). Coping with the unexpected: Perceptions of and responses to a relational expectancy violation as a function of intimacy and commitment. *Conference Papers – International Communication Association*, 1–33.

Appendix A

Typology and Description of Hurtful Events

1. **Partner Made a Negative Evaluation of Worth, Value, or Quality**: any action that is perceived as degrading to the relationship, disrespectful of the nature of the relationship, or reduces a partner's self-worth.
2. **Partner Violated a Confidence**: behavior that reveals personal information without prior approval or that creates mistrust.
3. **Partner Was Deceptive**: dishonest behavior, such as lying, omission, manipulation, or withholding relevant information.
4. **Partner Made a Unilateral Decision**: making a decision without communicating with or consulting the other relationship partner prior to that decision.
5. **Partner Spent Money We Did Not Have or Used Money in a Selfish Manner**: financial irresponsibility related to shared funds.
6. **Partner Dated or Flirted with Someone Else**: behavior that exhibits infidelity, perceived infidelity, or improper interaction with a person outside of the relationship.

7. **Partner Made an Unfair Accusation**: words or actions that place false blame on a partner in a relationship.
8. **Partner Changed or Forgot an Important Plan or Occasion**: a lack of integrity or commitment related to a shared plan or event.
9. **Partner Threatened Physical Harm**: words or actions that convey a likelihood of physical harm or aggression.
10. **Partner Abused Alcohol or Drug**: harmful behavior influenced by alcohol or drugs.
11. **Partner Broke Off the Relationship**: words or behaviors that terminate the relationship.

Appendix B

Frequency of Hurtful Events Experienced by Married and Dating Couples

Hurtful Event	**Married**	**Dating**
Partner made a negative evaluation of worth, value, or quality.	20	25
Partner violated a confidence	10	12
Partner was deceptive.	8	10
Partner made a unilateral decision.	10	3
Partner spent money we did not have or used money in a selfish manner	10	1
Partner dated or flirted with someone else.	2	7
Partner made an unfair accusation.	3	5
Partner changed or forgot an important plan or occasion.	2	4
Partner threatened physical harm.	4	1
Partner abused alcohol or drugs.	1	3
Partner broke off the relationship.	1	2

FIVE

Student-Sourced Verbal Aggression on Teaching Evaluations

Felecia F. Jordan Jackson

Verbal aggression is a communication construct that has been researched in many different contexts. Described as a destructive form of aggression, Infante (1987) defined verbal aggression as "the tendency to attack the self-concepts of individuals instead of, or in addition to, their positions on topics of communication" (p. 164). Its counterpart, argumentativeness, is described as a constructive form of aggression in which the issue, rather than the person, is the source of attack. The relatively vast amount of research done on these constructs has generally concluded that use of verbal aggression and *nonaffirming* communication (e.g., frowning, exhibiting an expressionless or tense countenance, little-to-no eye contact, sending messages of avoidance or unapproachability, etc.; Infante, Rancer, & Jordan, 1996) results in negative outcomes for the target of the messages and for the relationship between the interactants.

Much of the research on verbal aggression has focused on interpersonal contexts such as friendships, dating, familial, and marital relationships. For dating couples, self-reports of verbal aggression were found to be negatively related to relational satisfaction (Venable & Martin, 1997). Rancer and Avtgis (2006) assert that "within the family context, the consequences of verbal aggressiveness are most often destructive..." (p. 92). In marital relationships, marital satisfaction has been found to be inversely related to the husband's increased verbal aggression in husband and wife dyads (Payne & Chandler-Sabourin, 1990). Similar results were found among siblings. According to Martin, Anderson, Burant, and Weber (1997), siblings reported lower relational satisfaction when the use of verbal aggression was prevalent among them. In regard to the effects of parents' use of verbal aggression within the family and its effect on children, Martin et al. (1997) found children's use of verbal aggression and argumentativeness was positively related to their mother's use of these communication behaviors, regardless of the child's sex.

In instructional contexts, investigations have focused on perceptions and/or effects of the instructor as aggressor. Myers (2002) argued, for example, that studies in the instructional setting can be divided into four areas: (1) those that explore perceived instructor argumentativeness, (2) those that explore perceived instructor verbal aggression, (3) those that explore a combination of perceived instructor argumentativeness along with verbal aggressiveness, and (4) those that explore instructor use of messages that are verbally aggressive. Results of such research indicate that students of teachers/instructors who are perceived as using verbally aggressive messages report lower affective learning (Myers & Knox, 2000) and lower motivation, particularly when the aggression was in the form of character, competence, and background attacks along with threats, disparaging comments, and nonverbal expressions of aggression (Myers & Rocca, 2000). Perceptions of credibility were also lower when students perceived their instructors as being verbally aggressive (Tevan, 2001).

In the current examination, student-sourced verbal aggression toward the instructor in college classrooms was the focus. Specifically, the comment sections of student evaluations of instructors were reviewed to determine if students use this venue to aggress upon instructors, and if so, what form the verbal aggression takes (e.g., character attacks, competence attacks, profanity, etc.). Further, professor demographics were examined to determine who is more likely to be the recipient of these types of messages.

Verbal Aggression in the Classroom

Verbal and nonverbal immediacy are among the more highly researched variables associated with perceived instructor competence and student learning. Teacher immediacy behaviors are described as verbal and nonverbal messages that indicate warmth, involvement, and physical or psychological closeness (Andersen, 1999). While immediacy behaviors are more conducive to a positive classroom climate, verbal aggressiveness is significantly less so. Indeed, Rocca and McCroskey (1999) found a significant negative relationship between teacher immediacy and verbal aggression. Myers and Rocca (2000) found that instructors rated by students as high in use of verbally aggressive messages in the classroom were also perceived as facilitating a less supportive classroom climate than instructors rated moderate or low in verbal aggressiveness. Further, Myers (2002) reported that teachers perceived as exhibiting a combination of high argumentativeness and low verbal aggression "have students who are highly motivated, evaluate their instructors highly, report cognitive learning, and are highly satisfied" (p. 117). The fact that most of the studies on verbal aggression in the instructional setting take the perspective of students' perception of the instructor should not be ignored. One of the few studies that has examined student verbal aggression in the classroom was forwarded by Schrodt (2003) who reasoned that the finding that some students tend to see any verbal attack (even those which are argumentative and focus the attack on the issue) as a personal attack may be attributed to the students themselves being high in verbal aggression and low in perceived self-worth.

There is empirical evidence that humans use the cloak of anonymity to express personal, unpleasant, and/or dark communication messages. Joinson (2001) found, for example, that "[v]isually anonymous participants disclosed significantly more information about themselves than non-visually anonymous participants" (p. 177). In another study, Werner, Bumpus, and Rock (2010) suggested online anonymity impacts adolescents' aggressive behavior. In a vein related to instructional communication, many institutions of higher education afford students anonymity when requesting feedback and assessments of student perceptions of instruction. In some cases these evaluations are the sole measurement of teacher effectiveness. This process has been met with criticism for a number of reasons including the notion that "student evaluations promote sucking up to customers often at the expense of teaching effectiveness" and that they are biased, focusing on appearance (e.g., hair and beards), and their dislike for the subject (Schuman, 2014, para. 3).

The firsts step in this investigation was to examine student comments to their instructors as they pertain to the course; therefore, the following research questions were forwarded:

RQ1a: Do students use verbally aggressive comments in the qualitative section of teaching evaluations?

RQ1b: What types of verbally aggressive messages do students use?

Contextual Contributors to Student Verbally Aggressive Behavior

It is likely that few would argue that there are legitimate and mutable behaviors that may contribute to negative student perceptions of their classroom experience. In addition to verbal aggression and immediacy, a plethora of teacher character and classroom environmental factors have been found to contribute to students' perceived cognitive, affective, and behavioral intent learning, such as teacher enthusiasm (Mitchell, 2013) and classroom management (Boysen, 2012). In addition, immutable teacher variables may also contribute to students' dissatisfaction of teachers and yet lead to students' verbal aggression against the instructor during the course evaluation period. Some of these factors may include instructor sex, race/ethnicity, and course content (Hall & Sandler, 1982). Although much of the research in instructional communication settings and similar research by scholars in other disciplines has focused on how the classroom climate is unsupportive of female students, there is some evidence that (female) faculty may also experience the *chilly* (Hall & Sandler, 1982) and unsupportive climate that has been promoted. Interestingly, when asked, neither female nor minority students cited sex or race as a primary reason for the chilly and unsupportive classroom climate. It is a goal of this chapter to ascertain whether comments from students to and about instructors have any relationship to factors beyond the instructor's immediate control. To assess this behavior, the following research questions were posited:

RQ2: What is the race/ethnicity of the target teacher to whom the verbally aggressive comments are focused?

RQ3: What is the sex of the target teacher to whom the verbally aggressive comments are focused?

RQ4: What is the *nature* of the course being evaluated (e.g., course size, mode of presentation, and major/content area)?

One potential cause of verbal aggression, referred to as the skill deficiency explanation, is generally described as one's inability to form effective arguments due to lack of communication skill (or motivation) in arguing (Rancer & Avtgis, 2006). It may follow that if the students are unable (whether due to lack of skill on their part or the perceived lack of a positive classroom climate that facilitates open communication) and/or unwilling to express their frustrations with general or specific teacher behavior, they may see the evaluation as their only recourse or as a means of retaliation. This forum can be particularly scathing considering the weight these evaluations are often given in administrative decisions on merit, promotion, tenure, and retention. To ascertain whether these remarks target faculty during a particular point in their professional career (e.g., Teaching Assistant, Adjunct, Assistant Professor, Associate Professor, or Full Professor), the following research question was posited:

> **RQ5**: What is the professional status and/or rank of the recipients of student-sourced comments on the free-response portion of teacher evaluations?

Method

Demographic information of the instructor including, sex, race/ethnicity, rank/position/title, number of years in rank, age, along with contextual information such as course name and number, number enrolled, and type of course (e.g., STEM, Liberal studies, major/minor course, etc.) were requested of the instructor/participant. The written comments of two semesters (i.e., spring and fall) on teaching evaluations were obtained and examined in order to ascertain whether verbally aggressive messages were used. Each instance of these messages was categorized based on the types identified by scholars such as Infante and Wigley (1986), Infante (1987), and Kinney (1994). Rancer and Avtgis (2006) described the types of verbal aggressive (VA) messages as follows:

1. **Competence attacks** are "verbal attacks directed at another person's knowledge and/or ability to do something" (p. 21).
2. **Character attacks** are verbal messages which attack the person's character including their honor, nature, or integrity.
3. **Profanity** may "involve the use of obscene words, epithets, and vulgarities" (p. 22).
4. **Teasing** is considered to be a "playful form of verbal aggression" (p. 22) in which one is made fun of or is playfully mocked. It becomes verbal aggression when it "inflict[s] psychological harm and damage on the target" as a form of ridicule" (p. 22).
5. **Maledictions** are "verbally aggressive messages in which we wish someone harm" (p. 24).
6. **Threats** occur when one expresses intent to cause physical or psychological injury, pain, or hurt on another person.
7. **Nonverbal verbal aggression** "takes the form of speech-independent gestures" or emblems which are nonverbal behaviors used in place of verbal expression such as the American use of the peace sign or "displaying the middle finger" (p. 25).
8. **Personality attacks** are verbal attacks on another person's disposition, mood, temperament, temper, etc.
9. **Negative comparisons** are messages that compare the target negatively with someone or something else.
10. **Disconfirmations** are messages that negate another's existence or significance.

Sample

A convenience sample of faculty members from two predominantly White universities was targeted. The universities were both mid-to large-sized research universities, one in the South and one on the eastern coast of the United States. Both universities utilized the semester system to organize course offerings. Potential participants were asked to submit their student evaluations of teaching for any or all courses taught over the previous two years (i.e., 2012 through 2014). These years were chosen in order to make the data collection and analyses more manageable for this exploratory study. Voluntary participation via submission of the evaluations constituted informed consent. More than 50 potential male and female college instructors of various ranks, races/ethnicities, and ages received the request for participation, however, just 6 respondents provided the author with their evaluations for review and analyses. The 6 respondents yielded more than 1,500 teaching evaluations that were each examined for comments that could be classified as verbally aggressive. Because of the small participant sample size and subsequent low number of verbally aggressive comments that were identified, the results found here are presented as descriptive, are not generalizable, and should be considered anecdotal.

Data Analysis

To analyze the data, a content analysis based on a grounded theory approach was conducted. Specifically, the categories identified previously for the 10 types of verbally aggressive (VA) messages [see for example Infante & Wigley (1986), Infante (1987), Kinney (1994), Rancer & Avtgis (2006), and Infante, (1995)] were used to classify students' written comments for each professor. If there were comments that were not verbally aggressive in nature, those comments were not documented or categorized. If there was more than one verbally aggressive comment on one evaluation, each instance and type of verbally aggressive message was documented. Finally, comments were evaluated in order to ascertain whether they were directed toward the instructor or some other aspect of the course or institution that the student acknowledged was beyond the instructor's control, such as the time of day the course was taught, antiquated university facilities, or faulty equipment or software.

Results

The six participants provided a total of 1,538 individual teaching evaluations that fit the specified criteria for analysis (i.e., from students enrolled in courses taught over the previous two calendar years). The data represent 28 classes and the number of classes analyzed for each instructor ranged from 2 to 14 including courses taught during the spring, summer, and fall semesters. The sample was composed of one male (16.66%) and five female (83.33%) instructors. Two (33.33%) self-identified as White, three (50%) self-identified as Black, and one (16.66%) self identified as Hispanic. One (16.66%) was at the rank of Full Professor, four (66.66%) at Associate, and one (16.66%) at Assistant. All were tenured or tenure-earning full-time employees at their respective universities.

The courses represented ranged from sophomore (i.e., 200/2000 level) through graduate (i.e., 500/5000 to 600/6000) levels. The enrollment for the courses ranged from three to 197 students, with a median enrollment of 21 students. One course (3.57%) was offered exclusively online with the remaining 27 (96.42%) offered in the traditional face-to-face format.

The courses and instructors were representative of three Departments or Schools including English, Communication, and Recreation, Health, and Tourism.

Research Questions 1a

RQ1a was posed to determine whether students used teaching evaluations as a means to verbally attack instructors. Of the total number of student evaluations analyzed (i.e., 1,538), there were 45 instances of verbal aggression. Although the numbers were small (2.9%), there was evidence of verbal aggression in the evaluations of each of the six instructors represented in the sample.

Research Question 1b

RQ1b was asked to help determine the types of verbally aggressive messages students engaged in. In this sample, the most commonly used aggressive messages from among those forwarded by Infante (1987), Kinney (1994), and Rancer and Avtgis (2006) were: Competence Attacks (n = 14, 31.11%), Personality Attacks (n = 12, 26%), Character Attacks (n = 11, 24%), Disconfirmations (n = 2, 4.4%), Negative Comparisons (n = 5, 11%), and Profanity (n = 1, 2.2%). There were two additional instances of profanity written by students in the free response section, but they were clearly referring to the course and/or the larger institution and were not directed specifically toward the instructor.

Research Questions 2 and 3

Research Questions 2 and 3 were proposed to determine if the race or ethnic background and sex of the instructors, respectively, were associated with the type and number of verbally aggressive messages students used. Because of the small sample size, statistical analyses were not employed. Rather, these questions were operationalized by counting the number and types/categories of verbally aggressive messages for each ethnic/racial group and both sexes represented in the sample.

The results indicated that of the six professors, female professors received the most student-sourced VA messages with African-American females being the recipient of the most, followed by the Hispanic female. Specifically, two of the six professors self-identified as Black females (33.33% of the sample) and received 12 (26.66%) of the 45 VA messages. Two of the six professors self-identified as White females (33.33%) and received 25 (55.5%) of the 45 VA messages. One professor self-identified as a Hispanic (Latina) female (16.66%) and received 6 VA messages (13%), and one professor self-identified as a Black male (16.66%) and received 1 (2.22%) of the 45 VA message.

In terms of types or categories of VA messages received for each sex and racial/ethnic group, White females received more personality attacks (n = 8, 32%), competence attacks (n = 7, 28%), negative comparison (n = 5, 20%), and character (n = 4, 16%) than any other types of VA messages. They also received the greatest number of VA messages of any other racial/ethnic group overall. Most of the VA messages attacking Black female professors were in the category of competence attacks (n = 5, 41.66%). The Latina professor received more VA messages attacking her character (n = 3, 50%), than any other type of attack. The Black male instructor had one VA message, which attacked his character.

Research Questions 4 and 5:

Research Questions 4 and 5 were asked in order to determine if the nature of the course (e.g., size, mode of presentation, content/major) and the rank of the professor (e.g., Assistant, Associate, Full) affected the number and types of VA messages students sourced. The sample consisted of six professors representing three majors: English, Communication, and Sports and Recreation, and ranged in student size and mode from 3 (online) to 197 (face-to-face). The content also varied from public speaking to technology/computer design courses.

The (Black female) English professor is at the Full Professor rank and taught two courses (one undergraduate and one graduate) for a total of 55 students (or 50% of the students taught by the Black female professors) in this sample. This professor received 3 of the 12 (25%) VA messages received by Black female professors. These VA messages came from three separate students, all from the undergraduate course. No comments were offered in the graduate level course. All three of the messages were categorized as personality attacks.

The other Black female professor was at the Associate Professor rank, taught two courses and received 9 of the 12 (75%) VA messages that the Black female professors received and 20% of the 45 VA messages overall. One of the courses was at the undergraduate level from which she received 4 VA messages and one at the graduate level from which she received 5 VA messages.

Both White female professors were at the Associate Professor rank and taught in the communication content area. One of the professors in this demographic taught 14 of the 28 (50%) courses represented in the sample, which included 919 (59.7%) of the students represented. She received 10 (22.22%) of the 45 VA messages used overall. Additionally, the other White female taught five (17.85%) of the 28 courses, four at the undergraduate and one at the graduate level, representing 430 (27.95%) of the students. She received 15 (33.33%) of the 45 VA messages.

The Hispanic female also taught communication content. She was at the Assistant Professor rank and taught 72 (4.6%) of the students represented over three undergraduate courses and received 6 (13.33%) of the 45 VA messages overall.

Finally, the Black male was at the Associate Professor rank, taught 42 students in two undergraduate courses and received 1 (2.22%) VA message.

Discussion

The focus of this investigation was to determine if students utilize the free response or comment portions of student evaluations of teaching to verbally aggress against their instructors. The biggest limitation of this study was the small sample size in regards to the number of instructors represented, although over 1,500 individual evaluations were reviewed for evidence of inclusion of verbal aggression. It is evident that the answer to the core question of this investigation is "yes." All six of the faculty members who provided their evaluations for analyses showed evidence of student-sourced verbal attack over a two-year period. There appears to be specific types of messages that are most commonly used by students including competence, character, and personality attacks. It also appears that no one is immune as targets regardless of professor rank, mode of presentation, or content area. It should be noted, however, that at least for this sample, women as a whole and women of color received a disproportionate number of aggressive messages considering the number of courses and the number of students they teach.

Although this study did not target any particular demographic group, (e.g., sex or race/ethnicity) the implications of student-sourced verbal aggressive messages as it pertains to traditionally socially oppressed groups should be considered. Results of research examining student evaluations of instructors who are women and/or racial minorities give some indication that negative evaluations of these groups are more prominent than for White male faculty (Dukes & Victoria, 1989; Fries & McNinch, 2003). Pittman's (2010) investigation on the experiences of women faculty of color with White male students at a large predominantly White Midwestern research institution found via in-depth interviews with the women faculty of color that "White male students (1) challenged their authority, (2) questioned their teaching competency, and (3) disrespected their scholarly expertise" and that White male students engaged in behavior that was "threatening" and "intimidating" (p. 187). Further, Pittman offered the comments of a Black female faculty member who stated the following regarding White male behavior toward her both face-to-face and via email:

> They're snide, they'll sit with their arms crossed and they doodle and they sit right up in the front so that is definite passive-aggressive behavior. The tone sometimes in the e-mails they send, and it's kind of funny because it's the kind of things you don't even know how to express to other people. But you're like, if I was a White male you wouldn't dare write to me in that tone. (p. 188)

Pittman (2010) concluded that, based on the comments given by the women faculty of color in her study, White male students demonstrated a sense of entitlement to tell the women faculty of color how to manage their classroom which the faculty viewed as inappropriate, and, in some cases, aggressive. It was also clear that the faculty members associated these behaviors with White and/or male privilege, and these aggressive acts were not experienced in the classrooms of [White] male professors. Further, the effects of these behaviors were not only seen as inappropriate, but possibly having more far-reaching consequences such as lower teaching evaluations, which could affect career advancements with promotion and tenure.

Regardless of our stance on the appropriateness or inappropriateness of the procedures used to obtain student feedback, instructors should always be aware that as long as student dissatisfaction exists, there will be some anonymous outlet for them to express those feelings. According to Bolkan and Goodboy (2013), "Students need to believe that they have the power to make a difference in the classroom and they will not be punished for making their contradictory sentiments known" (p. 294). Adjusting the timing of these evaluations, however, may be key in increasing effectiveness of the responses and at the same time decreasing the number and types of verbal attacks aimed at the instructor. For example, employing mid-term assessments (in addition to end-of-the-term assessments) could help the instructor gauge student sentiment and have time to make adjustments, if warranted, while improving teacher effectiveness. Students would still have the opportunity to voice their concerns anonymously but could reap the benefits of a climate more conducive to learning while still enrolled. Instructors could identify remedies for legitimate issues and concerns expressed by students and address them before the frustrations escalate to the point of student retaliation via verbal attack and lowered student perceptions of teaching at the end of the term. Mid-term assessments may lead to an overall more positive classroom climate that is conducive to learning for the students and potentially less psychologically and professionally detrimental to the instructor.

References

Andersen, P. A. (1999). *Nonverbal communication: Forms and functions.* Mountain View, CA: Mayfield.

Bolkan, S., & Goodboy, K. (2013). No complain, no gain: Students' organizational, relational, and personal reasons for withholding rhetorical dissent from their college instructors. *Communication Education, 62*(3), 278–300.

Boysen, G. (2012). Teacher responses to classroom incivility: Student perceptions of effectiveness. *Teaching of Psychology, 39 (4)*, 276–279.

Dukes, R. L., & Victoria, G. (1989). The effects of gender, status, and effective teaching on the evaluation of college instruction. *Teaching Sociology, 17*(4), 447–457.

Fries, C. J., & McNinch, R. J. (2003). Signed versus unsigned student evaluations of teaching: A comparison. *Teaching Sociology, 3*(3), 333–344.

Hall, R. M., & Sandler, B. R. (1982). The campus climate: A chilly one for women? Washington, DC: Association of American Colleges. *(Report of the Project on the Status and Education of Women).*

Infante, D. A. (1987). Aggressiveness. In J. C. McCroskey & J. A. Daly (Eds.), *Personality and interpersonal communication* (pp. 157–192). Newbury Park, CA: Sage.

Infante, D. A. (1995). Teaching students to understand and control verbal aggression. *Communication Education, 44*(1), 51–63

Infante, D. A., Rancer, A. S., & Jordan, F. F. (1996). Affirming and nonaffirming style, dyad sex, and the perception of argumentation and verbal aggression in an interpersonal dispute. *Human Communication Research, 22*(3), 315–334.

Infante, D. A., & Wigley, C. J. (1986). Verbal aggressiveness: An interpersonal model and measure. *Communication Monographs, 53* (1), 61–69.

Joinson, A. (2001). Self-disclosure in computer-mediated communication: The role of self-awareness and visual anonymity. *European Journal of Social Psychology, 31*, 177–192.

Kinney, T. A. (1994). An inductively derived typology of verbal aggression and its relationship to distress. *Human Communication Research, 21*(2), 183–222.

Martin, M. M., Anderson C. M., Burant, P. A., & Weber, K (1997). Verbal aggression in the sibling relationship. *Communication Quarterly, 45*, 304–317.

Mitchell, M. (2013). Teacher enthusiasm: Seeking student learning and avoiding apathy. *Journal of Physical Education, Recreation and Dance, 84*(6), 19–24.

Myers, S. A. (2002). Perceived aggressive instructor communication and student state motivation learning, and satisfaction. *Communication Reports 15*(2), 113–122.

Myers, S. A., & Knox, R. I. (2000). Perceived instructor argumentativeness and verbal aggressiveness and student outcomes. *Communication Research Reports, 17*(3), 299–209.

Myers, S. A., & Rocca, K. A. (2000). Students' state motivation and instructors' use of verbal aggressive messages. *Psychological Reports, 87*(1), 291–294.

Payne, M. J., & Chandler-Sabourin, T. (1990). Argumentative skill deficiency and its relationship to quality of marriage. *Communication Research Reports, 7*, 121–124.

Pittman, C. (2010). Race and Gender Oppression in the Classroom: The experiences of women faculty of color with White male students. *Teaching Sociology, 38*(3), 183–196.

Rancer, A. S., & Avtgis, T. A. (2006). *Argumentative and Aggressive Communication.* Thousand Oaks, CA: Sage.

Rocca, K. A., & McCroskey, J. C. (1999). The interrelationship of student ratings of instructors' immediacy, verbal aggressiveness, homophily, and interpersonal attraction. *Communication Education, 48*(4), 308–316.

Schrodt, P. (2003). Student perceptions of instructor verbal aggressiveness: The influence of student verbal aggressiveness and self-esteem. *Communication Research Reports, 20* (3), 240–250.

Schuman, R. (2014). Needs Improvement: Student evaluations of professors aren't just biased and absurd—they don't even work. Retrieved from http://www.slate.com/articles/life/education/2014/04/student_evaluations_of_college_professors_are_biased_and_worthless.html

Teven, J. J. (2001). The relationship among teacher characteristics and perceived caring. *Communication Education, 50*, 159–169

Venable, K. V., & Martin, M. M. (1997). Argumentativeness and verbal aggressiveness in dating relationships. *Journal of Social Behavior and Personality, 12*, 955–964.

Werner, N. E., & Bumpus, M. & Rock, D. (2010). Involvement in internet: Aggression during early adolescence. *Journal of Youth Adolescence,* 39(6), 607–619.

Context Two

The Dark Side of Organizational Communication

SIX

Workplace Bullying: U.S. Academic Managers' Intervention Strategies

Susan L. Theiss and Lynne M. Webb

Workplace bullying is a rapidly increasing complaint among American workers (Duncan, 2011). Recent research reveals that ongoing exposure to workplace bullying can lead to turnover, lowered productivity, group disputes, and negative health outcomes (e.g., Hollis, 2013; Klein, 2012; Nielsen & Einarsen, 2012). However, many organizations and their cultures allow bullying to thrive (Hegranes, 2012; Power et al., 2013), especially organizations that highly reward performance and productivity (Cooper-Thomas, Gardner, O'Driscoll, Catley, Bentley, & Trenberth, 2013; Samnani & Singh, 2014) as well as organizations in which resources are in short supply (Wheeler, Halbesleben, & Shanine, 2010) and intimidating managers control the purse strings (Armstrong, 2011). Finally, employee workload and role conflicts also affect workplace bullying (Balducci, Cecchin, & Fraccaroli, 2012).

Bullying research has focused largely on understanding bullies and their victims (e.g., Lester, 2012; Lutgen-Sandvik, 2013) rather than managers whose actions (or lack of action) play a pivotal role in workplace bullying. Lack of managerial intervention can be decoded as passively condoning bullying and thus perpetuating bullying (Namie & Lutgen–Sandvik, 2010). Conversely, managerial intervention can ameliorate the impact of bullying, especially when managers intervene to break the cycle of abuse and actively discourage bullying (e.g., Cooper-Thomas et al., 2013; Giorgi, 2012; Hegranes, 2012).

Our research project examined managers' perspectives of workplace bullying in a unique organizational setting, higher education, where power differentials between personnel can be substantial and relationships can be complex (e.g., colleagues may be former professors). Consistent with other studies examining workplace conflict (Keaveney, 2008; Leung, 2010), our research was grounded in Sillars and Parry's (1982) Conflict Attribution Theory and Bandura's (2001) Social Cognition Theory.

Bullying Behaviors

"Bullying occurs when someone [or group] is systematically subjected to aggressive behaviors" which lead, either intentionally or unintentionally, to a stigmatization and victimization of the target (Einarsen, 1999, p. 16). When bullied, victims describe being targeted with unwanted, ongoing, aggressive behaviors. Einarsen posited that abuse related to workplace bullying seems "mostly to be of a verbal nature and seldom includes physical violence" (1999, p. 18). While most forms of verbal aggression do not lead to physical violence, acts of physical aggression typically are preceded by unresolved verbal violence or indirect aggression (Infante & Rancer, 1996).

Einarsen (1999) introduced two types of workplace bullying. *Dispute-related* bullying involves an escalated conflict. The bully employs coercive or aggressive conflict resolution strategies to resolve an ongoing dispute. *Predatory* bullying occurs when the organizational culture allows bullying as an interaction style. For example, tenured physicians in medical schools may be verbally abusive with residents because this was how they were treated as residents, and this practice is commonly accepted (and therefore reinforced). Predatory bullying tends to be more prevalent in rule-oriented and bureaucratic organizations (Ferris, 2004). Namie and Namie (2009) identified four additional types of bullying: Chronic, Accidental, Substance-abusing, and Opportunist. Opportunist bullies, who believe careers are built with political gamesmanship, are willing to succeed at the expense of their targets.

Issues of Power

When an organizational structure permits one person to exert more influence over another than is necessary or appropriate within the scope of his/her position, then the likelihood of reaching mutually beneficial resolutions and repairing relationships decreases (Folger, Poole, & Stutman, 2005). When such problems are referred to processes and personnel who do not understand the power differences at play between bullies and victims (Gibson, Medeiros, Giorgini, Mecca, Devenport, Connelly, & Mumford, 2014; Lutgen-Sandvik, & McDermott, 2011), then bullies' power can be reinforced, further institutionalizing bullying behaviors.

Various aspects of power (i.e., age, rank per se, and power differentials) provide bases for bullying. In a study that examined target age, Lutgen-Sandvik (2007) determined that rates of abuse were higher for younger employees. Older workers reported colleague/peer bullying more frequently; young employees received more verbal abuse from supervisors, were less likely to report abuse, and more likely to leave organizations. Gibson et al. (2014) found that power differentials can inhibit people's ability to make sound ethical decisions in academia, and that poor ethical decisions can lead to abusive actions.

The 2003 Report on Abusive Workplaces (Namie, 2003) found that 71% of bullies had higher ranks in organizations than their targets. Survey respondents identified higher-level managers as exhibiting behaviors that assist bullies in 24% of the cases. For example, if a supervisor who has been bullying an employee convinces upper-level management that the employee (the target) has performance problems, an upper-level manager might assist the bully by supporting and enforcing disciplinary actions requested by the supervisor (the bully). Combined, aggressors and supporters often have numerous opportunities to abuse their positions of power over their targets. Thus, managerial intervention can positively or negatively affect outcomes in bullying situations.

The Stages of Bullying

Einarsen (1999) identifies four-stages of bullying: (a) aggressive behavior, (b) bullying (c) stigmatization, and (d) severe trauma. When unwanted behaviors towards the target become frequent, then *aggressive behavior* becomes *bullying*. Victims often experience difficulty defending themselves, bring coworkers into the cycle, and thus set the stage for the move from *bullying* to *stigmatization* of the target. The associated stress eventually may cause performance problems for the target, which the bully then brings to others' attention. By highlighting the victim's weaknesses and performance problems, the bully can mislead managers and bystanders, culminating in *stigmatization.*

At stigmatization, the manager typically is asked to intervene at the request of the bully, the target, and/or the bystanders. The victim's inabilities become the focus of the bully, and typically of the intervener. In some situations, a target might believe that his/her behavior instigated the actions of the bully or may worry that others perceive him/her as timid, thin-skinned, or deserving of the negative attention (Keashly & Harvey, 2005). In reality, many targets are well educated, refuse to be subservient, and/or exhibit exceptional knowledge, skills, and ability to establish effective relationships with others, which the bully finds threatening (Namie & Namie, 2009; Klein, 2012). "Upper management or personnel administration tend to accept the prejudices produced by the [bullies], thus blaming the victim" (Einarsen, 1999, p. 20). If managers cannot address the situation without further victimizing targets, then targets may experience escalating health consequences, including *severe trauma.*

Costs and Consequences of Workplace Bullying

Rayner and Keashly (2005) identified three areas of organizational costs related to bullying: replacing staff, the time associated with staff coping with bullying, and administrative costs of mistreatment, including litigation. Additional cost of mistreatment can include increased absences and illnesses, poor and lowered productivity, as well as low morale (Meares, Oetzel, Torres, Derkacs, & Ginossar, 2004; Namie, 2003; Porath & Pearson, 2010). Serious physical and acute mental health problems associated with bullying are well documented (Hallberg & Strandmark, 2006; Mikkelsen & Einarsen, 2002; Namie, 2003); they range from anxiety and depression to thoughts of suicide (Brousse et al., 2008; Hallberg & Strandmark, 2006; Nielsen, Matthiesen, & Einarsen, 2008). Recent studies also suggest extreme bullying can lead to trauma and Post Traumatic Stress Disorder (Hogh, Mikkelsen, & Hansen, 2011). While health effects can vary, a review of current research reflects a strong correlation between job stressors (such as bullying) and depression (Schwickerath & Zapf, 2011).

The Importance and Impact of Effective Managerial Intervention

Perhaps the greatest challenge to the organization occurs when those with the most power to facilitate changes in bullying patterns do not understand that bullying is occurring or how to address it. When recognized, it may be difficult for managers to act because tactics typically used by bullies are not illegal or clear violations of organizational policies. Furthermore, targets of bullying often find it difficult to translate their experiences effectively into words that allow managers to understand the full implications of the bullying (Tracy, Lutgen-Sandvik, &

Alberts, 2006; Lutgen-Sandvik & McDermott, 2011). Moreover, when aggressive employees bully others, they typically have accomplices, and bullying can progress quickly, becoming a team issue (Namie & Lutgen-Sandvik, 2010).

Institutional climate and management temperament can exacerbate occurrences of bullying (Cleary, Hunt, Walter, & Robertson, 2009; Garling, 2008). Organizational tolerance of bullying is communicated through policies, norms, values, and managerial responses. Conversely, successful intervention is often multifaceted and long-term. Many forms of reprisal cannot effectively be prevented or addressed through formal channels, as many forms of retaliation are covert and cannot be addressed through enforcement of policies or rights-based procedures (Rowe, 1996).

The Unique Environment of Higher Education

Institutions of higher education provide a unique organizational context for bullying. Universities and colleges value academic freedom, which encourages sharing of diverse perspectives and sound arguments to defend those ideas. Tenure confers recognition of high achievement and establishes positional power. Collegiality is emphasized to encourage collaboration in research and curriculum development, which can discourage negative feedback. Student employees are hired to both further their education and to provide applicable work experience. Substantial power differences between typically younger student employees and experienced older workers abound. These organizational characteristics promote academic excellence but prevent conflict resolution via conflict avoidance, abuses of power, and verbal aggression.

Applied Theory

Bandura (2001) noted that nearly all social behaviors are learned through interactions with others, observing others' actions, and observing the related consequences of the actions. Thus, witnessing bullying could influence perceptions of workplace conflict. Furthermore, how bullies and targets perceive the organizational culture could influence the conflict resolution strategies (CRS) they use. The cycle of abuse can be continued by creation of a social reality that allows bullying or views bullying as an acceptable means for resolving disputes. "Emotionally abusive behaviors are more likely to occur in a societal context that is either tolerant of such behavior or does not define it as problematic" (Keashly & Harvey, 2005, p. 212).

Sillars and Parry's Attribution Theory (1982) suggests that interactants develop individual theories about why they are in conflict with others, based on their interpretation of others' behaviors, which influence selections of CRS. Sillars & Parry (1982) identified three common CRS:

- avoidance (not addressing conflict or minimizing its importance);
- competition (attempting to win in conflicts); and
- cooperation (seeking mutual agreement).

Although managers want to effectively address employee conflicts (Brotheridge & Long, 2007), Folger et al. (2005) noted that managers may not use preferred conflict styles in cases of bullying, if organizations' structure and culture prevent that style from being effective. For example, structures with only rights-based systems, such as grievance procedures, require documented steps and designated individuals making final decisions, thus inhibiting managers from facilitating cooperation among coworkers.

Purpose and Research Questions

Our study examined bullying in the unique organizational context of higher education. Given the convention in higher education of calling managers "administrators," we hereafter use the term "administrator" to reference employees at colleges or universities who directly supervise other employees. The purpose of this study was to assess administrators' perspectives of bullying by gathering reports of their interactions with parties involved in cases of bullying. We posed three research questions:

RQ1: How do administrators define "bullying"?

RQ2: What CRSS do administrators report using when intervening in cases of bullying and why? Specifically, do they distinguish between types of bullies or reference their organizational climate, policies, or permitted CRS?

RQ3: How do managers' CRSS change across stages of the bullying process?

Method

Participants

We recruited 15 U.S. participants, 10 from universities and 5 from colleges, all outside of the authors' institutions. The sample included four types of administrators: program directors (13%), department chairs (27%), deans (47%), and vice-presidents (13%). Their employers included universities in Arizona, Colorado, Illinois, Kansas, and Washington as well as community colleges in Arkansas, Florida, and Kansas. Longevity in management positions ranged from 2.5 to 34 years. Tenure status was split fairly evenly (47% tenured; 53% non-tenured). A minority (33%) reported supervising only staff; most participants (87%) reported supervising both staff and faculty members.

The majority of participants were age 50 or older (73%), but others were 40–49 years old (27%). Participants reported holding doctoral (80%) or masters' degrees (20%). Twelve were male; three were female. Participants reported three ethnicities: 80% Caucasian/white, 13% African American/black, and 7% Asian/Pacific Islander.

Instruments

A brief questionnaire gathered demographic information used to describe the sample (e.g., biological sex, age, ethnicity). Additionally, we developed a 3-part, original interview protocol. Initial questions captured participants' definitions of bullying by asking them to describe a witnessed or hypothetical bullying situation as well as the CRS they used to intervene (or not intervene). Next, the interviewer provided scenarios that portrayed clear changes in stages of bullying, asking what CRS the participant might use to intervene (or not intervene) in the stage-based situations. Finally, we asked participants to describe their roles and goals in multiple bullying scenarios.[1]

Procedures

After receiving IRB approval, the interviewing process was pre-tested with three participants drawn from the research population. After minor changes in the interview protocol, we recruited research participants by asking professional contacts at numerous institutions to recommend colleagues who might serve as participants. Initial interviewees were asked to recommend colleagues as potential participants, thus "snow-balling" the sample. When a participant was identified through e-mailed correspondence, a formal recruitment letter was e-mailed as an attachment. The letter provided a project overview, assurance of confidentiality, and contact information to schedule the interview. If a potential participant agreed to participate, a consent form and demographic questionnaire were forwarded for completion and returned by fax before the interview.

The interviewer was a Caucasian female, in her 40's, with a bachelor's degree in Business Administration, who completed training in active listening and inquiry techniques. She asked questions contiguously to avoid interrupting the flow of dialogue. She asked probing questions, as needed, to encourage participants to share clear and detailed information. During interviews, she maintained rapport with neutrality and refrained from leading the participant to specific answers (Hirsch, Miller, & Kline, 1977).

Analysis

Recorded interviews were transcribed, yielding 85 typed, single-spaced pages, containing 3,505 lines of data. The interviewer employed a contact summary form to document and categorize interview responses. Using Miles and Huberman's (1994) pattern coding methods, she recorded descriptions of participants' definitions of bullying, summaries of their recounting of previous experiences with bullying, and their perceptions of their roles and goals during various phases of bullying represented in the scenarios described by the interviewer.

Next, the interviewer and a second coder (a female, Caucasian graduate student in her 40's, studying counselor education) subjected the data to thematic analysis using Owen's (1984) criteria for interpreting themes: recurrence, repetition, and forcefulness. Recurrence was operationalized as statements from at least two participants with the same meaning (but potentially using different words). Repetition occurred when at least two participants used the same words to convey the same meaning. Forcefulness occurred when at least two participants stressed an issue by using dramatic language, vivid imagery, or vocal inflection. Each coder highlighted key concepts found in each interview and then summarized patterns in separate notes. The coders met to compare observations and identify common themes. There were no disagreements about the common themes.

The interviewer then reviewed all transcripts for negative evidence of the initial findings. Next, a review of all transcripts and notes occurred to outline themes that supported attribution theory and social cognition theory. A final review of data, themes, and tables occurred to look for multi-dimensional patterns. Thus, six reviews of the 85-page transcript were completed.

Results

Administrators' definitions and perceptions of bullying. One participant defined bullying as "one person trying to impose their will on another person in direct contradiction to the interests and desires of that other person." Another said, "I see bullying as a form of harassment in a way—someone dominating or someone having power over someone else and exercising that power [to control] the other person." These definitions provide good summations of the effects of the bullying behaviors and conditions identified across the interviews.

Indeed, participants defined bullying in terms of *Behaviors* and *Conditions: Direct Behaviors* (Verbal assaults, Retaliation, Threats, Altering job assignments, Physical intimidation) and *Indirect Behaviors* (Inappropriately influencing others' actions, Denial of behavior and/or redirecting blame) as well as five *Conditions (Repeated aggressive behaviors, Collateral impact, Abuse of Power, Structure that inhibits resolution, and Lengthy resolution process)*. Tables 1 and 2 display descriptions and examples of these themes.

Table 1. Behaviors that Define Bullying.

Direct Behaviors	***N***	**Descriptions/Examples**
Verbal assaults	11	Temper tantrums, verbal harassment, derogatory comments about ethnic groups, criticism, quiets junior faculty, rudeness, threats, intimidating statements, embarrassing comments about victim, condescending, interrupting, silent treatment, speaking sternly, negative comments about victims in public, discredit or ignore target's ideas, disruptive comments, get aggressive with those who disagree with them, discredit victim's program, character defamation
Threats	8	Threaten to sue, threaten consequences, sabotage victims' plans, make demands, threaten to prevent from getting a job, bring attorneys to meetings, marshal powerful resources, do it their way "or else", apply pressure
Retaliation	6	Ended GA position, changed evaluation from positive to negative, filed grievance against the target, bullied administrators that tried to help, removed from grant, accused target of scientific misconduct, excluded target from the team, withheld letter of recommendation, withheld information
Altering job assignments	5	Assigned demeaning activities, removed from grant activity, excluded from others, harder on one employee than another, withholding assistance
Physical intimidation	3	Individual was visually and physically intimidating, used threatening stance, standing up during staff meetings, wagging finger in the face of others, raised voice, would not look you in the eye when speaking

Indirect Behaviors	*N*	Descriptions/Examples
Inappropriately influencing others' actions	12	Create fear of repercussion, initiate investigation of an assisting chair, soliciting letters demanding chair be fired, people leave their positions, limit another's ability to act independently, restrict target from talking to others/ inhibit dialogue, prevent junior faculty from development opportunities, members of committee influencing each other's behaviors against a junior faculty, using closed questions, supervisor not being available to subordinate, sabotage meetings, interrupt or talk out of turn, express own opinion as the group's view, recruit support from others, cause people to "cave in" along the way, attempting to take funding from target's program, undermine program, keep others from succeeding, exclusion, victims and administrators sometimes feel desire to "push back"
Denial of behavior and/or redirecting blame	3	Bully adamantly upholds his right in defense of actions, delay processes helping targets, discredit the victim to others, project self as more credible than others, file counter grievances against the target or administrators

Table 2. Conditions that Define Bullying.

Conditions	*N*	Descriptions/Examples
Repeated aggressive behaviors	15	Acts as aggressor towards everyone, aggressive towards the same individual multiple times, has a history of aggressive behavior, aggressive behaviors are repeated
Collateral impact	15	Others who view bully's acts are afraid to address them, people leave while administrators are trying to address long-term cases, aggression occurs in a public setting, intimidation of women in public, people taking sides, supervisor does not see or decode the bullying behaviors, history of behaviors seen as ruining careers
Abuse of power	14	Positional status (tenured, supervisory, length of employment), group bullying, male gender, senior age, alliances with administrators, power over bystanders (target's/bystander's spouse is a student of the bully), imbalance of power created, abuse of "free speech rights." protected minority status, presence of attorneys
Structure that inhibits resolution	7	Bully files formal grievances against victim or administrator trying to help, hard to remove faculty, HR not viewed as helpful at institutions where limited to acting on sexual or racial harassment, structure doesn't support addressing cases of bullying, use position within organization to get what they want, committee structure supported senior/tenured faculty aggression, intervention needed at multiple levels, structure supports tenured professor over junior faculty, bullying is more easily addressed/ experienced less) in community colleges without tenure

Conditions	N	Descriptions/Examples
Lengthy resolution process	6	Cases involving tenured faculty take one or more years to resolve

Administrators' preferred CRS. Participants' unanimity of responses was striking: All participants perceived bullying as a serious issue and reported experiencing a duty to intervene. All participants described *Collaborative* strategies to initially address cases of bullying. Virtually all participants identified the same six CRS displayed in Table 3:

Table 3. Reported Initial Strategies for Intervention.

Collaborative Strategies	N	Descriptions/Examples
1. Early intervention	15	Address problem when brought to administrator's attention, address before problem escalation, monitor from beginning, do not ignore or dismiss, reduce power imbalance, intervene before recruitment (prevention)
2. Create a safe environment	9	Address individually first, "keep everyone safe," do not re-harm the victim, provide conducive environment for employees
3. Initial Assessment		
a. Individuals	15	Talk to both individuals, ask questions, collect data/facts, move discussion from a "public to private" place, investigate, substantiate/verify information, obtain documentation
b. Team	11	Visit with affected team members, get support from team, use transparency, create civility policy with team input, consider those "taking sides," substantiate information with bystanders
4. Consider policies, unique circumstances, and resources	15	Address with higher administration when needed, use Ombuds or other neutral third party (i.e., HR), consider status of employees, know policies, consider history of employees, may use multiple strategies depending on circumstances, consider power issues, what is a fair process, consider what is been done/tried to date, review documentation, check employee references
5. Provide education and coaching	15	Try reasoning, encourage appropriate behavior, use situation as teaching opportunity, mediate/bring parties together, reiterate civility policy, transformative discussions, find common ground, give options, negotiate options, encourage direct communication between parties, address issues that affect group climate, provide workshops/ethics training, encourage documentation, hold aggressor responsible for actions, inform of policies

Collaborative Strategies	*N*	Descriptions/Examples
6. Provide structure and follow through	12	Establish boundaries to prevent situation from escalating, address affected group, clarify boundaries, create shared expectations for behaviors, give options, create a quality environment, monitor situation, inform of actions to be taken, establish group climate, discourage triangulation, prevent parties from meeting alone until issue is resolved, clearly state anti-bullying policies, establish remediation plan, schedule follow-up meetings, document through evaluation process, be a role model

Administrators reported no changes across stages of the bullying process. Participants demonstrated awareness of the stages of bullying, but they did *not* report engaging in additional interventions at particular stages. Instead participants reported engaging in a "second stage of intervention" when and if prompted to do so by the ineffectiveness of first interventions. When they perceived that initial collaborative strategies did not work, as the bullying continued to progress or escalate, then ten participants reported *Re-assessing the situation*, and, depending upon the current conditions and organizational climate, they either provided *Additional coaching* or moved toward more *Dominant strategies*. The remaining participants reported immediately moving to *Dominant Strategies* including *Use of Formal Processes, Re-Assignment, Dismissal, or other Consequence.* Table 4 provides descriptions and examples of the four CRS themes.

Table 4. Next-Step Strategies for Intervention.

Collaborative Strategies	*N*	Descriptions/Examples
Re-Assess the situation	10	Continue dialogue, consider group effect on long-term environment, consider fair process, assess along the way; if changes do not occur, move towards progressive discipline; if informal collaboration doesn't work, move towards structured university process
Additional coaching	8	Continue coaching with individuals and/or team, monitor through planned meetings, create "feedback loop," prevent future conflicts from arising, give warning, try to negotiate solution, utilize third party resources, refer to hotline
Dominant Strategies	N	**Descriptions/Examples**
Use formal processes	9	Use progressive discipline if situation does not improve, process through HR, use structured university process if informal options do not work; may refer to Affirmative Action, get support at the institutional level, invoke probationary contract
Re-assignment, dismissal, or other consequence	5	Move the target or aggressor to another project or department, dismiss, or impose other consequences

Discussion

Summary of Findings

Participants recognized bullying via behaviors and conditions. They report intervening initially using six collaborative strategies. When and if bullying continued, participants reported reassessing the situation and then either providing additional coaching or employing dominant strategies such as progressive discipline. Figure 1 provides a pictorial representation of the findings.

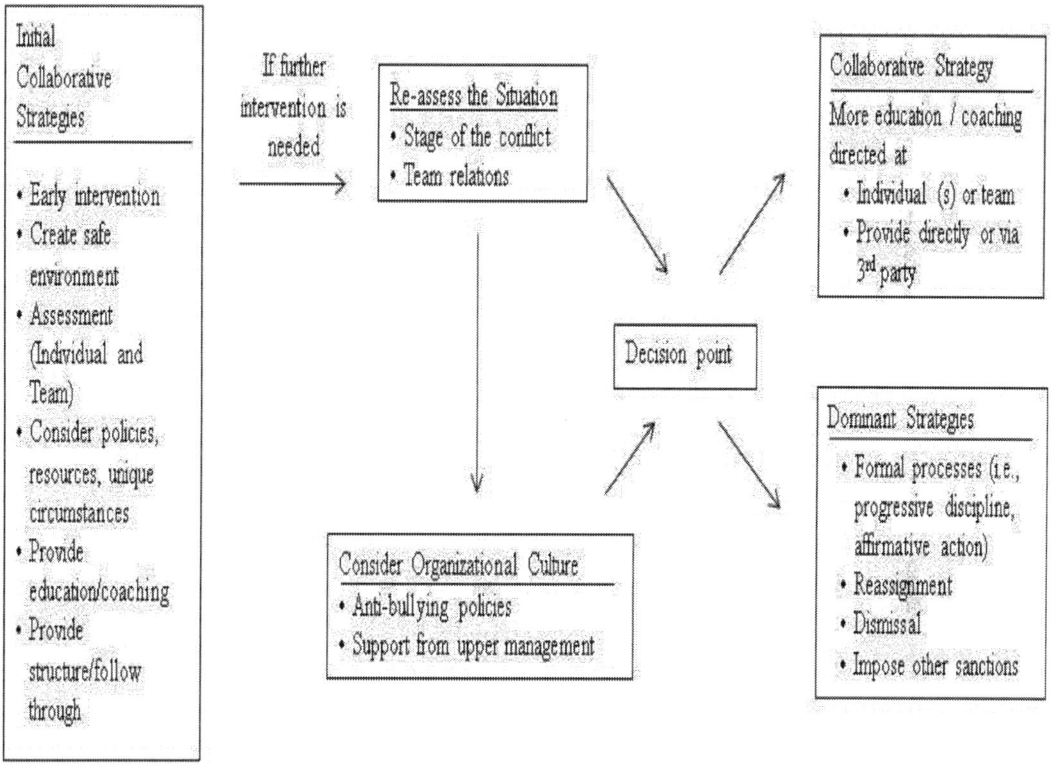

Figure 1. Progress of Administrators' Strategies in Cases of Bullying.

Interpretation of Findings

Bullying defined. Participants' descriptions of bullying behaviors they had observed provided further support for previously published definitions of bullying behaviors (Einarsen, 1999; Namie & Namie, 2009; Rayner & Keashly, 2005; Tracy, Lutgen-Sandvik, & Alberts, 2006). However, participants' views differed from previously reported findings in three important ways described below.

First, while Einarsen's (1999) definition offers repetition as the basis for bullying, our data indicates that individuals and groups can perceive themselves as bullied when targeted with *only one* act of aggression. If a victim previously witnessed a bully acting aggressively towards others, then he or she may perceive bullying during the first act of aggression directed at him or her. This finding was consistent with Lutgen-Sandvik's (2006) observation that "communication at work, including workplace bullying, is always social and public" (p. 426). In sum, observation of repeated acts of aggression, whether personally received or witnessed towards others, can facilitate identification of conflict as bullying as opposed to simply conflict or aggression.

Second, most published definitions of bullying rely on descriptions of aggressive behaviors. For example, Einarsen (1999) stated "Bullying occurs when someone [or group] is systematically subjected to aggressive behaviors" (p. 16). Our participants defined bullying in terms of the bully's behaviors *and the conditions surrounding the conflicts.* The conditions they identified provide insight into their view of bullying as a group or team phenomenon and offer evidence of complex perceptions of bullying, describing it in terms of collateral damage as well as its obvious behavioral manifestations.

Third, participants revealed their complex perceptions by identifying bullying behaviors as *direct, indirect, or both.* For example, if tenured Professor A files a grievance against untenured Professor B, Professor B will undergo an investigation, which will impact Professor B's time and allocation of duties, and likely cause stress. Indirectly, this grievance will communicate to administrators and coworkers that Professor B's performance or ethics are being questioned and may affect a review committee's ability to provide a fair and unbiased tenure evaluation, as well as potentially limit Professor B's equal access to institutional support.

Additionally, colleagues may be affected. For example, another professor who witnessed Professor B's experience may become fearful that others in positions of power may retaliate in the same manner towards him or her. This repercussion phenomenon further extends the impact of bullying, as viewers' perceptions of bullies' actions can influence viewers' subsequent behaviors and decisions. Such influence exemplifies social cognition theory, in that the social environment influences how individuals process observed social behavior (Bandura, 2001), and in turn the choices they make.

Administrators' conflict resolution strategies. Our results indicate that administrators' unanimous first-choice CRSS were collaborative strategies. Furthermore, our participants from various institutions in various states at various career stages and in various administrative posts achieved unanimous agreement on four CRSS and near-unanimous agreement on two CRSS (see Table 3). Such wide-spread agreement among diverse participants is rare in social science and may indicate that a finite set of effective initial strategies exists for addressing bullying in academic workplaces.

In *dispute-related* or *accidental* bullying, participants' reports suggested that early coaching and establishing behavioral expectations may end bullies' aggression. Cases of *predatory* or *opportunist* bullying prove more difficult. When and if initial collaborative intervention proved ineffective, participants reported re-assessing the situation and then deciding between two primary courses of action: either additional coaching or moving directly to dominant strategies. Participants stated that, at this point, the choice of collaborative versus dominant strategies depended on three factors: stage of the bullying process, relevant team conditions (e.g., history, power differences), as well as effectiveness of available institutional policy structure.

Given that our participants understood bullying and took steps to end it, why did bullying persist in most cases? Indeed, why does bullying exist today in academic workplaces? Power and effectiveness of third-party resources within an organization's structure influence administrators' ability to employ dominant strategies and force bullies to alter their behaviors when necessary. Most *university* participants reported that they do not have adequate policies in place to address bullying. Furthermore, they said that Human Resource offices (HR) can provide coaching, but often have no final authority to impose consequences (as in cases involving tenured faculty). In such cases, HR is viewed as powerless and largely ineffective. Participants reported that cases involving tenured faculty were either not resolved or took years to resolve. Long-term solutions only came when and if administrators at the highest levels (including deans, provosts, and if necessary, presidents, and chancellors) signed-off on solutions and agreed to not intervene in implementation. In such cases, participants reported that tenured faculty continued to use defensive behaviors (Ashforth & Lee, 1990), including counter-grievances and attorneys to defend themselves against accusations of bullying. Participants reported that enormous organizational costs (time, turnover, absenteeism) accompany long-term cases of bullying. In sum, the data indicate that administrators at the highest level must be willing to support actions of intervening deans, chairs, and directors for intervention to prove successful.

Theoretical Implications

Two theories guided our thinking: (1) Attribution Theory and (2) Social Cognitive Theory. Findings of this study are consistent with both theories. As Sillars and Parry's Attribution Theory (1982) suggested individuals, including administrators, develop theories about why employees are in conflict with each other. Participants' stories of bullying demonstrated that administrators collect information upon which they base their theories of attribution, and then choose their CRS accordingly.

As suggested by Social Cognitive Theory, organizational culture influenced administrators' choice of strategies, which, in turn, influenced organizational culture—thus, creating a system of reciprocal influence in reported cases of bullying. Moreover, participants reported that power structures within organizations and amongst employees influenced the dynamics of bullying processes and resolutions.

Limitations and Recommendations for Future Research

Although our study's participants were diverse in many ways and consistent themes emerged, sample size was small, all participants worked at U.S. institutions, and we did not employ random sampling. Thus, representativeness of the sample remains unknown. Recruitment efforts resulted in a sample of administrators who viewed bullying as an important issue. Such a purposeful sample enabled identification of intervention strategies. However, the sample may not capture perceptions of all higher education administrators. Nonetheless, we documented the unique challenges facing administrators in academic setting as they attempt to address bullying. Our results provide evidence that academic institutions provide fertile ground for research on bullying.

Conclusions

We offer the first study examining academic managers' CRSS in cases of bullying. The findings contribute to research on workplace bullying in four ways: (1) The study clarifies the definition of bullying provided by Einarsen (1999), allowing targets to receive *only one* act of aggression and perceive the act as bullying, *if* they have observed the bully repeatedly act aggressively towards others. (2) The study revealed, among at least one sample of administrators, complex understandings of bullying as involving both conditions and behaviors (direct and indirect).

(3) The study identified wide-spread agreement across administrators about six initial CRSS used to intervene in cases of bullying as a required duty of their role. (4) When and if initial intervention fails, administrators tend to reassess. Then, depending on their assessment and support they are likely to receive from upper-level management, they either provide additional coaching or employ dominant strategies such as formal processes, re-assignment, dismissal, or other consequences.

Notes

1 A copy of the interview protocol is available, upon request, from the first author at Sue.Theiss@oregonstate.edu.

References

Armstrong, P. (2011). Budgetary bullying. *Critical Perspectives on Accounting, 22*, 632–643. doi:10.1016/j.cpa.2011.01.011

Ashforth, B. E., & Lee, R. T. (1990). Defensive behaviors in organizations: A preliminary model. *Human Relations, 43,* 621–648. doi: 10.1177/001872679004300702

Balducci, C., Cecchin, M., & Fraccaroli, F. (2012). The impact of role stressors on workplace bullying in both victims and perpetrators, controlling for personal vulnerability factors: A longitudinal analysis. *Work & Stress, 26*, 195–212. doi: 10.1080/02678373.2012.714543

Bandura, A. (2001). Social cognitive theory of mass communication. *Media Psychology, 3,* 265–299. doi: 10.1207/S1532785XMEP0303_03

Brotheridge, C. M., & Long, S. (2007). The 'real-world' challenges of managers: Implications for management education. *Journal of Management Development, 26*, 832–842. doi: 10.1108/02621710710819320

Brousse, G., Fontana, K., Ouchchane, L., Boisson, C., Gerbaud, L., Bourguet, D., Schmitt, A., Llorca, P.M., & Chamoux, A. (2008). Psychopathological features of a patient population of targets of workplace bullying. *Occupational Medicine, 58,* 122–128. doi: 10.1093/occmed/kqm148

Bullying and harassment in the workplace: Developments in theory, research, and practice. (2nd ed.). CRC Press. Retrieved from http://www.crcpress.com/product/isbn/9781439804896

Cleary, M., Hunt, G. E., Walter, G., & Robertson, M. (2009). Dealing with bullying in the workplace: Toward zero tolerance. *Journal of Psychosocial Nursing, 47*(12), 34–41. doi: 10.3928/02793695-20091103-03

Cooper-Thomas, H., Gardner, D., O'Driscoll, M. P., Catley, B., Bentley, T., & Trenberth, L. (2013). Neutralizing workplace bullying: The buffering effects of contextual factors. *Journal of Managerial Psychology*, *28*, 384–407. doi: 10.1108/JMP-12-2012-0399

Duncan, S. H. (2011). Workplace bullying and the role restorative practices can play in preventing and addressing the problem. *Industrial Law Journal, 32*, 2331–2366. Retrieved from http://papers.ssrn.com/sol3/papers.cfm?abstract_id=1916138

Einarsen, S., Hoel, H., Zapf, D., & Cooper, C.(Eds.) (2010). *Bullying and harassment in the workplace: Developments in theory, research, and practice, second edition.* CRC Press.Retrieved from http://www.crcpress.com/product/isbn/9781439804896

Einarsen, S. (1999). The nature and causes of bullying at work. *International Journal of Manpower, Norway, 20,* 16–27. doi: 10.1108/01437729910268588

Ferris, P. (2004). A preliminary typology of organizational response to allegations of workplace bullying: See no evil, hear no evil, speak no evil. *British Journal of Guidance & Counseling, 32,* 389–395. doi: 10.1080/03069880410001723576

Folger, J. P., Poole, M. S., & Stutman, R. K. (2005). *Working through conflict: Strategies for relationships, groups, and organizations*, Boston, MA: Pearson.

Garling, P. (2008). Bullying & workplace culture: Final report of the Special Commission of Inquiry into acute care services in NSW public hospitals, 399–425. Retrieved from http://www.lawlink.nsw.gov.au/acsinquiry

Gibson, C., Medeiros, K. E, Giorgini, V., Mecca, J. T., Devenport, L. D., Connelly, S., & Mumford, M. D. (2014). A qualitative analysis of power differentials in ethical situations in academia. *Ethics & Behavior, 24*, 311–325. doi: 10.1080/10508422.2013.858605

Giorgi, G. (2012). Workplace bullying in academia creates a negative work environment: An Italian study. *Employee Responsibilities and Rights Journal, 24,* 261–275. doi: 10.1007/s10672-012-9193-7

Hallberg, L. R. M., & Strandmark, M. K. (2006). Health consequences of workplace bullying: Experiences from the perspective of employees in the public service sector. *International Journal of Qualitative Studies on Health and Well-being, 1*(2), 109–119. doi: 10.1080/17482620600555664

Hegranes, C. A. (2012). Leadership response to workplace bullying in academe: A collective case study. *Education Doctoral Dissertations in Organization Development.* University of St. Thomas (Minnesota). Retrieved from http://ir.stthomas.edu/caps_ed_orgdev_docdiss/17/

Hirsch, P. M., Miller, P. V., & Kline, F. B. (1977). Strategies for communication research. *Annual Review of Communication Research, 6*, 127–151.

Hogh, A., Mikkelsen, E. G., & Hansen, A.M. (2011). Individual consequences of workplace bullying/mobbing. In S. Einarsen, H. Hoel, D. Zapf, & C. Cooper (Eds.), *Bullying and harassment in the workplace* (pp. 107–128). Boca Raton, FL: CRC Press.

Hollis, L. P. (2013). *Bully in the ivory tower: How aggression and incivility erode American higher education.* Wilmington, DE: Patricia Berkly LLC. Retrieved from http://www.diversitytrainingconsultants.com/bully-in-ivory-tower

Infante, D. A., & Rancer, A. S. (1996). Argumentativeness and verbal aggressiveness: A review of recent theory and research. *Communication Yearbook, 19*, 319–351.

Keashly, L., & Harvey, S. (2005). Emotional abuse in the workplace. In S. Fox (Ed.), *Counterproductive work behavior: Investigations of actors and targets* (pp. 201–236). Washington, DC: American Psychological Association.

Keaveney, S. M. (2008). The blame game: An attributional theory approach to marketer-engineer conflict in high-technology companies. *Industrial Marketing Management, 37*, 653–663. doi: 10.1016/j.indmarman.2008.04.013

Klein, A. M. H. (2012). *Does workplace bullying matter? A descriptive study of their lived experience of the female professional target* (Unpublished doctoral dissertation). Capella University. Retrieved from http://www.anatomicaladvisors.com/documents/AnneKlein-Dissertation.pdf

Lester, J. (Ed.) (2012). *Workplace bullying in higher education.* Hoboken, NJ: Routledge. Retrieved from http://www.stanford.edu/group/cubberley/node/27437

Leung, Y. F. (2010). *Conflict management and emotional intelligence.* (Published doctoral dissertation) Southern Cross University, Australia. Retrieved from http://epubs.scu.edu.au/cgi/viewcontent.cgi?article=1121&context=theses

Lutgen-Sandvik, P. (2013). *Adult bullying—a nasty piece of work: A decade of research on non-sexual harassment, psychological terror, and emotional abuse on the job.* St Louis, MO: ORCM Academic Press. Retrieved from http://www.amazon.com/Adult-Bullying-A-Nasty-Piece-Work-ebook/dp/B00GM58GXK

Lutgen-Sandvik, P. (2007). "...But words will never hurt me," Abuse and bullying at work: A comparison between two worker samples. *Ohio Communication Journal, 45*, 81–105.

Lutgen-Sandvik, P. (2006). Take this job and...: Quitting and other forms of resistance to workplace bullying. *Communication Monographs, 73*, 406–433. doi: 10.1080/03637750601024156

Lutgen-Sandvik, P., & McDermott, V. (2011). Making sense of supervisory bullying: Perceived powerlessness, empowered possibilities. *Southern Communication Journal, 76*, 342–368. doi: 10.1080/10417941003725307

Meares, M. M., Oetzel, J. G., Torres, A., Derkacs, D., & Ginossar, T. (2004). Employee mistreatment and muted voices in the culturally diverse workplace. *Journal of Applied Communication Research, 32*, 4–27. doi: 10.1080/0090988042000178121

Mikkelsen, E. G., & Einarsen, S. (2002). Relationships between exposure to bullying at work and psychological and psychosomatic health complaints: The role of state negative affectivity and generalized self-efficacy. *Scandinavian Journal of Psychology, 43*, 397–405. doi: 10.1111/1467-9450.00307

Miles, M. B., & Huberman, A. M. (1994). *Qualitative data analysis.* Thousand Oaks, CA: Sage.

Namie, G. (2003). 2003 Report on abusive workplaces: The Workplace Bullying & Trauma Institute. Retrieved from http://www.bullyinginstitute.org

Namie, G., & Lutgen-Sandvik, P.E. (2010). Active and passive accomplices: The communal character of workplace bullying. *International Journal of Communication, 4*, 343–373. Retrieved from http://ijoc.org/ojs/index.php/ijoc

Namie G., & Namie, R., (2009). *The bully at work: What you can do to stop the hurt and reclaim your dignity on the job* (2nd ed.), Sourcebooks, Naperville, IL.

Nielsen, M. B., & Einarsen, S. (2012). Outcomes of exposure to workplace bullying: A meta-analytic review. *Work & Stress, 26*, 309–332. doi: 10.1080/02678373.2012.734709

Nielsen, M. B., Matthiesen, S.B., & Einarsen, S. (2008). Sense of coherence as a protective mechanism among targets of workplace bullying. *Journal of Occupational Health Psychology, 13*, 128–136. doi; 10.1037/1076-8998.13.2.128

Owen, W. F. (1984). Interpretive themes in relational communication. *Quarterly Journal of Speech, 70*, 274–278. doi: 10.1080/00335638409383697

Porath, C. L., & Pearson, C. M. (2010). The cost of bad behavior. *Organizational Dynamics, 39*, 64–71. doi:10.1016/j.orgdyn.2009.10.006

Power, J. L., Brotheridge, C. M., Blenkinsopp, J., Bowes-Sperry, L., Bozionelos, N., Buzády, Z., & Nnedumm, A. U. O. (2013). Acceptability of workplace bullying: A comparative study on six continents. *Journal of Business Research, 66*, 374–380. doi: 10.1016/j.jbusres.2011.08.018

Rayner, C., & Keashly, L. (2005). Bullying at work: A perspective from Britain and North America. In S. Fox & P. E. Spector (Eds.), *Counterproductive Work Behavior: Investigations of Actors and Targets* (pp. 271–296. Washington, DC: American Psychological Association.

Rowe, M. (1996). Dealing with harassment: A systems approach. *Sexual Harassment in the Workplace: Perspective, Frontiers, and Response Strategies, Women and Work, 5*, 247–271. doi: 10.4135/9781483327280

Samnani, A-K., & Singh, P. (2014). Performance-enhancing compensation practices and employee productivity: The role of workplace bullying. *Human Resource Management Review, 24*, 5–16. doi: 10.1016/j.hrmr.2013.08.013

Schwickerath, J., & Zapf, D. (2011). Inpatient treatment of bullying victims. In S. Einsarsen, H. Hoel, D. Zapf, & C. Cooper (Eds.), *Bullying and harassment in the Workplace* (pp. 397–421). Boca Raton, FL: CRC Press.

Sillars, A. L., & Parry, D. (1982). Stress, cognition, and communication in interpersonal conflicts. *Communication Research, 9*, 201–226. doi: 10.1177/009365082009002002

Tracy, S. J., Lutgen-Sandvik, P., & Alberts J. K. (2006). Nightmares, demons, and slaves: Exploring the painful metaphors of workplace bullying. *Management Communication Quarterly, 20*, 148–185. doi: 10.1177/0893318906291980

Wheeler, A. R., Halbesleben, J. R. B., & Shanine, K. (2010). Eating their cake and everyone else's cake, too: Resources as the main ingredient to workplace bully. *Business Horizons, 53*, 553–560. doi: 10.1016/j.bushor.2010.06.002

SEVEN

Defamation, Public Persecution, and Death Threats: Characterizing Retaliation against Whistle-Blowers

Brian K. Richardson and Dianne Gravely

I was reassigned from director of extracurricular activities; I was reassigned and placed in a closet inside the library at [the] high school and they gave me a table and set me up a table inside this closet... to stuff envelopes. And I started stuffing envelopes and [she] wrote me up for the day... for not stuffing envelopes fast enough.

Public school whistle-blower

The hate mail, email harassment, and death threats were the most prominent attacks I received. When the death threats started mentioning my children, I went to the FBI.

Collegiate sports whistle-blower

Introduction

The whistle-blowers quoted above described examples of the types of retaliation they faced for reporting alleged wrongdoing at their respective organizations. Such retaliation is not unusual for whistle-blowers, those individuals who speak out against illegal or unethical practices within organizations to which they belong or are associated with (Alford, 2001). Whistle-blowers often threaten the reputation or existence of the organization, particularly when the focal organization is highly dependent upon the alleged wrongdoing or accused wrongdoer (Miceli, Near, & Dworkin, 2008). Further, whistle-blowers may be perceived as violating organizational or group norms expecting conformity to ongoing business practices (Greenberger, Miceli, & Cohen, 1987). For these reasons, other organizational members or stakeholders may utilize retaliation in order to silence or discredit whistle-blowers, intimidate other organizational members into silence, and allow the alleged wrongdoing to continue. Based upon their nationwide survey which included 394 whistle-blowers, Rothschild and Miethe (1999) found "organizational retaliation against whistle-blowers (both internal and external reporters) is severe and common" (p. 120). Rates of retaliation are especially high for external whistle-blowers, or those who go outside the organization to report wrongdoing (Mesmer-Magnus & Viswevaran, 2005).

Retaliation is communicative in nature; whether verbal or nonverbal, written or oral, anonymous or identified, communicative behavior must be present before retaliation can occur. We further contend retaliation is representative of the "dark side of communication"; it is characterized by negative, unpleasant, manipulative forms of expression that are rarely discussed openly within organizational contexts. In fact, much retaliation is carried out in anonymous, hidden, or disguised manners, including writing derogatory comments about whistle-blowers on their workspaces, using social media formats that do not require identification to slander them, or using group processes to alienate them (Alford, 2001; Richardson & McGlynn, 2011). Though some research is conducted on whistle-blowers' experiences, including retaliation, we know relatively little about this particular form of "dark communication," particularly from those who experience it. Thus, the purpose of this chapter is to explore retaliation as a communicative form, the ways it is expressed, attributions of why it occurs, and its outcomes for whistle-blowers, retaliators, and organizations. In order to achieve this objective, we rely on our own research efforts, which include in-depth interviews with 21 whistle-blowers in collegiate sports and public K-12 education. Next, we review the literature on whistle-blowing, with particular attention paid to studies addressing retaliation.

Conceptualizing and Predicting Retaliation

Miceli et al. (2008) conceptualize retaliation as "undesirable action taken against a whistle-blower—and in direct response to the whistle-blower" (p. 11). Other perspectives of retaliation suggest it can include undesirable actions that were taken, and desirable actions, e.g. job promotions, that were withheld from whistle-blowers (Keenan, 2002). Determining what percentage of whistle-blowers encounter retaliation is challenging and may depend upon sampling procedures. Studies using nonrandom sampling procedures (Richardson & McGlynn, 2011; Rothschild & Miethe, 1999), which likely privilege high-profile whistle-blowing cases, found the vast majority of whistle-blowers face retaliation. However, random sampling procedures suggest smaller numbers; for example, some research suggests 16–38 percent of federal whistle-blowers, encounter retaliation (Near & Miceli, 1996). Still, even if the numbers skew to the lower percentages, a large number of whistle-blowers are encountering retaliatory behaviors by their coworkers and managers.

Much whistle-blowing research locates retaliation as internal to the organization; in other words, the agents carrying out retaliation are fellow organizational members, including peers and managers (Miceli et al., 2008). However, more recent scholarship suggests retaliation can originate from external agents who have a stake in the organization's success. Richardson and McGlynn (2011) found boosters, fans, and some media outlets, participated in retaliation against whistle-blowers in the collegiate sports industry. In categorizing retaliation, scholars have distinguished work-related retaliation, which is formal and documented, from social retaliation, which includes antisocial, often informal, behaviors that are undocumented (Cortina & Magley, 2003). Still, a typology of communicative forms of retaliation against whistle-blowers is lacking. Most research uses an *a priori* approach which privileges researchers' views while limiting the perspective and voice of those who actually faced retaliation. Developing a typology of retaliatory behaviors could spur research into factors predicting the various types of retaliation. Thus, we offer the following research question:

RQ1: How is retaliation against whistle-blowers expressed?

Scholars have utilized myriad frameworks for explaining why whistle-blowers encounter retaliation. Within the small group lexicon, whistle-blowers are akin to deviants or non-comformists who threaten the existence or purpose of the group (Bocchiaro & Zamperini, 2012). Organizational members frequently use retaliation in order to expel the deviant or bring him or her back in line with group norms (Schachter, 1959). From a group preservation standpoint, Alford (1999) suggests the whistle-blower is scapegoated and suffers retaliation so the remaining organizational members can persist in an unethical environment. Rothschild and Miethe (1999) found retaliation is increasingly likely and harsh when the alleged wrong-doing is most systemic and vital to the operations of the organization. Post-positivist research has found whistle-blowers are also at increased risk of retaliation if they use external channels (rather than internal channels only) and report on serious transgressions, while retaliation is less likely if the whistle-lower possesses convincing evidence of their allegations and enjoys support from their supervisors (Mesmer-Magnus-Viswesvaran, 2005). Though informative, such research rarely engages whistle-blowers in attempts to understand why they believe retaliation was used against them. Therefore, we posed the following research question to understand whistle-blowers' attributions of retaliation.

RQ2: What are the reasons retaliation is used against whistle-blowers?

Outcomes of Retaliation

Retaliation has consequences for the whistle-blower, the retaliators, and the organization. Interestingly, some research indicates retaliation emboldens the whistle-blower, leading many to increase their efforts and take their accusations external to the organization if necessary (Miceli et al., 2008). Scholars found outcomes of whistle-blowing can include job loss and blacklisting, severe depression and anxiety, feelings of isolation, distrust of others, bouts with alcoholism, declines in physical health and financial status, family relationship problems, and stigmatization (Alford, 1999; McGlynn & Richardson, 2014; Rothschild & Miethe, 1999). Rothschild and Miethe suggested the whistle-blowing process, particularly retaliation, impacted the "master status," or core personal identity, of the whistle-blower. The individual's identity becomes permanently linked to whistle-blowing; they see the world through the prism of their whistle-blower experience, which overwhelms their lives. They also found 90 percent of the whistle-blowers they studied would report again, while other research suggests the opposite (Alford, 1999). Research on outcomes for the retaliators and the organization is less prevalent. Obviously, in some cases, retaliators are also wrongdoers who have committed criminal behavior and are eventually punished. Other studies find retaliation serves its purpose for the organization and wrongdoers, as the whistle-blower is expelled and the status quo is preserved

(Alford, 1999). We sought to understand the outcomes of retaliation from whistle-blowers' perspectives. Thus, we posed the following research question:

> RQ3: What are the outcomes of retaliation for whistle-blowers, the retaliators, and the organization?

Method

In order to address these research questions, we combined data from two qualitative studies of organizational whistle-blowers. One study included 13 whistle-blowers in the collegiate sports industry (Richardson & McGlynn, 2011), while the other examined eleven whistle-blowers in the public school system (Gravley, Richardson, & Allison, 2015). These studies utilized interview survey methodologies. Each study received approval from our university's Institutional Review Board before data collection commenced.

Data Collection

In order to identify whistle-blowers, we conducted Internet searches and relied on personal networks. Between the two studies, we identified and requested participation from 43 whistle-blowers. Twenty-two whistle-blowers did not return our messages or declined to participate, while 24 agreed to participate (N = 24). The interview protocols varied slightly across the two studies, but each included open-ended questions of the individual's background at their respective organizations, the types of wrongdoing they encountered, their whistle-blowing decision-making processes, the retaliation/responses they encountered, and the long-term effects on their careers and lives. Interviews ranged from 45–110 minutes in length. All interviews were transcribed and, combined with the typed responses, resulted in 185 pages of single-spaced data. Twenty of the participants were interviewed, while four preferred to complete the interview protocol and email it back to us. Participants, who included thirteen women and eleven men, were located in every region of the U.S. Their ages ranged from the upper 20s to the mid-60s.

Data Analysis

For the present study, we focused our analysis on those statements which related to "retaliation." In the first stage of coding, we copied and pasted all references to retaliation in a separate Word document. Next, we read each of these statements and considered whether they addressed any of the three research questions. Those statements that did address one of the research questions were copied and pasted into separate documents. We then utilized open coding to identify particular statements that could be sorted and labeled into the respective research questions (Lindlof & Taylor, 2002). Specifically, we read the participants' quotes, discussed their meanings, compared them to related statements, and coded them into categories using an electronic copy-and-paste technique (Charmaz, 2006). Open coding was followed by axial coding which involved identifying relationships within and between categories, a process which led to the formation of coherent themes. Each unit of conversation, or complete statement, fell into a discrete category and was not used to support more than one theme. Next, we present these themes in the Findings section.

Findings

RQ1 asked "How is retaliation against whistle-blowers expressed?" Our analysis revealed four forms of retaliation expression from the perspective of participants: labeling, isolation (alienation and ostracism), surveillance, and fear tactics.

Labeling. We characterize labeling as delimiting a whistle-blower through false accusations, rumors, and name calling. Our analysis reveals three prominent contexts in which labeling was used to communicate retaliation: professional, social, and gendered/sexuality. In the professional context, Lane, a long-time member of the community and local school board trustee, discussed the rumors about his job performance, including the administration's attempt to label him as a trouble-maker. He said, "The superintendent's wife, who was implicated in some of the wrongdoings would actually go around telling blatant lies about me…They were…saying I was bad for the school." Whistle-blower Brad noted he was called "Chicken Little" by an administrator as a derogatory way of claiming he was blowing things out of proportion. Another whistle-blower, Nichole, described how the school administrators labeled her, a local business owner, as a "thief."

While many of the whistle-blowers in this present study experienced retaliation through labeling in a professional context, others experienced it on a more personal level. For example, whistle-blower Laura described the affects that labeling had on her friendships within the school.

> You know, teachers talk. And so I think in order to get along so that (my son's teacher) wasn't marked, she joined in the fray, saying, 'Oh, that mother is so weird, Oh my God, she asks so many questions'—so that's when the gossip portion started….And they would flat out tell you 'you're crazy, you're an idiot.'

Additionally, the use of personal labeling took on a form of gendered retaliation, causing some participants to describe their gender as a liability. For example, Asa explained the retaliation she received in this way:

> In some ways, my gender obviously worked against me. In the south, where the good old boy network runs the (athletic conference) and the state government, it was easy for the athletics interest to retaliate against me…and the local media soon followed suit, disenfranchising me as a professor and referring to me merely as '*Mrs.*'

Finally, Alice stated, "How many times are male whistle-blowers called 'sluts' or are accused of sleeping with the entire town? Some of the negativity I got from fans was based on my physical appearance or my sex life."

The act of labelling, or naming, someone cannot be underestimated for its symbolic import. Wood (1992) argues "naming is perhaps the fundamental symbolic act" (p. 352) as it isolates that which is named for attention. Further, those in power typically are allowed to name and, in doing so, control our perceptions about the object or person. Thus, the power of labelling a whistle-blower as "crazy," "thief" or "a slut" can be substantial. Narrowly defining the whistle-blower with a problematic label limits the ways potential allies or supportive stakeholders might view the stakeholder. Further, this form of retaliation can become a stigma that is difficult for the whistle-blower to shake (McGlynn & Richardson, 2014).

Isolation. By *isolation*, we refer to the attempt to alienate or ostracize an individual, in any symbolic or literal means. Treating individuals as if they were invisible (ignoring whistle-blowers or cutting off communication), reassigning or firing the whistle-blower, and intimidating potential allies of whistle-blowers into withdrawing support, were all isolation strategies utilized by those who retaliated. Whistle-blower Maci gave a powerful description of the first moment she realized she was about to face alienation.

> I drove up to the building that first morning—drove up to work because I had to—and I will never forget sitting there in the parking lot looking at the door…And I prayed to give me the strength to go in there. So I went in the office and…said hello to everybody. Everybody looked at me. No one spoke…all the way down the hallway…no one responded. So, the next day—no one ever spoke to me.

In addition to being cut off from communication, the whistle-blowers experienced alienation through loss of power, e.g., reassignment of job duties, or complete termination. For example, whistle-blower Bob was reassigned from his job as athletic compliance officer to another position on campus with less power and oversight over athletics.

Retaliation also took the form of social isolation when other members of the organization became agents of retaliation against the whistle-blower. Whistle-blower Alicia reported, "I was treated like a pariah. I was nauseous going to the office. One person that could have helped me would not be seen with me on campus. People wouldn't even eat lunch with me in town." By its very nature, whistle-blowing can be a lonely experience; it often involves a lone individual exposing the secrets of the group or organization. Making matters worse, retaliators further endeavor to make the whistle-blower feel increasingly alone through a number of isolating, ostracizing, and alienating strategies.

Surveillance. Some whistle-blowers experienced retaliation in the form of intensified scrutiny or surveillance, which emerged when members of the organization began seeking "documentation" or evidence against the whistle-blower. Cari described the school's use of her children in their attempt to seek evidence in which to discredit her.

> I started noticing all of a sudden that my kids were being zeroed in on…There was one day… they had sent [my son] down to the nurse's station because…the principal…said he stank… Later when I talked to the school nurse, she said she didn't smell anything…They were definitely…looking for something for me to get in trouble on.

While some surveillance was visible, other forms of this retaliation were less obvious.

For example, some whistle-blowers' narratives included accounts of invasion of privacy during these surveillance events. Millie discovered that the General Counsel had "bugged" her phone, while whistle-blower Maci described the moment she realized the administration had been reading her e-mail. "I went in one day to my computer, and I always turn…everything down, off at night. I booted up my computer and there was a window open on my email, and I don't do that…The email was open…So I figured out they were reading my e-mails."

Through direct monitoring, such as intently examining their children in order to allege abuse, or through more covert means of surveillance, retaliators attempted to confirm to the whistle-blower that their every move was being watched. The next form of retaliation was more direct in attempting to instill fear in the whistle-blower.

Fear Tactics. This final theme captures the most direct retaliation used against whistle-blowers and includes intimidation, physical violence, threats against families, and even death threats. While intimidation, and at times physical violence, were not widespread, they were used against several participants. The use of intimidation by anonymous agents emerged as Lucy shared a story about the morning she stepped outside to retrieve her morning paper and found a large rat trap—spring set—placed on her front porch. "I got the message, loud and clear," she said. Meanwhile, Alicia commented, "I had people tell me they were going to send me bombs...and then I realized people knew where my parents were—that kind of started freaking me out."

Forms of physical violence were also used against whistle-blowers. Alison reported that the tires on her daughter's car were slashed, and on another occasion, "we were slammed into the wall by the (restaurant) door and threatened (so we) would drop the case." Additionally, many of the whistle-blowers in this study noted the retaliators' targeting of family members as a fear tactic. Perhaps the most profound use of this strategy involved using CPS against whistle-blowers as described in Laura's account. She was told by an administrative assistant that

> [The principal said,] 'Well, if that woman does not shut the fuck up, I'm going to be calling Children's Protective Service on speed dial every time she so much as looks at me or comes on this campus. I'm going to take care of her.'

Finally, several participants reported receiving death threats. Alice, received an anonymous phone call with the message "we're going to kill you, you fucking bitch." Finally, whistle-blower Bill reported the following case:

> I had death threats against me immediately...The chief of police stopped me and said, 'Do you know where your wife and kids are?' and I said, 'Yes, I do,' and he said 'We need to get them home as quickly as possible—the death threat has been extended to them.'

The use of death threats seems to reveal how strongly retaliators feel they have been betrayed by the whistle-blower. Overall, the types of retaliation included within this theme reveal just how far wrongdoers and their allies will go in order to coerce silence and preserve the status quo. Next, we discuss whistle-blowers' perspectives on why retaliation was used against them.

Attributions of Retaliation

RQ2 asked "What are the reasons retaliation is used against whistle-blowers?" In our analysis we found that participants perceived several short-term, or micro, motives for retaliation, which we categorized into two themes: intimidation and discrediting of whistle-blower/eroding his or her power base. Additionally, the participants perceived broader, macro reasons for the retaliation they received, namely, financial motivations and preservation of a myth or identity.

Intimidation. When individuals are perceived as violating organizational norms by engaging in disclosure of wrongdoing, members or stakeholders of the organization may react to the threat in a variety of techniques. Our whistle-blowers attributed the organization's need to intimidate them as one motivation for retaliation. Whistle-blower Alison illustrated the connection between retaliation and the need to intimidate, saying, "I got threats all the time, my daughter got threats, we had calls to our home for a time...I think definitely some pressure was applied there to frighten me." Other participants perceived the use of intimidation in retaliation as an utter abuse of power. Maci explained how the members of the organization, specifically the public school administrators, utilized the power associated with their positions to "keep control" of employees and those in the community, and to intimidate them into behaving within organizational group norms:

> I remember that night being scared...They had power. They had been in the district for so long that they had so many favors out...Like, I've got your kid out of trouble one time, so you will never question what I do or if I need support...It was presumed power...So things like that created fear.

Considering scholars have identified intimidation as a cultural characteristic that prohibits dissent (Milliken, Morrison, & Hewlin, 2003), it makes sense that retaliation against whistle-blowers is used for similar purposes.

Discredit reputation/erode Power base. While some participants viewed intimidation as a motivation for retaliation, others viewed the act of retaliation as a means in which to discredit their reputation or erode their power bases. Maci described the administrators from her school as "extremely threatened" by a new candidate for school board, due to the large following this candidate received from the community. "It was probably the first real threat of anybody coming onto the board that they hadn't hand-picked, and they went ballistic. They started campaigning, had T-shirts made...they blatantly told people that [she was] evil ...". Similarly, whistle-blower Nichole reported the administration's desire to not only separate the whistle-blower from his or her power-base, but, upon doing so, to administer absolute seclusion upon the individual: "And of course...you can't be seen with me...because they're going to start giving you a hard time." As Near and Miceli (1995) argue, whistle-blowers are more likely to be effective when they are perceived as credible and as possessing power. The tactics discussed in the section above all serve to damage credibility and wear away power. Next, we discuss financial reasons and preservation of some myth or identity, as perceived reasons for retaliation.

Financial interests. When telling the truth threatens the financial well-being of an organization, retaliation may occur in order to silence the whistle-blower and protect the financial interests of the retaliators (Miceli et al., 2008). Brian explained his experience, saying "There is so much money involved—it's all about money," while another participant, Alison, reported how she witnessed others "placating to the money people," adding, "it's [about the] big bucks." Similarly, Nichole perceived the "big bucks" as one of the primary reasons for the alleged wrong-doing and the retaliation she received: "What I discovered was that they were dishonest. They were manipulating the numbers—that they were using tactics that bordered on being bullies."

Many of our participants expressed surprise and disappointment in the financial motivation for retaliation and wrong-doing. Millie explained, saying, "I never expected the administration that I had been a part of would turn against me, publicly disparaging me, slandering my character and my integrity to protect revenues in the athletic program." It is no surprise that whistle-blowers linked the retaliation they encountered with the organization's need to protect its financial resources. Research suggests retaliation will increase along with the seriousness of wrongdoing (Mesmer-Magnus & Viswevaran, 2005); it seems logical that as financial stakes of wrongdoing increase, so too does it seriousness. Whistle-blowers in the present study also attributed the retaliation they encountered to preservation of myth.

Preservation of myth. In addition to economic motivations, participants perceived the need for the retaliators to preserve an idealized image, sometimes a mythical identity, important to the organization or individuals within the organization, as a motivator for retaliation. For example, Lane explained the motivations of the board members and administrators within his school district, for the reactions he received when he, as a school board member, reported wrong-doing:

> You know, you're on the board with a bunch of people, and they're continually saying that they've got the kids' best interest in mind, and they've got the tax payers' best interest in mind, and I used to really believe that. Now I don't as much.

Lane believed the public school's mission of doing "what is best for the kids," was purely a myth that the school attempted to perpetuate. Millie contributed additional insight into the motivation for retaliation among stakeholders within college athletic organizations, attributing retaliation to the preservation of the myth and sanctity of collegiate athletics. She said,

> In a state and region where the population is undereducated, college football is a religion. Citizens live through difficult work weeks for the spiritual experience of a college game day. To attack their one source of identity and solace is to undermine their raison d'etre.

Some participants highlighted the retaliators' desire to protect the idealized image of the organization as a powerful motivation for retaliation. For example, Tracy, a whistle-blower in the same region as another famous whistle-blowing case, explained how powerful the motivation to protect one's identity can be:

> The principal, he was all about making everything positive. He came up with all of these programs to promote the kids...but then he was also afraid of anything bad being publicized about the school...or about him...because (of a previous case). But this was the principal that was hired after (that) incident and so he wanted the school to have a stellar image and for him to have a stellar image...and I think that he felt by me doing my job, then that was going to maybe scar the school's image—or his image.

Organizations project and desire images of ethical compliance and behavior; whistle-blowers publicly question that image and are frequently punished in return. Next, we discuss outcomes associated with retaliation.

Retaliation Outcomes

Our third research question asked "What are the outcomes of retaliation for whistle-blowers, the retaliators, and the organization?" For whistle-blowers, two identity-related themes emerged revealing the outcomes associated with retaliation: sustaining their identities or experience of a fractured identity.

Sustained/fractured identities. As mentioned above, researchers have identified a "master status" associated with whistle-blowing, revealing the individual's identity as sharply defined by this act. While Rothschild and Miethe (1999) observed a master status resulting in a "battled and embittered" whistle-blower, other research suggests that some whistle-blowers' who experienced retaliation, may develop a master status which sustains or even strengthens a favorable identity for the whistle-blower (Gravley et al., 2015). Consider Laura, who viewed herself as a crusader for good, specifically in the context of the public school system. Laura explained how she continued to 'crusade' for educational reform, even after surviving extreme retaliation.

> I've continued to show up at school board meetings and work with the community to elect other school board members... [and] volunteer...If I can give a voice to a problem that is in the system, then maybe that's my sole mission in life as a parent.

While retaliation resulted in Laura's sustainment of her identity, other participates experience a fractured identity as a result of experiencing retaliation. For example, Cari, the participant who was viewed as a "good mom" among those in her community, confessed that even several years after the whistle-blowing experience, she must "worry about appearance on everything." Cari explained that even after moving her children to a different school, her fear remained, saying, "I want (the new school) to think I'm a good mom—I'm afraid they don't think I'm a good mom—I'm afraid someone from (the former school) has talked to them and they already have it out for me." Even though Cari's identity was fractured, she expressed her desire to restore her previous identity, (i.e., "I want them to think I'm a good mom). However, other participants in our study admitted that they had abandoned their previous identities. For example, when asked if she would blow the whistle again, Alicia said "No, absolutely not, would have packed and up and left. I'll never be the same person and I'll never get back that time."

Retaliator advancement. While the participants viewed the outcomes associated with retaliation for themselves in terms of their personal identities, they viewed the outcomes for the retaliators as stipulated by their personal advancements; indeed several whistle-blowers noted those who retaliated against them the most vociferously were rewarded by the organization. For example, Millie explained the manner in which the people who carried out the retaliation against her were rewarded for their efforts. "These women (retaliators)...were clearly playing the ...game for their own advancement. It was quite disheartening to find these colleagues betray...our friendship...for their own personal advancement." Another whistle-blower, Paul, expressed similar frustration with the retaliator's personal advancement, lamenting the promotion the retaliator received. Paul said "He got rewarded for it, that's for sure," an outcome he associated with the act of retaliation.

Finally, we examined the outcomes of retaliation against the whistle-blower for the focal organization. Many participants reported a "chilling effect" on the organization subsequent to retaliation, resulting in the ability for the organization to continue in the acts of wrongdoing. For example, Nichole evoked the "Nazi" metaphor when describing the culture of her school district subsequent to the whistle-blowing and retaliation incident. "It's kind of—it's almost... as bad as living in Hitler's Germany, but I understand how people are afraid to say anything." To illustrate the chilling effect retaliation may have on other potential whistle-blowers, when asked if she would come out against misconduct again, if she had the opportunity, Maci clearly stated, "Hell no," while Brian advised, "never, ever, ever, never report anything."

Other whistle-blowers received the satisfaction and reassurance when the organization was forced to stop the misconduct. For example, Bill, even though he no longer works in the same school district, reported that the community and district are better off because of his actions.

> It was mishandled completely by the superintendent and it could have been handled better, [but] I had the opportunity to teach people about ethics, morals, values, character, and all those things...This sends a strong message to everybody...sends a strong message to the community that violations are not going to be tolerated.

In essence, some participants in this study expressed a sense of fulfillment when the outcomes for the organization resulted in a transformation of organizational behavior. Retaliation has consequences for all participants in the whistleblowing process: the whistle-blower, the retaliator, and the organization.

Discussion

The purpose of this study was to explore retaliation against whistle-blowers as a form of the "dark side of organizational communication." A key realization from this study was the extent retaliation against whistle-blowers is indeed "dark." Whistle-blowers were slandered, labeled with pejorative terms, physically and psychologically cut off from their jobs and support structures, physically intimidated, covertly monitored, and subjected to death threats. When we consider the dramatic ways retaliation was doled out, it is surprising there is a paucity of communication research into whistle-blowing in general, and these specific forms of retaliatory expression in particular. There are few experiences that impact organizational members as much as systematic retaliation. Thus, this form of communication warrants additional research.

First, scholars could examine retaliation as a form of organizational feedback. Specifically, we view retaliation as a form of feedback utilized by organizational members to correct or alter a whistle-blower's behavior (e.g., by encouraging silence). From the whistle-blower's perspective, this form of feedback possesses a negative valence and "consists of messages that refer unfavorably to the recipient's behavior" (p. 632). Geddes and Linnehan (1996) further deconstructed the negative valence dimension of feedback, noting that it was destructive in nature. As they explained, such feedback was characterized by negative language and a harsh delivery, conditions which certainly align with the retaliation we identified above. This connection to feedback led to us developing a definition of retaliation. We define retaliation against whistle-blowers as *negatively-valenced feedback directed toward organizational whistle-blowers in order to harm their reputations and undermine their power, while preserving the wrongdoer or organization's ability to carry out unethical behavior.* Conceptualizing retaliation in this way can foster research into this phenomenon. Specifically, scholars could measure the negative valence of various forms of retaliation, and consider which factors lead to more (or less) of this type of negative feedback.

Next, we should investigate perspectives of management and coworkers as they make decisions about whether to support or retaliate against whistle-blowers. Social information processing may be especially valuable in understanding how groups reach decisions on whether to support a whistle-blower. This theory proposes that "job attitudes are a function of the communicative activities of employees" (Miller & Monge, 1985, p. 365–366.) Pfeffer and Salancik (1978) argued that individuals' attitudes and needs revolving around an issue are influenced by the information available to them at the time those issues are discussed. Social information is critical as group members interpret complex cues from the environment and make judgments about which environmental cues are salient. Scholars should examine decisions to support/retaliation against whistle-blowers as a produce of social information processing.

Conclusion

Based upon the findings of the present study, we are confident retaliation against whistle-blowers is indeed representative of the "dark side of organizational communication." Communication that serves to alienate, ostracize, engender fear, and intimidate organizational members who are attempting to address unethical behavior is not routinely discussed within scholarly and practitioner articles. However, considering the impact on whistle-blowers' lives *and* the climate of the organization, it is imperative we better understand retaliatory communication. If history is any indication, organizational wrongdoing will persist; thus, it is likely an increasing number of people will face the decision of whether to blow the whistle.

References

Alford, C. F. (1999). Whistle-blowers: How much can we learn from them depends on how much we can give up. *American Behavioral Scientist, 43*, 264–277.

Alford, C. F. (2001). Whistle-blowers and the narrative of ethics. *Journal of Social Philosophy, 32*, 402–418.

Forgiveness, apology, and communicative responses to hurtful events. *Communication Reports, 19*, 45–56.

Bocchiaro, P., & Zamperini, A. (2012). Conformity, obedience, disobedience: The power of the situation. In Gina Rossi (Ed.), *Psychology—Selected Papers*. Rijeka, Croatia: InTech

Charmaz, K. (2006). *Constructing grounded theory: A practical guide through qualitative analysis*. London: Sage.

Cortina, L. M., & Magley, V. J. (2003). Raising voice, risking retaliation: Events following interpersonal mistreatment in the workplace. *Journal of Occupational Health Psychology, 8*, 247–265.

Geddes, D., & Linnehan, F. (1996). Exploring the dimensionality of positive and negative performance feedback. *Communication Quarterly, 44*, 326–344.

Gravley, D., Richardson, B. K., & Allison, J. M. (2015). Navigating the narrative "abyss": Using narrative analysis to explore relationships among whistle-blowing, retaliation, and identity. *Management Communication Quarterly, 29*, 171–197.

Greenberger, D. B., Miceli, M. P., & Cohen, D. J. (1987). Oppositionists and group norms: The reciprocal influence of whistle-blowers and co-workers. *Journal of Business Ethics 6*, 527–542.

Keenan, J. P. (2002). Comparing Indian and American managers on whistleblowing. *Employee Responsibilities & Rights Journal, 14*, 79–89.

The effects of interpersonal closeness and issue seriousness on blowing the whistle. *Journal of Business Communication, 34*, 419–436.

Lindlof, T. R., & Taylor, B. C. 2002. *Qualitative communication research methods* (2nd ed.). Thousand Oaks, CA: Sage.

McGlynn, J., & Richardson, B. K. (2014). Public support, private alienation: Whistle-blowers and the paradox of social support. *Western Journal of Communication, 78(2)*, 213–237.

Mesmer-Magnus, J. R., & Viswesvaran, C. (2005). Whistleblowing in organizations: An examination of correlates of whistleblowing intentions, actions, and retaliation. *Journal of Business Ethics, 62*, 277–297.

Miceli, M. P., Near, J. P., & Dworkin, T. M. (2008). *Whistle-blowing in Organizations*. New York: Routledge.

Miller, K. I., & Monge, P. R. (1985). Social information and employee anxiety about organizational change. *Human Communication Research, 11*, 365–386.

Milliken, F. J., Morrison, E. W., & Hewlin, P. F. (2003). An exploratory study of employee silence: Issues that employees don't communicate upward and why. *Journal of Management Studies, 40*, 1453–1476.

Near, J. P., & Miceli, M. P. (1996). Whistle-blowing: Myth and reality. *Journal of Management, 22*, 507–526.

Pfeffer, J., & Salancik, G. R. (1978). *The external control of organizations*. New York: Harper & Row.

Richardson, B, K., & McGlynn, J. (2011). Rabid fans, death threats, and dysfunctional stakeholders: The influence of organizational and industry contexts on whistle-blowing cases. *Management Communication Quarterly, 25*, 121–150.

Rothschild, J., & Miethe, T. D. (1999). Whistle-blower disclosures and management retaliation. *Work and Occupations, 26*, 107–128.

Schachter, S. (1959). *The Psychology of Affiliation*. Palo Alto, CA: Stanford University Press.

Wood, J. T. (1992). Telling our stories: Narratives as a basis for theorizing sexual harassment. *Journal of Applied Communication Research, 20*, 349–362.

EIGHT

"A Cat Fight in the Office": The Use of Gossip as a Means of Resource Control

Falon Kartch and Kathleen S. Valde

Destructive workplace communication interferes with people's ability to work, hurts information flow, and damages trust and cooperation (Lutgen-Sandvik & Sypher, 2009). Destructive workplace communication behaviors include bullying, incivility, ostracism, sexual harassment, racial harassment, and hostile workplace relationships (Lutgen-Sandvik & Sypher). The dark side of workplace gossip, with its destructive potential as an aggressive form of communication, should be included among these behaviors.

Gossip is common in the workplace (Bartunek, Kolb, & Lewicki, 1992; Bergmann, 1987/1993). Gossip is "informal and evaluative talk in an organization, usually among no more than a few individuals, about another member of that organization who is not present" (Kurland & Pelled, 2000, p. 429). Gossip is a social activity occurring within a larger, existing system of relationships (Archer & Coyne, 2005; Van Vleet, 2003). Employees use gossip to vent about bothersome workplace issues and receive support from colleagues (Bartunek et al.).

Gossip has a "chaotic quality" that disrupts the workplace (Bergman, 1987/1993, p. 134) and harms relationships critical to a successful workplace (Bartunek et al.; Hafen, 2004). Thus, organizational gossip fits into the paradigm of dark side communication. This chapter will focus on the dark elements of gossip and the ways in which gossip is an aggressive, destructive form of workplace communication.

Gossip as Aggressive Communication

Gossip is labeled a form of aggression (Archer & Coyne, 2005; Baron, 2004). A variety of types of aggression exist including physical, verbal, indirect, social, and relational (Willer & Cupach, 2011).

Gossip can be viewed as both relational and social aggression. Relational aggression refers to attempts to harm another by manipulating their relationships with others (Crick & Grotpeter, 1995; Willer & Cupach, 2011). Social aggression refers to the attempt to bring social harm to another by threatening their self-esteem and/or social status (Galen & Underwood, 1997). Because gossip can be used to manipulate relationships and alter others' self-esteem and social status, gossip may be an attractive tactic for individuals seeking social power.

Resource Control Theory

This study employs resource control theory (RCT) (Hawley, 1999, 2007) to explore gossip as a socially and relationally aggressive tactic within organizations. According to RCT, various forms of aggression are used to exert power over others, thus gaining resources within a group (Willer & Cupach, 2011). Based on evolutionary views, RCT explains aggression as an adaptive strategy (Hawley, 1999). Historically, humans relied on access to resources such as food to stay alive. Relationships were another important resource, as access to a larger group meant hunting in groups and protecting one another from harm. In today's organization, relationships remain a valuable resource as they provide access to information and various forms of support.

RCT assumes individuals are motivated to access and control resources (Hawley, 2007; Willer & Cupach, 2011). One way an individual obtains control over resources is through membership and/or a leadership position within a social group (Willer & Cupach, 2011). Individuals achieve power, dominance, and status within a social group using prosocial or coercive strategies (Hawley, 2007). Prosocial strategies involve using "socially acceptable behavior" to obtain access to resources (Hawley, 2007, p. 12). People also obtain resource control through coercive strategies (Hawley, 2007), such as aggression and threats (Willer & Cupach).

Gossip as a Resource Control Tactic

Organizational members use gossip to obtain social status and power (Baumeister, Zhang, & Vohs, 2004; Bergmann, 1987/1993; Kurland & Pelled, 2000). Guendouzi's (2001) study of women in conversation found bitching was "power-driven, enabling its participants to compete for social capital" (p. 47). DiFonzo and Bordia (2007) depict gossip as a comparison process of putting other people down in order to build up self. Noon and Delbridge (1993) described gossip as a power-gaining strategy.

Three ways people might use gossip to control resources include gossiping to manipulate relationships, build cliques, and ostracize organizational members. First, people use gossip to manipulate workplace relationships among individuals and groups (Michelson et al., 2010). Using French and Raven's (1959) types of power, Kurland and Pelled (2000) argue gossipers enhance their power in organizations through gossip. Two types of power related to relationship manipulation are coercive power and expert power. Gossipers' coercive power results from recipients' concerns about becoming targets of gossip, which can lead the recipient to cooperate with the gossiper (Kurland & Pelled). The relationship between gossiper and recipient is manipulated because fear motivates the recipient's actions. Expert power results from other people perceiving one has expert knowledge or information (French & Raven). When the recipient views gossip as expert information, the gossiper gains expert power (Kurland & Pelled). If a gossiper provides gossip (information about people) with the intention of increasing his or her power, the gossip potentially manipulates the relationship.

Second, gossip used to build cliques is a form of resource control. Gossipers tend to develop a network of trusted people with whom they share gossip. For recipients, gossip might create a sense of belonging. Sharing of gossip with the same group creates and reproduces insider/outsider relationships (Michelsen et al., 2010). Thus, gossip can result in the development of "cliques," or groups of individuals who know one another well, typically spend a considerable amount of time together, and share common characteristics (Ennett & Bauman, 1996). Membership in a workplace clique provides people with power in two ways (Lamertz & Aquino, 2004; Sias, 2009). First, cliques enable people to obtain and share information. Second, clique membership protects individuals from social ostracism and provides opportunities to influence fellow clique members.

Third, gossip used to control and influence people through exclusion is a form of resource control. For targets, gossip damages relationships and emphasizes their outsider status (Michelsen et al., 2010) or ostracism. Sias (2009) indicates ostracism occurs when workplace cliques intentionally exclude organizational outcasts from groups.

Gossip used to ostracize can be a means for gaining power. Ostracism is an influence tactic used to bring people into compliance with group norms and beliefs (Williams, 2001). Because ostracism removes one's feelings of connection and belonging and damages self-esteem, those who are ostracized seek to restore these lost needs (Williams). In seeking to restore needs, ostracized individuals may comply with the wishes of the clique from which they have been excluded.

By chronicling a real-life organizational gossip situation, this study seeks to address how organizational members use gossip as a means of resource control. This chapter explores the dark side of organizational communication by addressing the following research question: How can gossip be used as a means of resource control in the workplace?

Methods

Data Collection[1]

With the university's institutional review board approval, data were drawn from participant-observer ethnographic fieldwork done at Gateway, and ethnographic interviewing with organizational members. Gateway is a local living history museum in a mid-sized, upper-Midwestern city. Gateway employs 14 full-time staff members, five part-time staff members, and 20 living history interpreters. They rely heavily on over 300 volunteers to perform special events and daily operations.

As a participant-observer, the first author collected data over a 10-week period by working alongside administrative staff in two different roles (Emerson, Fretz, & Shaw, 1995). For the first five weeks, the researcher worked in the administrative offices as the Marketing Assistant. Located in these offices is the largest copy machine in the building, the fax machine, the laminator, all employee mailboxes, and the supply storeroom, making it a prime location to observe and interact with staff.

For the subsequent five weeks the researcher worked with the Museum Store Supervisor at the museum's front desk. Tasks included working as a cashier and answering phones. The front desk is positioned by the main entrance and employee timesheets are located at the desk providing ample opportunities to interact with employees and volunteers that work in the museum.

Ethnographers must consider the degree to which they will participate in the setting and how much they plan to observe (Bishop, 1999). Sometimes the researcher stayed at her desk and observed the actions and conversations of employees. Other times the researcher actively participated in conversations and activities by asking questions to enhance the clarity and detail of the data. Having the Marketing Assistant position allowed a great deal of access to the Director of Marketing, Marie, which created both an opportunity and a limitation. Despite trying to remain neutral, organizational members viewed the researcher as being part of Marie's clique. This perception made it difficult to access information from certain employees due to their negative feelings toward Marie. These data would have made for a more well-rounded analysis and offered additional insight into the escalation of the central conflict and use of gossip within the other clique. However, being seen by organizational members as part of Marie's clique allowed privileged access to that clique. Data were enriched by thick description of gossip conversations and events within Marie's clique that an outsider would not have been able to access.

The researcher took extensive fieldnotes after every visit to ensure maximum accuracy and detail. Fieldnotes contained a written, chronological, re-creation of the time in the setting (Emerson, Fretz, & Shaw, 1995) and included tasks done and all interactions with and overheard between members. Conversations were recorded with as much verbatim text as possible. Fieldnotes focused on describing what took place in the setting by privileging member meanings. To accomplish this, the researcher withheld all analysis and value judgments. Fieldnotes and transcribed interview data combined resulted in 171 single-spaced pages of data.

Ethnographic interviews with four employees were also conducted. An interview protocol provided an outline of questions and follow up questions were asked when appropriate to maximize the amount of data collected, (Lofland & Lofland, 1995). All interviews were audio-recorded with written consent from participants and transcribed for analysis.

Data Analysis

The data were analyzed using Spradley's (1979) semantic relationships to identify domains of meaning. Spradley proposed semantic relationships help specify relationships between data and domains of meaning. The means to end semantic relationship was used to identify ways in which gossip was used to gain control of relational resources. During data analysis, we noticed cause-effect relationships and began coding to better understand the effects of gossip.

Results

The research question asked how gossip can be used as a means of resource control in the workplace. Results indicate gossip was prevalent in the organization and used as relationally and socially aggressive tactics to manipulate workplace relationships, build cliques, and exclude people. Data analysis also identified two destructive outcomes from the gossip.

The target of the gossip was Marie (pseudonyms were used), who was head of the Marketing Department. Jane, a member of the Education Department, was the primary gossiper.

Marie was uncertain what had caused Jane to begin gossiping about her. However, as demonstrated in a conversation with the researcher, Marie was aware of the gossip and believed Jane used gossip strategically to gain power.

> Researcher: Really? You think she has that much power?
> Marie: Oh, she is good. What is that phrase?
> There is a pause.
> Marie: Tickle his ear—that's what the Bible says—tickle their ears, people like gossip and she knows how to spin it.

Jane's ability to "spin" gossip enabled her and her network to control important workplace resources; thus, through gossip, Jane and her insider network gained power.

Gossip and Relationship Manipulation

One tactic employed to control important resources was using gossip to manipulate relationships. In the following fieldnote excerpt, Marie talks about the indirect way Jane caused harm through gossip:

> I would never hear it directly because it was about me so it was usually my closer friends that we talked about earlier who would tell me over lunch, when we would sit in at lunch together, they wouldn't always tell me stuff. I would just say, well you know I'm going through blah blah blah you know this situation, and then reluctantly they would say "well we kind of overheard this" or "we overheard that" or maybe "that explains why you know this other supervisor was treating this other gal a certain way."

This excerpt illustrates Baron's (2004) claim that gossip, as a form of aggression, indirectly harms the target. Jane did not confront Marie directly; instead, she indirectly harmed Marie by altering her relationships with her friends. Her friends' hesitations in sharing the stories they heard demonstrate how gossip, as a relational form of aggression, manipulates and damages relationships. Jane's gossip puts Marie's friends in an awkward position. If they speak against Jane and her gossip and it gets back to her, they risk becoming targets of her gossip (Hafen, 2004). However, friends, typically, share gossip with each other. By not telling Marie the gossip, her friends violate friendship norms and weaken the relationships.

Jane and other employees also used gossip to influence the boss, John, president of the museum. In an interview, Marie talked about other employees going to John with gossip. "I don't know if it's just because of the department I was in, because sometimes there's strength in numbers if you share similar opinions then the boss goes 'ooh whooh, three of my staff thinks this.' So you know it can be persuasive." Marie believed John was persuaded by Jane's gossip. This is further illustrated in a fieldnote excerpt where Marie discusses John's response to gossip he heard from Jane:

> ...when I was meeting with John for my review he said that he hears things, probably about me from them I think, and that he is going to believe them because "they came to him first." She throws her hands up in the air and shakes her head... Instead of coming to me and talking to me about it, asking me my side of the story he just believes them.

The excerpt demonstrates the effectiveness of gossip as a tactic for obtaining social power in the workplace by manipulating relationships. Hafen (2004) explains when gossip is shared with someone in authority, the possibility exists the recipient will deem the gossip important and convert it to information to be acted on. John's response suggests he is going to treat Jane's gossip as information. By viewing the gossip as valuable information, John validates Jane's group as having more social power. Jane gains coercive power over John and is able to manipulate his relationship with Marie (Kurland & Pelled, 2004).

In an interview, Marie's comments about the ways gossip eroded trust and made her feel betrayed further evidenced the manipulation of organizational relationships:

> I just think gossip in an office just destroys the morale. I just think it's probably one of the number one things to make a place bad to work at. I mean, you know, gossip and then of course lying, stealing, and I mean lying, that's probably number one, when you can't trust. But you know, gossip's right up there because you know if that's happening then you don't know about coworkers and you can't trust them. Who can you trust?

When speaking about Jane, Jane's clique, and John, Marie stated, "I really felt betrayed by them." Gossip created a climate within Gateway where Marie distrusted some of her coworkers and felt betrayed by them. Scholars have conceptualized betrayal as a vehicle for relational devaluation, which occurs when an individual's behavior makes the target question how much the other actually values the relationship (Kowalski, Walker, Wilkinson, Queen, & Sharpe, 2003). Jane's gossiping enabled her to manipulate and damage relationships between Marie and other organizational members. This manipulation exemplifies relational aggression (Crick & Grotpeter, 1995). The resulting damage to Marie's reputation and organizational status illustrates how gossip operates as social aggression (Galen & Underwood, 1997).

Gossip and the Formation of Clique

A second tactic for controlling relational resources was gossip used to build cliques. As gossip permeated Gateway, coworkers took sides in the conflict and two cliques formed. In the following fieldnote excerpt Marie describes this phenomenon:

> ...it has become a "cat fight in the office. It is me, Cindy, Amanda, and Joe." The researcher interjects: "What about Ashley?" She replies that Ashley is with her too. She goes on to say that she thinks Julie is on her side too, but does not talk about it with her much because "I don't want to drag her in the middle." Then she says the other side consists of Adam, Jane, and everyone in Education. "Even Emily?" The researcher asks. She replies affirmative.

Marie is explaining who is a part of the two cliques: hers and Jane's. In an interview, Marie talked about how gossip created a clique against her:

> ...she [Jane] was a part of another department by then and I soon gathered that her opinions became common conversations in her new department. So then there were at least two other people in her department then who started to share her same similar talk and it got to be like a gossip thing or something...I knew that this person was responsible for influencing at least three other people on staff and maybe a fourth in their attitudes towards me. I just know that she influenced a lot of bad blood...

From Marie's perspective, Jane actively created a clique that shared a similar, negative opinion about her. By turning coworkers against Marie, support for Jane's position grew and she gained social power.

Gossip, Cliques, and Ostracism

Third, gossip was used as a tactic for gaining relational resources through ostracism. In an interview, Cindy talked about the creation of cliques and Marie's exclusion from Jane's clique:

> ...I think the climate is, it's getting, areas are getting subdivided and I just think it's a shame. I think everyone needs to be on the same page, be a team... I see one department being not very nice to Marketing and that bothers me, because I feel that our Marketing Director hasn't done anything wrong and I just don't understand. They are just on her on her all the time... there is one big division going on right now and I guess that is what I am speaking of.

Cindy believed other coworkers were intentionally excluding Marie and did not believe that behavior was warranted. Williams's (2001) research on ostracism suggests exclusion damages one's sense of belonging, which leads one to try to re-establish connections and a sense of belonging in the workplace. Marie's attempt to rebuild a sense of connection and belonging is evidenced in this fieldnote excerpt where she talked to Jason, a coworker and clique member, to find out how the boss heard about a gossip story:

> Marie: Jason left me a message last night around 8. I wonder if it had anything to do with this. She picks up her office phone and dials a number. There is a pause. Marie begins talking into the phone: hey. You called me last night. There is a long pause. Marie laughs. There is another long pause. Marie: I got called into John's office today because he found out I was talking to Cindy about going to [the] managers' [meeting]. There is a long pause. Marie: do you know how he knows? Another pause. Marie: okay, well I just wanted to see if that's why you called. "Just another day in paradise." Bye. Marie hangs up the phone and turns to me, he doesn't know anything.

Jane and her clique used gossip as social and relational aggression against Marie. By forming her own clique, Marie attempted to protect herself from Jane's and her clique's gossip.

In an interview, Marie described how her clique kept her informed of Jane and her clique: "I knew things that had happened. I mean I had my own little informants who had told me so and so said this and so and so said that. And so I don't think they knew that I was finding out about those behind-doors conversations..." By forming her own clique, Marie was able to re-establish a sense of connection and belonging, and she also used the clique defensively to take back some of her social power.

Ultimately, Jane's use of gossip to gain control of resources prevailed. In an interview, Marie explained John terminated her because of continued gossip:

> ...she [Jane] ended up winning. I didn't think that she would but, I mean I shouldn't say that she won, but I don't know if it was just they didn't stop and think about the damage that was being done and how far it would. But I think that someone of her position and age [laughs] should have been a little bit wiser in really what she was doing. This is just not potshot time with Marie, you know, this is serious and when you go to the boss with stuff like this what are you hoping to accomplish? So I don't know if she really had a goal in it, but it certainly wasn't to see me succeed.

Gossip was used as an aggressive tactic to control resources—relationships—in this workplace. The more coworkers Jane brought into her clique, the more powerful she became. The final move was to get John on her side. It is unknown whether this was Jane's goal; however, the data illustrate how gossip was used as a tool to control and gain power.

Destructive Outcomes of Resource Control Gossip

Data analysis suggests two destructive outcomes resulted from Jane's use of gossip to gain control of relational resources. One destructive outcome was social isolation of employee Cindy, who refused to give up her association with Marie. Cindy's supervisor, Liz, joined Jane's clique and made it clear she did not want Cindy interacting with Marie. In the following fieldnote excerpt Marie describes the situation:

> The researcher says to Marie: what is going on with Cindy? She does not seem like herself today.
>
> Marie: she is having problems with Liz. Liz told her she needs to stay behind the desk all the time and told her not to come in here and talk to me. She said, "stay away from Marie she is in trouble."
>
> Researcher: why would Liz say something like that? I thought you and Liz got along well.
>
> Marie: We did. Now "Jane has been bending her ear. And Jane can be very convincing. She gossips in a way that does not come across as gossip, but as fact."

When Liz viewed Jane's gossip as important workplace information (Hafen, 2004), she sought to decrease the interaction between Cindy and Marie. However, as discussed in an interview, Cindy refused to stop talking to Marie and doubted the gossip: "I'm not going to disown somebody just because she's [Liz] hearing something that probably isn't true. That I know it is not true. But for whatever reason Jane has it out for Marie and it's scary."

This situation between Cindy and Liz continued to escalate. In the following excerpt Cindy explains a rule created to prevent her from talking with Marie: "I was talking to the Marketing Director and I got in trouble for that and then it just somehow happened to come up then at [the] manager's meeting a new rule that I couldn't leave the store." The new rule isolated Cindy and interfered with her ability to do her job responsibilities, which included running the museum store, answering phones, and coordinating food and merchandise purchasing. While she spent much time in the museum store, she still had to move about the museum to make copies, check the mail, send faxes, and communicate with coworkers about events and online museum store purchases. On a typical day, Cindy communicated with a variety of volunteers and employees, including Marie. The new rule was stressful for Cindy: "It is a very caged feeling when you can't, you feel like, you can't go to the bathroom, you can't make a copy, you can't do a fax."

Liz's demands that Cindy stop talking with Marie and the rule about not leaving the museum store resulted in the Cindy's ostracism, which was essentially a compliance move. Williams (2001) argues ostracism can be used to discipline people and get them to comply with norms and beliefs. Williams contends targets will comply in order to end feelings of loss of control and connection that result from ostracism. Liz and other managers in Jane's clique created the rule in order to influence (coerce) Cindy. However, their compliance attempt was unsuccessful. Instead, the physical isolation resulted in a decline in Cindy's ability to do her job and in her organizational satisfaction.

The second destructive outcome of gossip used to control resources was damage to the organizational climate. During an interview, Cindy described how she saw the gossip affecting the workplace: "There's a lot of people whispering and a lot of tenseness going on around me." Her comment makes it clear gossip was contributing to a tense workplace climate. Marie echoed a similar sentiment in an interview when she discussed how the organization had been through a similar situation prior to her working there:

> Well there were a couple of people that were working the museum when the first cloud was there and then obviously during my time there when it was good and then when this kind of stuff started to resurface again. So then they were the ones that said: "this is just like when so and so worked here" and "this is not good" and "we've been through this before" and "why are they doing this" and "it's not fun to work here anymore" and "we used to be such a team." I mean those were exact quotes that I heard from people who had been through it, out of it, and now going back into it.

Marie was not the only person in the organization who noticed and was affected by the gossip. Employees in different departments and at different levels were concerned about the gossip, and their job satisfaction was negatively affected by it.

Discussion

This study indicated gossip was used as a means of social and relational aggression through manipulation of workplace relationships, development of cliques, and isolation of coworkers. Certain employees used workplace relationships as a means of gaining power and status within the organization and used gossip as a tactic to control and manipulate relationships. These results also illustrate how clique formation, relationship manipulation, and isolation can occur simultaneously. Results highlight the destructive, dark elements of workplace gossip and the adaptive nature of this behavior.

Gossip to Manipulate Relationships

Gossip was used to perpetrate relational aggression in this workplace. Crick and Grotpeter (1995) define relational aggression as attempts to harm another by manipulating relationships. Gossip altered Marie's relationship with friends who knew of the gossip but were reluctant to share the gossip with Marie. Gossip also destroyed Marie's relationship with John. Bringing gossip to the boss led to further workplace destruction. RCT would argue Jane's use of gossip to gain power in her relationship with John and to decrease Marie's power with him was adaptive and effective (Hawley, 1999).

Gossip to Form Cliques

Gossip was also used to build cliques. According to RCT, coworkers can be seen as resources because they provide support and assistance (Willer & Cupach, 2011). As Marie said in an interview, "There's strength in numbers." Jane and Marie were attempting to develop strength by forming cliques. The creation of these cliques exemplifies social aggression, which involves attempts to bring social harm by threatening another's self-esteem and/or their social status (Galen & Underwoood, 1997). By using gossip to establish a clique, Jane was able to exclude Marie and obtain social power.

Gossip to Ostracize

Finally, gossip was used to isolate coworkers. Jane used gossip to form a clique that ostracized Marie. Sias (2009) argues some cliques intentionally exclude. Jane's gossip established Marie as an organizational outcast, which damaged Marie's sense of belonging. In response, Marie sought to establish relationships with her coworkers, but unfortunately was unsuccessful in re-establishing her reputation and social status.

Destructive Outcomes of Gossip

Scholars have described the destructive implications of gossip in the workplace (Bartunek et al., 1992; Baumeister et al., 2004; Michelsen et al., 2010). Our results support arguments about the destructive nature of gossip. Destructive outcomes included erosion of the organizational climate, decreased trust, no longer feeling like a "team," and an overall "tenseness" at work. While RCT provides a lens for understanding effective use of gossip to obtain power, it is important not to lose sight of the destructive implications of using gossip to perpetrate aggressive acts. Organizational gossip that erodes relationships and organizational climate should be viewed as a dark side behavior.

Resource Control Theory and Communication Competence

Through the lens of RCT, Jane's socially and relationally aggressive use of gossip was successful and adaptive (Hawley, 2007; Willer & Cupach, 2011). She adapted to her workplace and used her coworkers to obtain and utilize social dominance over Marie. However, this situation was also destructive to Marie, other employees, and the organizational climate. Scholars using RCT have argued aggression should be viewed as competent to the extent it can be used to successfully achieve desired outcomes (e.g., Farmer, Xie, Cairns, & Hutchins, 2007). Therefore, it is important to consider the extent to which Jane's use of gossip was competent.

Interpersonal communication competence consists of three characteristics: effectiveness, appropriateness, and ethics (Rubin & Martin, 1994). Effectiveness refers to the degree an individual was successful in achieving the interaction goal(s) (McCroskey, 1982). RCT emphasizes this characteristic (Hawley, 2007; Willer & Cupach, 2011).

Appropriateness refers to the degree an individual's actions fit cultural, situational, and relational norms (Spitzberg, 1983). An organizational outsider would likely view Jane's gossip as inappropriate. However, in this organization some members apparently viewed gossip as appropriate. Hafen's (2004) argument about the revolving door between gossip and information provides insight on why some people did not view Jane's gossip as inappropriate. When John and Liz heard gossip from Jane, they viewed the content as important and shifted it from gossip to information.

Finally, interpersonal competence requires ethical communication behaviors (Cupach & Canary, 2000). According to Redding (1996), coercive and destructive are two forms of unethical communication. Coercive communication intimidates or threatens in order to limit the hearer's autonomy. Jane's use of gossip to alter John's relationship with Marie exemplifies coercive gossip. Destructive communication attacks the target's self-esteem, reputation, and feelings. Jane's use of gossip indirectly reduced Marie's self-esteem and status. Because Jane's use of gossip to gain resource control is unethical, it cannot be viewed as entirely competent. Future research should explore the ways an organizational culture may deem an inappropriate behavior, like gossip, appropriate.

Notes

1 Data were originally collected as part of the first author's Master's thesis under the direction of the second author. Findings from this project were previously presented at the Chicago Ethnography Conference, the Midwest Popular Culture Association Annual Conference, and as part of a panel discussion at a National Communication Association Convention, but the theoretical framework, analysis, and results presented here are new.

References

Archer, J., & Coyne, S. M. (2005). An integrated view of indirect, relational, and social aggression. *Personality and Social Psychology Review, 9,* 212–230. doi: 10.1207/s15327957pspr0903_2

Baron, R. A. (2004). Workplace aggression and violence: Insights from basic research. In R. W. Griffin & A. O'Leary-Kelley (Eds.), *The dark side of organizational behavior* (pp. 23–61). Hoboken, NJ: Wiley.

Bartunek, J. M., Kolb, D. M., & Lewicki, R. J. (1992). Bringing conflict out from behind the scenes. In D. M. Kolb & J. M. Bartunek (Eds.), *Hidden conflict in organizations: Uncovering behind-the-scenes disputes* (pp. 209–228). Newbury Park, CA: Sage.

Baumeister, R. F., Zhang, L., & Vohs, K. D. (2004). Gossip as cultural learning. *Review of General Psychology, 8,* 111–121. doi: 10.1037/1089-2680.8.2.111

Bergmann, J. R. (1987/1993). *Discrete indiscretions: The social organization of gossip* (J. B. Bednarz, Jr. & E. K. Barron, Trans.) New York: Aldine De Gruyter. (original work published in 1987).

Bishop, W. (1999). *Writing ethnographic research.* Portsmouth, NH: Boynton/Cook.

Crick, N. R., & Grotpeter, J. K. (1995). Relational aggression, gender, and social-psychological adjustment. *Child Development, 66,* 710–722.

Cupach, W. R., & Canary, D. J. (2000). *Competence in interpersonal conflict.* Prospect Heights, IL: Waveland.

DiFonzo, N., & Bordia, P. (2007). *Rumor psychology: Social organizational approaches.* Washington D.C.: American Psychological Association.

Emerson, R. M., Fretz, R. L., & Shaw, L. L. (1995). *Writing ethnographic fieldnotes.* Chicago, IL: The University of Chicago Press.

Ennett, S. T., & Bauman, K. E. (1996). Adolescent social networks: School, demographic, and longitudinal considerations. *Journal of Adolescent Research, 11,* 194–215.

Farmer, T. W., Xie, H., Cairns, B. D., & Hutchins, B. C. (2007). Social synchrony, peer networks, and aggression in schools. In P. H. Hawley, T. D. Little, & P. C. Rodkin (Eds.), *Aggression and adaption: The bright side to bad behavior* (pp. 209–233). Mahwah, NJ: Lawrence Erlbaum Associates.

French, J. R. P., & Raven, B. (1959). The bases of social power. In D. Cartwrite (Ed.), *Studies in social power* (pp. 150–167). Ann Arbor: University of Michigan Institute for Social Research.

Galen, B. R., & Underwood, M. K. (1997). A developmental investigation of social aggression among children. *Developmental Psychology, 3,* 589–600. doi: 10.1037/0012-1649.33.4.589

Guendouzi, J. (2001). 'You'll think we're always bitching': The functions of cooperativity and competition in women's gossip. *Discourse Studies 3*(1): 29–51.

Hafen, S. (2004). Organizational gossip: A revolving door of regulation and resistance. *Southern Communication Journal, 69,* 223–240. doi: 10.1080/10417940409373294

Hawley, P. H. (1999). The ontogenesis of social dominance: A strategy-based evolutionary perspective. *Developmental Review, 19,* 97–132. doi: 10.1006/drev.1998.0470

Hawley, P. H. (2007). Social dominance in childhood and adolescence: Why social competence and aggression may go hand in hand. In P. H. Hawley, T. D. Little, & P. C. Rodkin (Eds.), *Aggression and adaption: The bright side to bad behavior* (pp. 1–29). Mahwah, NJ: Lawrence Erlbaum Associates.

Kowalski, R. M., Walker, S., Wilkinson, R., Queen, A., & Sharpe, B. (2003). Lying, cheating, complaining, and other aversive interpersonal behaviors: A narrative examination of the darker side of relationships. *Journal of Social and Personal Relationships, 20,* 471–490. doi: 10.1177/02654075030204003

Kurland, N. B., & Pelled L. H. (2000). Passing the word: Toward a model of gossip and power in the workplace. *The Academy of Management Review, 25,* 428–438. doi: 10.5465/AMR.2000.3312928

Lamertz, K., & Aquino, K. (2004). Social power, social status and perceptual similarity of workplace victimization; A social network analysis of stratification. *Human Relations, 57,* 795–822. doi: 10.1177/0018726704045766

Lofland, J., & Lofland, L. H. (1995). *Analyzing social settings: A guide to qualitative observation and analysis.* (3rd ed.). Belmont, CA: Wadsworth.

Lutgen-Sandvik, P., & Sypher, B. D. (Eds.). (2009). *Destructive organizational communication: Processes, consequences, and constructive ways of organizing.* New York, NY: Routledge.

McCroskey, J. C. (1982). Communication competence and performance: A research and pedagogical perspective. *Communication Education, 31,* 1–7.

Michelson, G., van Iterson, A., & Waddington, K. (2010). Gossip in organizations: Contexts, consequences, and controversies. *Group & Organization Management, 35,* 371–390. doi: 10.1177/1059601109360389

Noon, M., & Delbridge, R. (1993). News from behind my hand: Gossip in organizations. *Organization Studies, 14,* 23–36.

Redding, W. C. (1996). Ethics and the study of organizational communication: When will we wake up? In J. A. Jaksa & M. S. Pritchard (Eds.) *Responsible communication: Ethical issues in business, industry, and the professions* (pp. 17–40). Cresskill, NJ: Hampton.

Rubin, R. B., & Martin, M. M. (1994). Development of a measure of interpersonal communication competence. *Communication Research Reports, 11,* 33–44.

Sias, P. M. (2009). Social ostracism, cliques, and outcasts. In P. Lutgen-Sandvik & B. D. Sypher (Eds), *Destructive organizational communication: Processes, consequences, & constructive ways of organizing* (pp. 145–163). New York, NY: Routledge.

Spitzberg, B. H. (1983). Communication competence as knowledge, skill, and impression. *Communication Education, 32,* 323–329.

Spradley, J. P. (1979) *The ethnographic interview.* Belmont, CA: Wadsworth.

Van Vleet, K. (2003). Partial theories: On gossip, envy, and ethnography in the Andes. *Ethnography, 4,* 491–519. doi: 10.1177/146613810344001

Williams, K. D. (2001). *Ostracism: The power of silence.* New York, NY: Guilford.

Willer, E. K., & Cupach, W. R. (2011). The meaning of girls' social aggression: Nasty or Mastery?. In W. R. Cupach & B. H. Spitzberg (Eds.), *The dark side of close relationships II* (pp. 297–326). New York, NY: Routledge.

Nine

Dispelling Darkness through Dialogue in Discrimination Crises: Learning Diversity Lessons the Hard Way

Donyale R. Griffin Padgett, Melvin Gupton, and Idrissa N. Snider

"Where there is no difference, there can be no dialogue."

Heather M. Zoller (2000)

Modaff, Butler, and DeWine (2012) note that outside the explosion of information technology in the workplace, the issue of corporate diversity is perhaps the second most significant change in the modern day organization. Historically, conversations on diversity have focused on diversifying the workforce. However, the issue of diversification alone does not speak to the need "to *integrate* workers of many cultures, backgrounds and ideologies" into the domains of organizational life (Modaff, Butler & DeWine, 2012, p. 7). We are all too familiar with statistics related to dramatic shifts in the U.S. population that were predicted to significantly change the makeup of our society. Johnston and Packer's (1987) widely cited *Workforce 2000* report predicted, by the year 2000, almost one in every three persons in the U.S. would be African American, Hispanic, Asian, or Native American. According to the U.S. Census Bureau (2012), minority populations make up over a third of the U.S. population at 37 percent; however, by 2043, for the first time minorities will become the majority and comprise 57 percent of the U.S. population (U.S. Census Bureau, 2012).

These and other statistics have become so commonplace that we have even coined terms to document this new reality—the "browning of America" and the "majority-minority." The effects of population growth have undoubtedly permeated the modern-day workforce, particularly related to the numbers of women, people of color, and individuals with a disability. Although we have seen much truth among the various predictions on population growth, this "browning" of America has presented more challenges for organizations, primarily in the area of work life.

Since the late '80s and early' 90s, documented cases of workplace discrimination have resulted in class action lawsuits with some of the largest payouts in our nation's history. Four companies stand out: Denny's, Texaco, Coca-Cola, and Wal-Mart. Among them, cases involving Denny's and Texaco are perhaps most notable. During 1993 and 1994, Denny's settled two class action discrimination lawsuits for a total of $54 million after the U.S. Justice Department moved in and imposed a consent decree. Texaco's case, which stemmed from a *New York Times* article citing evidence that company leaders had planned to sabotage an ongoing discrimination case involving black and female employees, resulted in a record settlement of $176 million. The 21st century is marked by two stand-out cases involving systemic discriminatory practices at Coca-Cola and Wal-Mart. In 2000, Coca-Cola paid out the single largest settlement of a race discrimination suit. At a record $192.5 million, the company's settlement involved salary adjustments, oversight programs and sweeping changes to its internal system of promotions, and internal evaluations of employees.

Then there is retail giant Walmart, which has been buffeted by accusations of discrimination for decades, but it has successfully avoided large settlements. One of the earliest cases, dating back to 1993, involved a religious discrimination lawsuit. Another 2010 case involved disability discrimination for failure to accommodate a deaf employee. Since then, the largest retailer has battled accusations of sexual discrimination, including a 2014 suit filed by the Equal Employment Opportunity Commission on behalf of a pregnant, maintenance worker who fell ill while using harsh chemicals to clean the store's bathrooms. On one occasion, she fainted at the bus stop and was rushed to emergency. At 20-weeks pregnant, she provided a doctor's note to her supervisor and asked to be reassigned, to no avail. After using her sick time, she was subsequently fired and felt she was forced to choose between a paycheck and a healthy pregnancy (Schulte, 2014). Another recent filing against Wal-Mart is the unprecedented gender discrimination case in which 1.5 million women plaintiffs filed suit for unequal pay. The largest gender discrimination case in U.S. history, it went all the way to the Supreme Court, which ruled in 2010 that the plaintiffs did not qualify as a uniformed class and failed to prove the company operated under a company-wide policy of discrimination (Adams, 2013). Since that time, regional gender discrimination cases are still pending in the lower courts.

These and other cases demonstrate that instances of discrimination in organizations are not isolated, but systemic. When discrimination is normative in any organization, it poses two concerns. Externally, it increases the likelihood of organizational crises. When an organization is forced to respond to actions for which it is culpable, it jeopardizes the mission, impedes its ability to reach high-priority goals and forces the organization to make strategic attempts to regain legitimacy among stakeholders (Fearn-Banks, 2011; Padgett, Cheng, & Parekh, 2013; Ulmer, Sellnow, & Seeger, 2011). Internally, these environments are evidence of "bad behavior", or what has been theorized as the "dark side" of organizational life. Despite efforts to crusade diversity initiatives, numerous organizations continue to be plagued by "serious overt and subtle discrimination" (Combs, 2002; Dipboye & Halberson, 2004).

Allen includes a call to action for communication scholars to explore communication as a centering point in the study of diversity and difference in workspaces (Ashcraft & Allen, 2003). System theorists have long documented the centrality of communication in creating and sustaining organizational life (Everett, 1989; Weick, 1979), but we must continue to explore communication strategies that lead to more inclusive organizational environments.

In this chapter, we embrace Allen's call to action. We use three landmark crisis cases involving systemic discrimination to advance two pertinent arguments. First, we argue that it is not just communication or even "communication effectiveness" that is needed to confront dark side behaviors. We highlight the need for a dialogic approach to communication, which includes a generative dialogue about difference and diversity broadly, and race, more specifically. Second, we echo Allen's (2007) point that discussions about race and privilege are critical to dialogue about organizational diversity. Race is at the very core of our perceptions of others—not only because we are racialized beings, but also because our identities are cloaked in racial histories that privilege some, while marginalizing others. Race is not the only aspect of diversity, but as Allen points out, it is "inextricably connected to other salient facets of social identity, such as gender, social class, age, ability, religion, and sexuality" (p. 263). As challenging as it is to address issues of race, discussions on privilege remain taboo in and outside of workspaces. However, it is this failure to acknowledge the taken-for-granted assumptions of privilege that impede true dialogue.

The cases that follow are not representative of all aspects of diversity, but they highlight two of the most prominent forms of discrimination—racial and gender inequality. At the close of the chapter, we explore ways to facilitate more dialogue in organizations that promotes both integration and equality in workspaces.

Diversity and Dialogue

The reality of population growth has produced both challenges and opportunities for organizations, especially those with a focus on social responsibility. As Sadri and Tran (2002) purport, although most organizations accept diversity as a glaring reality, "there has been disagreement over how to deal with such diversity" (p. 228). From extensive diversity programming intended to eradicate discrimination and create civility to making a "business case" for diversity, approaches to the topic are vast (Jayne & Dipboye, 2004). Over the last three decades, scholarship has chronicled diversity training and corporate retreats, affinity groups, and benefits programs for domestic partners and veterans as ways to create systemic culture change (Combs, 2002; Jayne & Dipboye, 2004; Vertovec, 2012). One of the more contemporary arguments for engaging corporate diversity has to do with creating a corporate climate characterized by characterized by "openness, trust and mutual respect." (Sadri & Tran, 2002, p. 229).

While these strategies demonstrate a move in the right direction, we argue that a focus on communication merely reaches the surface of engaging difference. The cases outlined in this chapter signal the need for a more dialogic approach to diversity efforts.

In her article on theorizing communication and race, Allen (2007) encourages us to see organizations as "sites of identity construction" where people from different racial and ethnic backgrounds interact. In the quest to value diversity, organizational leaders must not ignore the opportunities that "difference" creates (Allen, 2007, p. 262).

We turn to the work of dialogue scholars in order to explore the type of communication that is necessary to create felt change in workspaces. Dialogue, in the broadest sense, moves beyond a mere, one-way communication activity that sees others as a means to reach a goal. At its roots, it is predicated on relationship and fosters mutual respect and understanding, and involves ongoing communication (Kent & Taylor, 2002). Martin Buber, regarded as the progenitor of the contemporary use of dialogue, conceived dialogue as "an effort to recognize the value of the other" within the context of a communication exchange "based on reciprocity, mutuality, involvement, and openness" (Kent & Taylor, 2002, p. 22). This view of dialogue casts it as essentially relationship building through authentic interaction with others. Dark side behaviors such as racial slurs, epithets, harassment, discrimination and the like, violate the tenants of mutual respect and valuing the other as an equal. Dark side behaviors also do not conform to Johannesen's five characteristics of dialogue: "genuine, accurate, empathetic understanding; unconditional positive regard; 'presentness'; spirit of mutual equality; and a supportive psychological climate" (Kent, & Taylor, 2002, p. 22). To be truly dialogic means to "risk one's position in order to arrive at new understandings and make a commitment to use communication to keep the conversation going" (Zoller, 2000, p. 193).

Review of the Dark Side

The term *dark side* in popular culture was made mainstream in the late '70s with the release of George Lucas's initial installment of the epic *Star Wars* trilogy. It refers to the destructive side of the universal force influencing the film's leading antagonist—Darth Vader. When applied to organization and management studies, dark side fundamentally represents divergence from the bright/light or ethical side of organizational behavior. It is the antithesis of appropriate conduct, and we have learned through several landmark cases (Exon-Valdez, Enron, Lehman Brothers, etc.) that such behavior can be harmful to individual employees, organizations, and an entire industry. Vaughan (1999) associates dark side behavior with "organizational deviance" that leads to unanticipated, negative consequences (p. 283). The dark side of organizational behavior is further theorized to be motivated, intentional behavior (Griffin & Lopez, 2005) that is costly, injurious, and unethical. Expressly, it is demonstrated as workplace aggression, workplace politics, workplace violence, discrimination, mistreatment, sexual harassment, incivility, and even intimate partner violence, among others (Griffin & O'Leary-Kelly, 2004).

Several researchers argue that dark side conduct occurs at both the individual and organizational levels (Griffin & O'Leary-Kelly, 2004; Vaughn, 1999). Dark side behaviors injurious to individuals are: sexual harassment, unsafe work practices, sabotage, theft, as well as verbal, psychological, and even physical violence. Similarly, behaviors injurious to the organization include incidents such as proprietary information leaks, gender bias, leader bias, code violations, member underperformance, theft and destruction of organization assets (Dipboye & Halverson, 2004).

With an increase in the scholarship aimed at describing dark side behavior, at the turn of the century Griffin and Lopez (2005) developed a five-part organizing framework. They use the term "*bad behavior*" to identify any form of conduct that is or potentially is injurious to the organization and/or individuals (Griffin & Lopez, 2005, p. 988). Their typology of bad behavior includes: (1) *dysfunctional behavior* (e.g., workplace bullying, incivility, workplace revenge and retaliation); (2) *workplace deviance* (i.e., noncompliant behavior or those actions that violate operational standards and depart from social norms); (3) *workplace aggression* (e.g., assault or threat of assault, including aggressive nonphysical behavior toward a person or thing); (4) *workplace violence* (e.g., battery or touching with hostile intent, including aggressive physical behavior); and (5) *antisocial behavior* (i.e., conduct equivalent to dysfunctional behavior but more specific). Excluded from this framework, however, were behaviors the authors believed to be "best addressed in terms of legality" such as criminal negligence, discrimination, and sexual harassment among others (Griffin & Lopez, 2005, p. 989).

This omission makes the case for an expansion of the typology to include those everyday dark side behaviors that undermine employee effectiveness such as discrimination. Even the excessive use of the outdated term "minority" is potentially harmful and can be construed as demeaning, as it designates one as an "other" instead of being included as a part of the collective (Edmondson et al., 2009).

Other researchers have chosen to include workplace discrimination as a particularized form of dark side behavior (Griffin & O'Leary-Kelly, 2004). However, these destructive behaviors are not limited to racial discrimination. Instead we note that other groups have been singled out as targets of these unfair practices. Among the more visible ones are the LGBTQ community, individuals with religious affiliations, and those with disabilities (Rosaik, 2012). According to a *Washington Times* article, cases of disability discrimination are quickly rising. In the last fiscal year, "disability-related complaints lodged with the EEOC also rose to their highest level, at 26,000, and payouts to complainants through that process nearly doubled to $103 million compared with the figure from 2007. That does not include money paid out to those who took their complaints to court" (Rosaik, 2012).

Without genuine dialogue and left unchecked, dark side behaviors undermine the organizational culture, erode employee morale and thwart one's sense of organizational belonging, which jeopardizes institutional legitimacy. Indeed, the business corpus is replete with such cautionary tales, three of which are examined in the next section.

Denny's Class Action Lawsuit

On March 25, 1993 the U.S. Justice Department filed suit against Denny's restaurant chain after African American patrons in northern California cited claims of discrimination (Chin et al., 1998). On April 1, 1993 in Annapolis, Maryland, six African American service men reported that a Denny's waitress refused to serve them breakfast.

After a series of incidents involving African American and Asian American patrons, Denny's restaurant employees began to speak in the press about the company's "unofficial 'blackout' policy" (Kanso, Levitt, & Nelson, 2012). According to the employees, when a "blackout" – characterized by too many black patrons in the restaurant – went into effect "staff was instructed to deny admittance to African American customers or require them to pay a cover charge and prepay for their meals" (Kanso et al., 2012, p. 362).

The Justice Department imposed a strict consent decree in conjunction with the National Association for the Advancement of Colored People (NAACP), which mandated sweeping policy changes within the then 1,600-unit restaurant chain. The massive inequities that existed within the company, proved discriminatory practices were systemic. Denny's was ordered to initiate companywide changes that included appointing an independent taskforce to review its current policies, redesigning its programs for advancement and supplier programs, and creating a strategic diversity plan. The company was also mandated to settle the discrimination suit, and eventually agreed to a payout in excess of $54 million.

Denny's credibility began to weaken under the weight of negative media attention. It became apparent that remedying its discriminatory practices internally was only half the battle (Kanso et al., 2012). After a series of failed public relations attempts to restore the company's image, Flagstar, Denny's parent company, agreed to spend $1 billion to promote business ventures for people of color.

Denny's case is unique because the extent of dark side behaviors was not confined to the company's internal structure. The company's discriminatory practices were far-reaching. In addition to a lack of diversity in the company's hiring and advancement practices, there were also issues with supplier diversity and a lack of inclusion among franchise owners. In the aftermath of its crisis, Denny's took steps to salvage its reputation and to change the company's culture, but despite the reconciliation period brought on by new initiatives, the dark side behaviors inflicted during this crisis haunted the company for years.

Coca-Cola Class Action Lawsuit

One of the most contemporary workplace discrimination cases occurred at the turn of the century. In 1999, Coca-Cola, the world's largest soft drink company was faced with charges of racial bias discrimination. Up until this point, a prior case involving discrimination at Texaco had netted the largest award for a class action lawsuit in U.S. history. That would change after four Coco-Cola employees located in Atlanta filed a suit based on the premise of pay discrimination.

The global brand that is Coca-Cola was severely damaged after the discovery of widespread instances of grouping African American employees at the bottom of the pay scale. *The New York Times* reported that Coca-Cola paid its black employees up to $26,000 less annually than its white employees (Winter, 2000). Although pay issues were cited as the sole complaint from the victims, other dark side behaviors were levied against the company as well. Employees stated that they had been spied on, made to work in a hostile work environment, and eventually wrongfully fired.

The case damaged the brand's image globally. A *New York Times* article reported that during interviews, Coke officials "did not deny that black employees had often been paid less than they should have or failed to get the promotions they deserved" (Winter, 2000). However, Coke officials admitted there was evidence of this among non-black employees throughout the company as well (Winter, 2000). Perhaps the most damaging discourse during the case was Coke officials characterizing the company's actions toward minority employees as "benign neglect" (Winter, 2000).

Although there is not as much information available to detail Coca-Cola's racial crisis and pursuit of post-crisis legitimacy, the evidence of dark side behaviors is most damaging among the three cases. News reports revealed complainants were "humiliated, ignored, overlooked or unacknowledged" and showed evidence of stress-related illnesses like depression (Winter, 2000).

Likewise, Coca-Cola ultimately agreed to a $156 million payout. In addition to restitution, the settlement agreement included further concessions. Companywide changes were mandated. Coca-Cola agreed to use an independent taskforce external to the company to monitor its behaviors (Maharaj, 2000). They also agreed to provide funding for black-owned businesses and non-profit organizations located within traditionally marginalized communities. These additional measures resulted in an additional $36 million, bringing the settlement total to $192.5 million.

In a news report, the lead attorney for the plaintiffs noted, "this settlement sets a new standard for corporate diversity" (Maharaj, 2000, para. 12). Company officials said the settlement "will help the company build a more talented workforce by bringing total transparency to how employees are hired" (Maharaj, 2000, para. 12).

Wal-Mart's Class Action Lawsuit

In 2001, fifty-four-year-old Betty Dukes accused Wal-Mart, the world's largest retailer, of discrimination. In her suit, she alleged unequal pay and promotion rates for the company's female employees, which comprise 57% of the retailer's U.S. workforce. Her case sparked what continues to be over a decade of sex discrimination litigation against the company. Dukes began as a part-time cashier at Wal-Mart's Pittsburg, California, store in 1994. She quickly moved to full-time status on the strength of her performance appraisals and three years later, in 1997, was promoted to Customer Service Manager (Adams, 2013). Then, in the fall of 1997, she filed a complaint with her district manager citing gender discrimination.

On the heels of her complaint to management, Dukes's pay and hours worked were summarily cut; she was subsequently demoted to her former cashier status and dissuaded from pursuing any open management positions. Dukes even received disciplinary action for behavior she alleged her male counterparts were given a pass on, and she was consistently not notified of any managerial openings (Adams, 2013). In June 2001, five other female employees of Wal-Mart joined Dukes and filed a formal complaint in a California district court seeking class action certification against Wal-Mart for discriminatory practices.

In June 2004, the U.S. District Court for the Northern District of California concluded there was merit to the *Dukes et al.* case and believed the statutory conditions were sufficiently sound to proceed with the largest class action lawsuit in U.S. history. The certification represented 1.5 million women employed over five years. Had they been successful, it could have cost the mega-retailer as much as $11 billion to compensate those female workers dating back to 1998. As evidence, complainants cited less pay than their male counterparts for comparable work, fewer promotions to management, and longer wait periods for advancement (Adams, 2013, p. 258).

Anecdotal support for claims identified by Adams (2013) came from 110 supervisors who testified that "female managers were required to go to Hooters sports bars as well as strip clubs for meetings and office outings" (p. 257). Additionally, female workers were subjected to demeaning tags such as "girls" and one worker was told to "doll up" to be promoted (p. 256). Plaintiffs characterized the Wal-Mart culture as consistently chauvinistic as reflected by statements such as: "retail is 'tough' and not 'appropriate' for women," "men need to be paid more than women because men have families to support," and "God made Adam first, so women would always be second to men" (Adam, 2013, p. 256). An expert witness for *Dukes et al.*, William Bielby, corroborated these attitudes and testified that the company, from his perspective, sustained a "sophisticated system of centralized coordination, reinforced by a strong organizational culture" with "subjective and discretionary features of the company's personnel policy" that make "decisions about compensation and promotion vulnerable to gender bias" (p. 256).

The company argued that the scope of the filing was too large and "unmanageable" and appealed its case in 2005 to the U.S. Ninth Circuit Court of Appeals in San Francisco. The higher court affirmed the class action certification, but limited the number of plaintiffs to 1 million women. Wal-Mart then appealed to the Supreme Court, which on June 20, 2011, reversed the appellate courts' decision citing the plaintiff's failure to prove that Wal-Mart engaged in the uniform and general practice of employment discrimination. The Justices also ruled that the plaintiffs' evidence that Wal-Mart's policy affords local supervisors discretion in making employment decisions was insufficient proof that these supervisors do so in a uniformed fashion.

Despite the ruling, this case is far from resolved. The highest court's decision to throw out the class action suit after over a decade of legal battles did not rule on individual discriminatory practices. The global conglomerate, which operates 11,000 stores in 27 countries and employs 2.2 million workers worldwide, continues to receive gender discrimination claims filed by its female employees. The perception of unequal treatment by its female workers persists, and the company's claim that its discrimination policy with penalties shields its workers from unfair treatment remains at odds. This is evident in a gender bias case in London, Kentucky, where Wal-Mart reached a court settlement of $11.7 million in damages with its female workers who were denied jobs because of their sex (Lynch, 2010).

Themes between the Cases

Dipboye and Halverson (2004) remind us that "not so long ago discrimination in the workplace was open, tolerated, and even encouraged" (p.131). Where legislative policy was once the vehicle for many individuals fighting against workplace discrimination, today, class action lawsuits force companies to make policy changes and create plans of action for more inclusive workspaces. The three cases presented in this chapter involved some of the world's most recognizable companies. These cases are shining examples of what happens when dark side behaviors are allowed to operate without regard for impact on organizational members. The cases demonstrate that acknowledgment of wrong doing alone is not sufficient to rectify systemic discriminatory practices. Companies must present compelling diversity strategic plans and demonstrate a long-term commitment to implementation.

Among the cases presented here three specific errors occurred. First, these cases involved discriminatory practices, which were not random. As evidenced in the case analyses, dark side behaviors were targeted and widespread. For example, the Denny's case demonstrates through the use of its "blackout" policy that discrimination against certain groups was the accepted norm. Second, our analysis confirms an inability by company leaders to identify dark side behaviors as adverse and engage in necessary steps toward company-wide dialogue about diversity and difference. For instance, based on statements and observations obtained from numerous supervisors, Wal-Mart's culture was characterized by sexually offensive language (Adams, 2013). At one point, a senior official at Coca-Cola communicated to then-chairman M. Douglas Ivester that the two were on different ends of a visibility spectrum. Ivester, she wrote, had the problem of being "too visible," while she had the problem of being too "invisible," speaking to the invisibility often associated with people of color and other traditionally marginalized groups (Maharaj, 2000, para. 12). What is perhaps most telling is that in her communication, the female executive attributed her invisibility to "chauvinism, power… and absolute disrespect" (Maharaj, 2000, para. 12).

A final error is that each of the three companies was forced to create broad-based diversity plans that outlined clear steps toward a culture shift. This points to the need for company leaders caught in the snare of discrimination crises to use dialogue about diversity and difference as a proactive tool to bring those groups traditionally on the margins in conversation with mainstream groups. This kind of dialogue may not eradicate instances of marginalization, but it has much potential to aid in the healing process brought on by the trauma of such discrimination and support the recovery process necessary for organizations to make peace with the past.

Dispelling the Darkness: Recommendations for Praxis

Our examination of these cases demonstrates the need for new approaches to corporate diversity that emphasizes dialogue as a core feature of systemic change and not court-ordered fixes. Specifically, our discussion highlights three strategies for making this change possible.

1. *Make room at the table for different perspectives*—Diversity is not just an issue for diverse groups. It is a mission to be taken up by the entire organization. Conversations on equity and inclusion in any workspace must include members from traditionally marginalized groups and members from dominant groups. In order to create a more dialogic environment, organizational leaders must cast a wider net around diversity issues and make room at the table for generative dialogue that shapes and is shaped by people from different social positions and cultural experiences. Often, individuals who are traditionally marginalized are uniquely poised to see instances of marginalization more clearly because of their social positioning on the margins. Likewise, members from dominant groups often fail to see their own privilege because it is normative. In many organizations, the dominant culture is the standard by which all others are measured. We ought not to see these two perspectives as dialectically opposed. They are in fact different. Dialogue creates opportunities to bring all organizational members into conversation with one another about how best to engage difference and facilitate inclusivity. Failure to acknowledge our social position impedes true dialogue and can potentially impede the shift that is necessary for systemic cultural change.

2. *Engage in generative dialogue*—True dialogue involves a "collaborative orientation" in which those engaged in the conversation have a unique perspective that they are willing to "advocate for…vigorously" with an understanding that no party to the conversation has absolute truth (Kent & Taylor, 2002, p. 25). Collaborative partners in the conversation are willing to hear, understand and be heard and give the same courtesy to others. Presupposed in dialogic exchanges is a recognition of equality that avoids the one-sidedness that fosters feelings of superiority and privilege. Likewise, notions of inferiority and disadvantage are also abandoned in light of true authentic engagement. Another feature of dialogic engagement is the investment of the total self to the conversation and a sense of abandonment to the immediate conversation. Kent and Taylor (2002) note that parties to the conversation are "communicating in the present about issues, rather than after" decisions are made and they do so with all deliberateness as they "seek to construct a future for participants that is both equitable and acceptable to all" (p. 26).

3. *Create system-level change* – The road to creating systemic cultural change is fraught with numerous barriers. Adopting dialogic communication in the organization as a matter of practice, as is being recommended here, is subject to uncertainties. The risks that come from sharing one's views, beliefs, and desires makes space for individual and organizational growth, but it also make those sharing "vulnerable to manipulation or ridicule" (Kent & Taylor, 2002, p. 28). Besides vulnerability, the spontaneous nature of dialogic communication makes it subject to unpredictable exchanges that may not go according to the organizational script. Kent and Taylor (2002) refer to this risk as "emergent unanticipated consequences" (p. 28).

At the center of this discussion is the debate over whose responsibility it is to create this change. Is the responsibility for change at the system/organizational level? Or does it rest with the individuals who lodge complaints of dark side behaviors? The final principle of dialogic communication answers these questions—it is a threefold commitment. Dialogue is predicated on a commitment to honesty and "genuiness" and placing "the good of the relationship above the good of the self or organization" (Kent & Taylor, 2002, p. 29). The commitment to honesty and genuineness is closely aligned with a commitment to the conversation "for the purposes of mutual benefit and understanding" (p. 29). Finally, there is a commitment by all parties to really "work at dialogue to understand often-diverse positions" (p. 29).

Final Thoughts

The cases we examined present clear evidence of employee mistreatment and discrimination that can stem directly from individual prejudices, fear, and the inherent discomfort with cultural differences we addressed earlier in the chapter. However, as these cases demonstrate, no organization is absolved from its responsibility to provide workspaces where people can thrive regardless of their unit of difference. As Kanter (1977) pointed out decades ago, because of the evidence of managers mistreating employees and other micro-level stressors related to diversity issues, the responsibility for change must be at the system level (p. 152). This statement is even more true today than it was decades ago.

Despite Dipboye and Halverson's (2004) cogent argument for future researchers "to understand the cause of unfair discrimination and ways to eliminate it" (p. 136), there remains a dearth of literature on the dark side behavior that is discrimination. We argue, in the foregoing discussion, for acknowledgment and expansion of existing typologies to incorporate discrimination as a categorical form of dark side dysfunction worthy of scholarly investigation. If left unaddressed, discriminatory practices harm individuals, work groups, and the organization itself by exacerbating difference as an oddity to be contained. As we have seen, this can be a costly proposition. Organizations that deliberately address diversity initiatives through dialogue are poised to dispel dark side behaviors such as discrimination. By being more dialogic, diversity leaders can ultimately create more inclusive environments where individuals can thrive, not in spite of their differences, but because of them.

References

Adams, R. J. (2013). David v. Goliath: A brief assessment of the U.S. Supreme Court's 2011 ruling denying class certification in Dukes v. Wal-Mart. *Business and Society Review, 118*(2), 253–270.

Allen, B. J. (2007). Theorizing communication and race. *Communication Monographs, 74*(2), 259–264.

Ashcraft, K. L., & Allen, B. J. (2003). The racial foundation of organizational communication. *Communication Theory, 13*, 5–38.

Chin, T., Naidu, S., Ringel, J., Snipes, W., Bienvenu, S. K., & DeSilva, J. (1998). Denny's: Communicating amidst a discrimination case. *Business Communication Quarterly. 61*(1), 180–197.

Combs, G. M. (2002). Meeting the leadership challenge of a diverse and pluralistic workplace: Implications of self-efficacy for diversity training. *The Journal of Leadership Studies, 8*(4), 1–16.

Dipboye, R. L., & Halverson, S. K. (2004). Subtle (and not so subtle) discrimination in organizations. In R. W. Griffin, & A. M. O'Leary-Kelly (Eds.), *The dark side of organizational behavior* (pp. 131–158). San Francisco, CA: Jossey-Bass.

Edmondson, V. C., Gupte, G., Draman, R. H., & Oliver, N. (2009). Focusing on communication strategy to enhance diversity climates. *Journal of Communication Management, 13*(1), 6–20.

Everett, J. L. (1989). Communication and sociocultural evolution in organizations and organizational populations. *Communication Theory, 4,* 93–110.

Fearn-Banks, K. (2011). *Crisis communications: A casebook approach.* Mahwah, NJ: Erlbaum.

Griffin, R. W., & Lopez, Y. P. (2005). "Bad behavior" in organizations: A review and typology for future research. *Journal of Management, 31,* 988–1005.

Griffin, R. W., & O'Leary-Kelly, A. M. (2004). *The dark side of organizational behavior.* San Francisco, CA: Jossey-Bass.

Griffin, R. W., & O'Leary-Kelly, A. M. (2004). An introduction to the dark side. In R. W. Griffin, & A. M. O'Leary-Kelly (Eds.), *The dark side of organizational behavior* (pp. 1–19). San Francisco, CA: Jossey-Bass.

Jayne, M. E. A., & Dipboye, R. L. (2004). Leveraging diversity to improve business performance: Research findings and recommendations for organizations. *Human Resource Management, 43*(4), 409–424.

Johnston, W., & Packer, A. 1987. *Workforce 2000: Work and Workers for the 21st Century.* Indianapolis, IN: Hudson Institute.

Kanso, A. M., Levitt, S. R., & Nelson, R. A. (2012). Public relations and reputation management in a crisis situation: How Denny's restaurants reinvigorated the firm's corporate identity. In W. T. Coombs, & S. J. Holladay (Eds.), *The Handbook of Crisis Communication* (pp. 359–377). Malden, MA: Wiley-Blackwell.

Kanter, R. M. (1977). *Men and women of the corporation.* New York, NY: Basic Books.

Kent, M. L., & Taylor, M. (2002). Toward a dialogic theory of public relations. *Public Relations Review, 28,* 21–37.

Lynch, M. (2010). Gender suit against Wal-Mart advances. *WWD: Women's Wear Daily, 199*(89), 2–1.

Maharaj, D. (2000, November 17). Coca-Cola to settle racial bias lawsuit. *Los Angeles Times.* Retrieved from http://articles.latimes.com/2000/nov/17/news/mn-53405

Modaff, D. P., Butler, J. A., & DeWine, S. (2012). *Organizational communication: Foundations, challenges and misunderstandings* (3rd ed., pp. 1–25). Glenview, IL: Pearson.

Padgett, D. R. G, Cheng, S. S., & Parekh, V. (2013). The quest for transparency and accountability: Corporate social responsibility and misconduct cases. *Asian Social Science, 9*(9), 31–44.

Rosaik, L. (2012, May 31). Discrimination lawsuits double as definition of "disability" expands. *Washington Times.* Retrieved from http://www.washingtontimes.com/news/2012/may/31/discrimination-lawsuits-double-as-definition-of-di/?page=all

Sadri, G., & Tran, H. (2002). Managing your diverse workforce through improved communication. *Journal of Management Development, 21*(3), 227–237.

Schulte, B. (2014, December 17). Wal-Mart faces new pregnancy discrimination charges. *The Washington Post.* Retrieved from http://www.washingtonpost.com/blogs/local/wp/2014/12/17/wal-mart-faces-new-pregnancy-discrimination-charges/

Ulmer, R. R., Sellnow, T. L. & Seeger, M. W. (2011). *Effective crisis communication: Moving from crisis to opportunity.* Thousand Oaks, CA: Sage.

U.S. Census Bureau, Newsroom Archive. (2012). *U.S. Census Bureau projections show a slower growing, older, more diverse nation a half century from now.* Retrieved from U.S. Census Bureau website: http://www.census.gov/newsroom/releases/archives/population/cb12-243.html

Vaughan, D. (1999). The dark side of organizations: Mistake, misconduct, and disaster. *Annual Review of Sociology, 25,* 271–305.

Vertovec, S. (2012). "Diversity" and the social imaginary. *Archives Européennes De Sociologie, 53*(3), 287–312. doi:http://dx.doi.org/10.1017/S000397561200015X

Weick, K. (1979). *The Social Psychology of Organizing* (2nd ed.). Reading, MA: Addison-Wesley.

Winter, G. (2000, November 17). Coca-Cola settles racial bias case. *The New York Times.* Retrieved from http://nytimes.com/2000/11/17/business/coca-cola-settles-racial-bias-case.html

Zoller, H. M. 2000. "A place you haven't visited before." Creating the conditions for community dialogue. *The Southern Communication Journal, 65,* 191–207.

TEN

Microaggressive Communication in Organizational Settings

Shawn D. Long, Haley Wonznyj, Marcus J. Coleman, Amin Makkawy, and Calvin Spivey

Diversity and inclusion efforts have emerged as a critical touchstone in American society, especially within formal organizations and institutions. As the U.S. demography shifts from a White majority to a more diverse and pluralistic populace (Toossi, 2012), there still remains persistent remnants of discriminatory and prejudicial attitudes and practices that undermine efforts toward equality and equal access to resources and opportunities. As policies, laws, and regulations are enacted to increase diverse representation in all aspects of American life, so too have counter-political movements and social forces seeking to marginalize, oppress, or mute diverse voices or bodies. Typically, the actions to maintain the racial, ethnic and gender status quo are overt and conscience. However, research suggests that the most insidious practices used to undermine interpersonal inequality are sub- or unconscious. These unwitting practices have been termed microaggressions (Sue et al., 2007). Microaggressions are emerging as both a scholarly and cultural phenomenon that affects and influences the interpersonal, group, and organizational contexts. Culturally, it is becoming a pervasive and recognizable action to the target; yet there still remains a deniable or frequently passive misunderstanding posture embraced by the perpetrator of this communication event. This chapter extends the scope of microaggressions to include microaggressive communication; as communication is the medium in which individuals enact microaggressive actions and individuals are microaggressed against in episodic or serial ways. We offer a background on microaggressions, a series of case studies illuminating this activity and finally offer theoretical and organizational implications associated with microaggressive communication.

Literature Review

The nature of discrimination has evolved (e.g., Jones, 1997; Shenoy-Packer, 2015; Thompson & Neville, 1999; Wang, Leu, & Shoda, 2011). Prior to legislation like the Civil Rights Act or the Equal Employment Act, marginalized groups in the United States were often the victims of overt prejudice and discrimination (King et al., 2011). Such regulations were designed to reduce the discrimination and prejudice that marginalized groups experienced. Though these legislative changes, combined with a shift in public opinion, have contributed to a decrease in overt discrimination, they have not, however, been successful at abating more covert and subtle discrimination that emerges in interpersonal interactions (Dovidio, Gaertner, Kawakami, & Hodson, 2002; King et al., 2011; Sue et al., 2007). Research suggests subtle forms of discrimination are more harmful to victims' physical and psychological well-being than more traditional, overt forms (Jones, Peddie, Gilrane, King, & Gray, 2013; Solorzano, Ceja, & Yosso, 2000).

Microaggressions capture this contemporary form of discrimination and emphasize the individual experience of everyday racism and discrimination (Sue et al., 2007). Formally, microaggressions refer to "brief and commonplace daily verbal, behavioral, or environmental indignities, whether intentional or unintentional, that communicate hostile, derogatory, or negative racial slights or insults toward the target person or group" (Sue et al., 2007, p. 273). An Asian American who was born in the United States, yet who receives a compliment for how well he or she speaks English is an example of a microaggression; the message communicates that the Asian American is a foreigner in his or her own country.

The definition of microaggressions connotes many pieces that comprise the experience of subtle discrimination. First, microaggressions are "brief and commonplace"; as such, they emerge frequently in everyday interactions (Deitch, Barsky, Butz, Chan, Brief, & Bradley, 2003; Sue et al., 2007; Wang et al., 2011). Second, microaggressive messages can be communicated through a variety of different mediums: verbally, behaviorally, or environmentally (Sue et al., 2007). The example given above with the Asian American speaking English well represents a verbal microaggression. In a team setting, an example of a behavioral microaggression would be a team member avoiding eye contact with their Black teammates. Finally, environmental microaggressions are less interpersonal in nature, but communicate offensive messages nonetheless. An example of an environmental microaggression could be an office that decorates for Christian holidays, ignoring the holidays that are important to other religious groups.

Third, the definition of microaggressions recognizes that subtle discrimination varies in its intentionality. People who engage in microaggressions often do not know that they have communicated offensive, derogatory messages; people generally believe themselves to be well-intentioned, moral and hold egalitarian beliefs without any negative stereotypes towards marginalized groups (Sue et al., 2007; Wang et al., 2011). Consequently, microaggressions are often unintentional and seem innocent to the microaggressor. For example, Burdsey (2011) found that perpetrators often communicated microaggressions in the form of jokes, though they denied responsibility for and intentionality of the offensive message. The potential unintentionality of microaggressions makes it difficult for the victims to interpret whether the message was malicious or not. In addition, it is similarly difficult to prove whether discrimination has actually occurred, making discrimination lawsuits (particularly with regard to employment discrimination cases) burdensome for the victim. Yet, because microaggressions are often invisible to the perpetrator, Sue et al. (2007) emphasizes the thoughts, feelings and experiences of the victim rather than the perpetrator.

The final piece of the definition of microaggressions is the most important. Microaggressions are offensive and denigrating to the victim. They demean the experiences and identity of the victim based on their membership of a particular group and suggests that they are a lesser human being (Dietch et al., 2003; Sue et al., 2007). It is important to note that microaggressions can be against a number of different groups, including, but not limited to, women, racial and ethnic minorities, people with disabilities, and lesbian, gay, bisexual, and transgender (LGBT) people. Such hostile and negative messages are detrimental to the victim's psychological well-being, regardless of which group they belong to. For example, microaggressions increase anger, depression, and anxiety and decrease perceptions of worthiness (a more detailed review of the consequences of microaggressions is provided below; e.g., Carter, 2007; Sue, Capodilupo, & Holder, 2008). In addition, microaggressions perpetuate the negative biases and stereotypes about particular marginalized groups on which they are based.

Types of Microaggressions

Scholars have identified three general types of microaggressions: microassaults, microinsults, and microinvalidations. The first type of microaggressions, *microassaults*, are similar to traditional forms of racism; they are the most overt type and are often intentional (Sue et al., 2007). In particular, microassaults are conscious and explicit derogations that are purposefully meant to offend and harm the intended victim through verbal attacks, avoidant behavior, and even physical violence. For example, yelling racial epithets or drawing a swastika on a bathroom stall are examples of microassaults. Microassaults are different from the other two types of microaggressions because perpetrators of microassaults are often aware of the negative stereotypes they hold about a group. Due to shifts in public opinion towards equality and away from tolerance of such behaviors, microassaults are generally communicated in private and/or when the microaggressor can remain anonymous (Sue et al., 2007). Although, perpetrators can communicate microassaults, perhaps unintentionally, in an emotional outrage.

Microinsults, the second form of microaggressions, are more subtle than microassaults. Microinsults are subtle communications that are characterized by rudeness and that demean a person's identity as a member of a particular group. It is often the case that the microaggressor is unaware that they have offended the victim. For example, microinsults are frequently in the form of a compliment (e.g., after a Black student passed an exam with flying colors, the teacher tells them they are a credit to their race). Although such compliments are intended to be positive, the hidden message that Blacks are not as intelligent or hard working as Whites is insulting and demeaning. Microinsults can also be nonverbal in nature; for example, women could be offended if a male counterpart hangs a swimsuit calendar in his office.

The final type of microaggressions, *microinvalidations*, are similar to microinsults in that they are usually subtle and invisible to the perpetrator. Microvalidations communicate messages that "exclude, negate, or nullify the psychological thoughts, feelings, or experiential reality of a [member of a particular group]" (Sue et al., 2007, p. 274). The example about the Asian American speaking English well that we first used to introduce microaggressions is an example of this form. Another widely cited example of microinvalidations is colorblindness, which refers to the disregard of racial characteristics when making decisions or interacting with others. Majority group members often celebrate colorblindness as an egalitarian belief.

Yet, colorblindness completely ignores the experiential reality that members of marginalized groups live each day. Consequently, colorblindness is an example of how microinvalidations are offensive to the victims.

Consequences of Microaggressions

The implications of microaggressions are widespread (Sue et al., 2008). In general, research findings suggest that victims of microaggressions experience feelings of discomfort anger, and humiliation (Nadal, 2013). Much of the focus on microaggressions has been within the clinical psychology field, focusing on how microaggressions in the therapist-client counseling relationship influence client outcome success. However, microaggressions are present in all interactions, not just those between therapist and client in a counseling context. Accordingly, scholars in other fields have begun to examine the consequences of microaggresions on issues such as health and well-being, education, and employment.

Counseling, health & well-being. Within the counseling psychology literature, scholars have focused a great deal on how subtle discrimination can influence the relationship between a therapist and client and client outcomes (Sue et al., 2007). For example, when a therapist microaggresses against their client, the client perceives a poorer relationship with their therapist (known as the working alliance; Constantine, 2007; Owen, Tao, & Rodolfa, 2010). Clients who have been microaggressed against also report lower satisfaction with their therapy (Constantine, 2007). Furthermore, Microaggressions can decrease the success of therapy (Owen et al., 2010), and can even drive clients to discontinue treatment (Sue et al., 2007) or discourage them from seeking treatment all together (Crawford, 2011).

In addition to the mental health outcomes associated with counseling, microaggressions have also been shown to have a substantial influence on general well-being outcomes. For instance, people who are the victims of microaggressions experience increased stress and anger and have more depressive symptoms (Sue et al., 2008; Torres, Driscoll, & Burrow, 2010).

Education. Research also shows that microaggressions can have implications for educational settings as well. Numerous qualitative and quantitative studies show that students experience microaggressions in school (e.g., Grier-Reed, 2010; Solorzano, 1998; Yooso, Smith, Ceja, & Soloranzo, 2009). For example, using critical race theory as a foundation, Soloranzo found that Latino/a graduate students experience microaggressions in education, which made them feel out of place in the academy. In addition, graduate students felt that their instructors had lower expectations of them because of their race and gender. The presence of macroaggressions in the classroom often leads to difficult dialogues on racism (Sue, Lin, Torino, Capodilupo, & Rivera, 2009b). However, teachers often do not have the skills and resources to adequately recognize microaggressions and facilitate such difficult dialogues, potentially reinforcing the negative stereotypes and beliefs on which microaggressions are based.

Employment. Microaggressions are also found in interpersonal interactions at work and can influence many different aspects of the work experience. For example, members of marginalized groups can experience microaggressions during the job application process (Sue, 2010; Sue, Lin, & Rivera, 2009a). More specifically, recruiters could argue that members of marginalized groups don't "fit in" to the organization, or organizations may signal through text and pictures that they value only certain types of people. Furthermore, microaggressions can manifest in many different interactions at work, including peer-to-peer or supervisor-to-subordinate, the latter of which may influence promotion decisions (Coleman, 2004).

Furthermore, victims of microaggressions have reported receiving inadequate mentoring and feeling invisible in the organization (Constantine, Smith, Redington, & Owens, 2008). Like microaggressions in counseling relationships, microaggressions at work drive employees to discontinue their employment at the organization, which has important organizational implications (e.g., replacement training costs, lost knowledge; Sue et al., 2009a).

Coping Strategies

In addition to the physical, psychological and behavioral implications of microaggressions, researchers have also begun to investigate how victims cope with the implications of microaggressions. Victims of microaggressions choose many different ways to cope with the negative feelings they experience as a result of the offensive messages. For example, some people choose to confront the perpetrator, questioning the intentions behind the microaggression (Nadal et al., 2011), while others may express their anger by vocalizing their frustrations concerning the experience (Nadal, Hamit, Lyons, Weinberg, & Corman, 2013). Moreover, studies have shown women and LGBT people to be relatively passive, choosing to accept prejudices and therefore to not react after experiencing a microaggression (Nadal et al., 2011; Nadal, 2013; Nadal et al., 2013). In educational settings, forming networks or support groups has been shown to increase students' ability to cope with the subtle discrimination (Grier-Reed, 2010; McCabe, 2009). Various communication approaches, mainly assertive communication, sense making, and discursive strategies (e.g., rationalization, creation of alternative selves) can also help victims manage the negative implications of microaggressions (Camara & Orbe, 2010; Shenoy-Packer, 2015). Finally, some groups engage in preemptive avoidant behaviors, attempting to protect themselves from microaggressions. For example, African Americans report modifying their body language and communication style to communicate a non-threatening behavior (Sue et al., 2008).

Taking the microaggressions literature as a whole, these subtle forms of discrimination are widespread and found in many different settings in which members of different groups interact with each other. It is often the case that the perpetrators of microaggressions are unaware of the negative biases they hold or that they offended another party. Because microaggressions communicate harmful, denigrating messages, victims of microaggressions often experience negative physical and attitudinal consequences that influence behaviors in a host of contexts. However, research has shown that various groups have developed a variety of ways to cope with more subtle forms of discrimination. The following section presents specific case studies of microaggressions to further illustrate the phenomena.

Case Studies

Microaggressions represent a specific tool of oppression. Yearwood (2013) recounts an incident with her faculty colleague that left her uneasy because there was a subversive attempt to deny race and racial experiences. Though similar tales of macroaggressions have been well documented and researched, they persist. Discrimination, as commonly stated, is alive and well in the United States in varying forms. As mentioned previously, contemporary forms of discrimination function covertly and evolved from more overt bigoted forms of racial mistreatment against people of color, e.g., slavery, Jim Crow, lynching, etc. (Sue et al., 2007). The aforementioned definitions of discrimination lead toward a ritualistic (i.e., cultural) characterization of communication, not merely transmission (Carey, 1989). Ritualistic communication is described as, "not toward the extension of messages in space but toward the maintenance of society in time; not the act of imparting information but the representation of shared beliefs (Carey, 1989, p. 239). The representation of shared beliefs is reflected in both our positive and negative brief commonplace interactions, conscious or not (Sue et al., 2007; Yearwood, 2013). Below we explore microaggressive communication—in particular, microinsult, microassault, and microinvalidation—with regard to sexual orientation, race, ethnicity, and gender.

Microinsult. As mentioned previously, Sue and colleagues (2007) describe microinsults as more subtle snubs that convey an insulting message. An exemplar of microinsults was reported in a *Washington Post* story about a peaceful protest at a high school in southwestern Virginia. Students protested the school's dress code by donning confederate apparel to express their dissatisfaction with their inability to display symbols that commemorate their southern heritage (Shapiro & Balingit, 2015). Because the protest happened in a racially charged time in American history with publicized racial violence (e.g., police brutality against black men and women and a fatal assault of black parishioners at their place of worship), the reemergence of confederate symbols as representations of southern heritage are highly contested. Nonetheless, the high school students exercised their first amendment right to pay homage to their heritage by wearing confederate symbols. But, what does that mean to those who do not share their heritage?

To illustrate, a black male student who recently graduated from the high school in question expressed frustration and historical pain. Shapiro and Balingit (2015) report that the alumnus saw the confederate symbols worn while he was a student at the school. He noted that some teachers upheld the dress code policy when students wore confederate flags, but others overlooked it. Though the alumnus felt uncomfortable when his classmates donned the flag, he withheld expressing his feelings to his white classmates. Thus, we see a clear distinction between the interpretations of confederate symbols; the pride displayed one side and the pain experienced on the other. Microinsults are commonplace and subtle, but rarely ever discussed with the perpetrator. As illustrated, microaggressive communication is indeed subversive and can silence its victims, but it can also occur overtly and seek to injure.

Microassault. As described earlier, microassaults are derogatory, often purposeful attacks meant to offend the victim (Sue et al., 2007). For example, as reported by the *Washington Post*, the Student Government Association (SGA) president at the University of Southern California, Rini Simpath, was recently verbally and physically assaulted while walking from a friend's apartment when a person shouted from a fraternity house window, "You Indian piece of s—!" and threw a drink at her (Bever, 2015). Simpath's experience was extremely unfortunate but it was not outside the realm of imagination regarding a historical view of how minorities have been treated in the United States. Simpath speaks to the pain that she felt as a result of incident, "It brought back all these memories of growing up as immigrant in America.…All the things people said started playing back in my head, over and over, like a broken record" (Bever, 2015). This is especially salient when our national discourse focuses on the supposed burden of immigration thrust upon American citizens. Microaggression functions at varying levels of consciousness and communicative expression, while inflicting pain on its victim(s), i.e., physical [emotional] and as we see next, the psychological aspect of microaggression, i.e., microvalidation.

Microvalidation. Microvalidations are communicative acts that exclude the thoughts, feelings, and reality of members of marginalized groups (Sue et al., 2007). Microaggressions are a common occurrence. Though microaggresive communication is often assumed to occur in interracial contexts (Harris, 2008), we must also be aware of microaggressions that occur with consideration to gender, ethnicity, sexual orientation, and physical and mental disabilities (Sue, 2010). Labels are important because they help shape our perception of others (Burke, 1966). With that in mind, professors at Washington State University were criticized due to their efforts to create learning environments where labels describing men and women of color, ethnic minorities, sexual orientation immigrants, and others are more equitable. As reported by the *Washington Post*, the interim president of the University, Dan Bernardo, was not supportive of the faculty members' efforts to penalize offensive language in their classrooms (Moyer, 2015).

The internal and external forces against the professors who were attempting to encourage and enforce more equitable labels for marginalized groups facilitated opportunities to invalidate, exclude, and negate the experiences of people of color, ethnic minorities, sexual orientation, immigrants, and the mentally and physically disabled via language. Thus, Washington State University's interim president and *The Daily Caller* both invalidated the use of perceived equitable labels in the classroom. Distinctions in sexual orientation, ethnicity, race, and gender can produce negative feelings of distrust and divisiveness due to perceived difference.

Theoretical and Practical Implications

According to the widely cited definition provided by Sue et al., (2007), microaggressions are at their essence communicative events. It is ironic that relatively little theoretical work has been conducted regarding microaggressions in the communication studies literature; instead the scholarly conversation surrounding our understanding of microaggressions has been centralized in the psychology literature. In the realm of communication research, the theoretical implications of microaggressions are plentiful. Two primary theoretical implications that merit further exploration are (1) the integration of the concept of microaggressions into critical and postmodern perspectives on communication, and (2) the use of a rhetorical lens to further define and explore the concept/types of microaggressions.

Critical perspectives on communication could benefit in examining microaggressions as events that act to reify societal relationships of power and subordination. At its crux, microaggressions are communicative events that rely on societal positioning of different social groups on a hierarchy of power. It is the use of this hierarchy of political relationships that ultimately produce and strengthen microaggressions. Postmodern communication scholars would benefit in examining microaggressions to further understand the communicative acts that support the scaffolding that constitutes social divisions (racial, gender, etc.). Microaggressions provide another concept that can be used in the field of communication to further understand how communication is key in understanding societal inequalities via everyday communicative acts. For example in the light of co-cultural theory (Orbe, 1998), microaggressions can be placed in the larger arena of co-cultural communication. In this case non-dominant group members attempt to gain power in the face of microaggressive behaviors that degrade their position in society via the reinforcement of cross group power differentials.

A deeper rhetorical understanding of microaggressions could be useful in gaining a unique, if not more refined definition, of the fundamental building blocks that constitute a microaggression. While the psychology literature provides a typology of different types of microaggressions, a rhetorical analysis of microaggressions could provide a more refined/ communication-centered typology of microaggressions. Providing this typology would ultimately assist in defining microaggressions from a perspective that is commensurate with the communicative nature of the concept. Appreciating microaggressions as rhetorical occurrences lends a hand to the previously discussed implications of the integration of microaggression research into critical communication research. Understanding the rhetorical bindings that define a microaggression, critical communication scholars can be a useful tool in understanding representations of societal cross group power differentials as embedded in language.

Practical implications of microaggression research for communication practitioners are centralized around understanding and staying cognizant of microaggressions. Training should be implemented to make organizational members knowledgeable of the relevant microaggressions that researchers have identified in the literature. Public relations specialists should take care not to mistakenly reify societal inequalities by using microaggressions, as people in such positions represent their organizations to various stakeholder groups and the public in general. Inside organizations, communication practitioners should be aware that the presence of anti-discrimination practices and the absence of racism does not preclude microaggressive incidents (Burdsey, 2011). For example, the claim made to an employee of Arab heritage that a thobe is not appropriate attire for a company party because employees are expected to wear their *best* clothing and therefore a suit and tie would be better is a microinvalidation. However, this does not constitute a racist incident and current anti-discrimination practices do not regulate such subtle discourse directed at an individual.

Microaggressive discourse and traditional racist incidents are not the same and should not be considered to necessarily accompany one another. For this reason, communication practitioners should make this distinction obvious and provide mechanisms and a language for expressing the subtleties characteristic of a microaggressive experience that may not be agreed upon or succinctly understood by all the actors involved in the microaggressive experience. Communication practitioners must manage microaggressive situations appropriately to demonstrate that even covert forms of discrimination still count as discrimination. In doing so, it is important for communication practitioners to provide a sense-making process for all of those involved and to be aware of the voices that are being muted.

The theoretical and practical implications illustrated above place communication scholars central to the future of microaggression research. The nature of microaggressions as communicative acts, the strength of research methodologies used by communication scholars, and the expertise regarding critical and rhetorical approaches in understanding human interaction that communication scholars bring to the table insure this central role. This is not to say that cross disciplinary and interdisciplinary endeavors in understanding microaggressions should be neglected, but instead that these endeavors have great potential especially when the expertise of communication scholars are utilized.

Conclusion

Microaggressions are both behavioral and communicative. We situate microaggressions as negative interpersonal and organizational communication that needs greater scholarly and pragmatic attention. We posit that macroaggressions, whether intended or not, are a learned and cumulative socialized process that seeks to marginalize the other, often historically marginalized, party.

Microaggressions have significant consequences on the perpetrator and the target of this communication breach. Interpersonal communication is vital to the human condition, yet when differences in background and diversity are introduced to the interaction, complicated communication often results. Ill-intended micro-communication consequences happen primarily because many individuals have not been socialized to exist in a pluralistic society and their latent privilege standpoint prevails. We assert that microaggressive communication is an attempt to maintain the power differential status quo via communicative acts, which are underlined by implicit bias.

There is fertile ground for future research in the area of microaggressive communication. This includes contexts such as the workplace, social media, interracial communication, disability research, and within the broad area of diversity and inclusion scholarship. Foundational and contemporary research methods are needed to help support the advancement of this emerging area of study. In short, microaggressive communication is an everyday interpersonal and organizational event that needs rigorous study and exploration as a negative communicative act.

References

Bever, L. (2015, September 21). The shocking racial epithet hurled at USC's student body president. *Washington Post.* Retrieved from: http://www.washingtonpost.com

Burdsey, D. (2011). That joke isn't funny anymore: Racial microaggressions, color-blind ideology and the mitigation of racism in English men's first-class cricket. *Sociology of Sport Journal, 28*, 261–283.

Burke, K. (1966). *Language as symbolic action: Essays on life, literature, and method.* Berkeley, CA: University of California Press.

Camara, S. K., & Orbe, M. P. (2010). Analyzing strategic responses to discriminatory acts: A co-cultural communicative investigation. *Journal of International and Intercultural Communication, 3*, 83–113.

Carey, J. W. (1989). Cultural approach to communication. In J. W. Carey (Ed.), *Communication as culture: Essays on media and society* (pp. 13–36). Boston, MA: Unwin Hyman.

Carter, R. T. (2007). Racism and psychological and emotional injury: Recognizing and assessing race-based traumatic stress. *The Counseling Psychologist, 35*, 13–105.

Coleman, H. S. M. (2014). *Portraits of four African-American women who earned doctoral degrees from a predominately White institution.* DePaul University: University Libraries. Retrieved from http://via.library.depaul.edu/cgi/viewcontent.cgi?article=1058&context=soe_etd.

Constantine, M. G. (2007). Racial microaggressions against African American clients in cross-racial counseling relationships. *Journal of Counseling Psychology, 54*, 1–16.

Constantine, M. G., Smith, L., Redington, R. M., & Owens, D. (2008). Racial microaggressions against Black counseling and counseling psychology faculty: A central challenge in the multicultural counseling movement. *Journal of Counseling & Development, 86*, 348–355.

Crawford, E. P. (2011). Stigma, racial microaggressions, and acculturation strategies as predictors of likelihood to seek counseling among black college students. (Doctoral dissertation). Retrieved from ProQuest LLC.

Deitch, E. A., Barsky, A., Butz, R. M., Chan, S., Brief, A. P., & Bradley, J. C. (2003). Subtle yet significant: The existence and impact of everyday racial discrimination in the workplace. *Human Relations, 56*, 1299–1324.

Dovidio, J. F., Gaertner, S. L., Kawakami, K., & Hodson, G. (2002). Why can't we all just get along? Interpersonal biases and interracial distrust. *Cultural Diversity and Ethnic Minority Psychology, 8*, 88–102.

Grier-Reed, T. L. (2010). The African American student network: Creating sanctuaries and counterspaces for coping with racial microaggressions in higher education settings. *The Journal of Humanistic Counseling, Education and Development, 49*, 181–188.

Harris, R. S. (2008). Racial microaggression? How do you know? *American Psychologist, 63*, 275–276.

Jones, J. M. (1997). *Prejudice and racism* (2nd ed.). Washington, DC: McGraw-Hill.

Jones, K. P., Peddie, C. I., Gilrane, V. L., King, E. B., & Gray, A. L. (2013). Not so subtle: A meta-analytic investigation of the correlates of subtle and overt discrimination. *Journal of Management*. Advance online publication. doi:10.1177/0149206313506466

King, E. B., Dunleavy, D. G., Dunleavy, E. M., Jaffer, S., Morgan, W. B., Elder, K., & Graebner, R. (2011). Discrimination in the 21st century: Are science and the law aligned? *Psychology, Public Policy, and Law, 17*, 54–75. doi: 10.1037/a0021673

McCabe, J. (2009). Racial and gender microaggressions on a predominantly-White campus: Experiences of Black, Latina/o and White undergraduates. *Race, Gender & Class, 16*, 133–151.

Moyer, J. W. (2015, September 2). Washington State University class bans 'offensive' terms like male, female, tranny, illegal alien. *Washington Post*. Retrieved from http://www.washingtonpost.com.

Nadal, K. L. (2013). *That's so gay! Microaggressions and the lesbian, gay, bisexual, and transgender community*. Washington, DC: American Psychological Association.

Nadal, K. L., Hamit, S., Lyons, O., Weinberg, A., & Corman, L. (2013). Gender microaggressions: Perceptions, processes, and coping mechanisms of women. In M. A. Paludi (Ed.), *The psychology of business success* (pp. 193–220). Santa Barbara, CA: Praeger.

Nadal, K. L., Wong, Y., Issa, M., Meterko, V., Leon, J., & Wideman, M. (2011). Sexual orientation micoraggressions: Processes and coping mechanisms for lesbian, gay, and bisexual individuals. *Journal of LGBT Issues in Counseling, 5*, 21–46. doi:10.1080/15538605.2011.554606

Orbe, M. P. (1998). *Constructing co-cultural theory: An explication of culture, power, and communication*. Thousand Oaks, CA: Sage.

Owen, J., Tao, K., & Rodolfa, E. (2010). Microaggressions and women in short-term psychotherapy: Initial evidence. *The Counseling Psychologist, 38*, 923–946.

Shapiro, T. R., & Balingit, M. (2015, September 17). Virginia high school students suspended for wearing Confederate flag apparel. *Washington Post*. Retrieved from: http://www.washingtonpost.com.

Shenoy-Packer, S. (2015). Immigrant professionals, microaggressions, and critical sensemaking in the U.S. workplace. *Management Communications Quarterly, 29*, 1–19.

Solórzano, D. G. (1998). Critical race theory, race and gender microaggressions, and the experience of Chicana and Chicano scholars. *International journal of qualitative studies in education, 11*, 121–136.

Solórzano, D., Ceja, M., & Yosso, T. (2000). Critical race theory, racial microaggressions, campus racial climate: The experiences of African American college students. *Journal of Negro Education, 69*, 60–73.

Sue, D. W. (2010). *Microaggressions in Everyday Life: Race, Gender, and Sexual Orientation*. Hoboken, NJ: John Wiley & Sons.

Sue, D. W., Capodilupo, C. M., & Holder, A. (2008). Racial microaggressions in the life experience of Black Americans. *Professional Psychology: Research and Practice, 39*, 329.

Sue, D. W., Capodilupo, C. M., Torino, G. C., Bucceri, J. M., Holder, A., Nadal, K. L., & Esquilin, M. (2007). Racial microaggressions in everyday life: Implications for clinical practice. *American Psychologist, 62*, 271–286.

Sue, D. W., Lin, A. I., & Rivera, D. P. (2009a). Racial microaggressions in the workplace: Manifestations and impact. In J. L. Chin (Ed.), *Diversity in mind and in action* (pp. 157–172). Santa Barbara, CA: ABC-CLIO, LLC.

Sue, D. W., Lin, A. I., Torino, G. C., Capodilupo, C. M., & Rivera, D. P. (2009b). Racial microaggressions and difficult dialogues on race in the classroom. *Cultural Diversity and Ethnic Minority Psychology, 15*, 183–190.

Thompson, C. E., & Neville, H. A. (1999). Racism, mental health, and mental health practice. *Counseling Psychologist, 27*, 155–223.

Toossi, M. (2012). Labor force projections to 2020: A more slowly growing workforce. *Monthly Labor Review, 135*, 43–64.

Torres, L., Driscoll, M. W., & Burrow, A. L. (2010). Racial microaggressions and psychological functioning among highly achieving African-Americans: A mixed-methods approach. *Journal of Social and Clinical Psychology, 29*, 1074–1099.

Wang, J., Leu, J., & Shoda, Y. (2011). When the seemingly innocuous "stings": Racial microaggressions and their emotional consequences. *Personality and Social Psychology Bulletin, 37*, 1666–1678. doi:10.1177/0146167211416130

Yearwood, E. L. (2013). Microaggression. *Journal of Child and Adolescent Psychiatric Nursing, 26*, 98–99.

Yosso, T. J., Smith, W. A., Ceja, M., & Solórzano, D. G. (2009). Critical race theory, racial microaggressions, and campus racial climate for Latina/o undergraduates. *Harvard Educational Review, 79*, 659–690.

Eleven

Hazing as a Tool of Destructive Organizational Identification and Loyalty

Creshema R. Murray and Kenon A. Brown

Organizations are communities that foster and promote culture and identity (Jablin & Putnam, 2001). Organizational life and the discourse promoted within organizations are critical in the development of relationships, individual identities, and organizational identities (Cheney & Christensen, 2001; Eisenberg & Riley, 2001). Through the use of symbols, language, and interactions, individual perspectives merge and work under the realm of one unifying organizational belief. Organizational power is necessary and when used in a constructive manner, power contributes to the success of many organizations (Howard & Geist, 1995; Mumby, 1996, 2001). Supervising job assignments, making personal network connections, and holding the title of leader in your field, all represent power in organizational life. However, when organizations and people within organizations misuse power it can cause problems of incivility, distrust, emotional abuse, and physical harm (Lutgen-Sandvik, Namie, & Namie, 2009). Sexual harassment, emotional tyranny, and workplace bullying are just a few examples of destructive organizational practices where the misuse of power was inherent in organizations (Gill & Sypher, 2009; Lutgen-Sandvik et al., 2009; Waldron, 2009).

In recent years, many organizational scholars have taken great interest in researching destructive practices in organizations and how those practices lead to a loss of productivity, decreased worker commitment and impoverished workforces (Allen, 2009; Doughtery, 2009; Keashly & Neuman, 2009; Lewis, 2009; Lutgen-Sandvik et al., 2009; Seibold, Kang, Gailliard, & Jahn, 2009). However, the destructive practice of organizational hazing has been a topic that many scholars have yet to fully explore. Historically, hazing has been categorized as a tool of power commonly practiced in tight-knit groups such as fraternities, sororities, the military, athletic organizations, and gangs (Finkel, 2002). Hazing represents a rite of passage used to determine the level of commitment that a member may have to the organization (Van Maanen & Schein, 1979). By participating in hazing activities members gain a sense of identity and a connection to the organization (Sweet, 2004).

Traditional research examining the hazing phenomenon does not explore hazing as either a communicative issue or as a destructive means of organizing. However, hazing should be examined as a communicative problem due to the nature in which messages of power and dominance are reinforced through verbal and nonverbal oppressive communication tactics (Lutgen-Sandvik et al., 2009). As a method of abusive and tyrannical social interactive strategies, hazing is constructed as a necessary organizational ritual deemed essential for organizational entry. The unwritten symbols and messages associated with the discourse around hazing create systems of meanings that preserve the desire to prolong the initiation practice for any organizational newcomer. Hazing sustains the tradition of organizing around an ideology that is necessary for organizational access and acceptance.

To understand the power hazing has in organizations, we examined the association between hazing and the creation of organizational identity and organizational loyalty. To this end, we first reviewed existing literature describing hazing as a destructive communication practice while considering the connection hazing has to the construction of members' organizational identity. Then, we analyzed the relationship that hazing plays in members, loyalty to an organization. Finally, we conducted an investigation that examined how hazing affects members' organizational identity and loyalty.

Hazing Defined

Hazing is considered a popular rite of passage practiced throughout much of the world (Butt-Thompson, 1908/2003; Hoover, 1999; Hoover & Pollard, 2000; Jeong, 2003; Lewis, 1992; McCarl Jr., 1976; Parks & Brown, 2005; Schlegel & Barry, 1979; Shaw, 1992). Leemon (1972) deemed hazing as a social mechanism used to enhance and revitalize relationships. Nuwer (1999) noted that hazing is an activity used by high-status organizational members to humble newcomers. In broad terms, hazing is "any activity expected of someone joining a group that humiliates, degrades, abuses or endangers, regardless of the person's willingness to participate" (Hoover, 1999, p. 8). Hazing is an action taken or a situation intentionally created that is intended to cause mental or physical harm (Crow & Rosner, 2005).

As a destructive organizational communication practice, hazing is used to demean, exploit, oppress, and abuse organizational members (Lutgen-Sandvik et al., 2009). Destructive communication is an interaction ranging from subtle gestures to physical assault, making organizational life unhealthy and unproductive (Lutgen-Sandvik et al., 2009). Increased pressure to participate in the historical practices of hazing, results in human and organizational damage (Gill & Syper, 2009). The hazing phenomenon is a painful cultural practice engendered with tactics that strips power away from novice participants. However, this practice of organizing is not deemed destructive by all.

Culturally, advocates for hazing indicate that there is a sense of unity and love for the organization and members performing the hazing (Sirhal, 2000). Parks and Brown (2005) argue hazing has a positive effect for members, indicating that people who participate in hazing have a higher level of organizational commitment than those members who did not participate in hazing. Additionally, members that were hazed upon organizational entry use that learned behavior to continue the hazing cycle with new organizational members (Montague, Zohra, Love, McGee, & Tsamis, 2008). Despite the destructive nature of hazing, we propose that:

H1: There is a positive correlation between the self-reported frequency of hazing in an organization by a member and that member's perception of hazing.

H2: There is a positive correlation between the self-reported importance of hazing to an organization by a member and that member's perception of hazing.

Construction of Identity

Due to the communicative nature of organizations, the messages, interactions, and relationships we form construct and reify the identity of organizational life. Organizations are sites of social interactions that assist in the construction of meaning; therefore, identity is constructed through meaning. Organizations must be reconceptualized as public spheres that shape identities (Mumby, 2001). Hazing is a practice that works to shape individual identities into group identities. The core values behind this practice generate an understanding of unity through human interaction. Organizational identity has a positive impact on member satisfaction and organizational effectiveness (Chaput, Brummans, & Cooren, 2011). Members gain a sense of self through organizational association which can either be positive or negative (Chaput et al., 2011; Miller, 2002). Organizational identification can be seen as the degree to which individuals use organizational attributes to not only define the organization but to also define themselves (Dutton & Dukerich, 1991). Not only is identity a rhetorical process (Burke, 1972; Cheney, 1983), but it is also a performed process (Barker, 1993). Organizational behaviors produce identity through performance (Chaput et al., 2011). This performance, or act, produces a sense of sameness generating a unifying identity. Organizational identity allows members to construct a sense of self that is often seen as necessary to create and maintain credibility within and outside of the organization.

Organizational members perform acts to not only serve the organization but to also enhance their sense of self (Rogers, 2007). Sweet (2004) indicated that organizational members participate in hazing activities because the act of hazing creates an organization centered identity tying self-concept to organizational identity. Thus, we predict:

H3: There is a positive correlation between the self-reported frequency of hazing in an organization by a member and the degree of identification that member has with the organization.

H4: There is a positive correlation between the self-reported importance of hazing to an organization by a member and the degree of identification that member has with the organization.

Organizational Loyalty

There is a growing need for organizational members to belong (Cheney & Christensen, 2001). Due to many of the changes in organizational culture, loyalty amongst members of organizations has also shifted in recent years, despite the growing desire from organizational members to belong (Cheney & Christensen, 2001). Members of organizations are searching for a connection with an entity larger than their individual selves. According to organizational socialization (Van Maanen & Schein, 1979), group members become integrated into the established culture of an organization learning patterns that are considered routine. For most new members of an organization, initial entry is a time to learn traditional organizational behaviors, beliefs, and values (Jablin, 2001). Much of this initial learning comes through observation and practice.

Through hazing, traditional organizational practices are used to determine the level of commitment a new member has to the organization (Rogers, 2007). Despite the damaging physical and psychological effects hazing can have on a person (Finkel, 2002), organizational members deem the practice as a natural form of organizational socialization (Sweet, 2004). With these ideas in mind, we project that:

> H5: There is a positive correlation between the self-reported frequency of hazing in an organization by a member and that member's degree of loyalty to the organization.
>
> H6: There is a positive correlation between the self-reported importance of hazing to an organization by a member and that member's degree of loyalty to the organization.

Although this study will look at the effects of hazing in college organizations as a whole, organizations that are more prone to hazing are worthy of closer examination. Greek organizations, marching bands, spirit squads, and sanctioned sports teams are organizations with histories of hazing incidents, and the following research question examines the effects of hazing within these organizations:

> RQ1: Are there perceptions of hazing, organizational identification, and organizational loyalty among members of organizations that have a history of hazing issues different than other organizations?

Method

To examine how the frequency and importance of hazing within an organization can affect a member's perceptions of hazing, identification with the organization, and loyalty to the organization, a survey was conducted via an online questionnaire to college undergraduates at two Southern public universities. The researchers recruited participants through the universities' research pools.

Participants

After obtaining Institutional Review Board approval, data were collected from 287 participants: 69 males (24%), 213 females (74.2%), and 5 that did not report his/her gender. The sample was predominately Caucasian (219 respondents, 76.3%) with a mean age of 19.9 years (*SD* = 1.89 years). Participants were asked to report their organization affiliations by answering a multiple-choice question prompting them to choose the type of organization ("Please indicate the type of organization whose membership experience you will complete this section about."). Descriptions of the types of organizations were derived from Messiah College's webpage on student involvement and leadership programs (http://www.messiah.edu/offices/student_affairs/student_programs/clubs/). Participants that were affiliated with more than one organization were allowed to report a second organization. The Appendix provides a list of the types of organizations described in the questionnaire. Participants were also asked to specify if they were members of specific organizations with a history of hazing (fraternity, sorority, marching band, spirit squad, or intercollegiate athletic team). Overall, the 287 participants provided information for 345 organizations. Table 11.1 provides a breakdown of the reported organizations.

Table 11.1. Frequency of Self-Reported Organization Affiliation.

Type of Organization	Reported Affiliation
Academic/Professional Organization	32 (9.3%)
Intramural or Club Sports Team	36 (10.4%)
Fine Art and Music Organization	21 (6.1%)
Cultural and Ethnic Organization	10 (2.9%)
Greek Organization	104 (30.1%)
Honor Society	14 (4.1%)
Political Organization	6 (1.7%)
Religious Organization	50 (14.5%)
Service Organization	24 (7%)
College or University-Sponsored Organization	24 (7%)
Special Interest Organization	22 (6.4%)
Organization Not Reported	2 (0.6%)

Overall, 287 participants provided information for 345 organization affiliations.

Variables

Frequency of hazing and importance of hazing were measured using respective one-item, seven-point Likert scale items. The item measuring frequency asked how often his/her organization hazes members compared to other organizations, and ranged from much less (1) to much more (7). The item measuring importance asked how important hazing is to his/her organization's initiation and membership process, and ranged from not at all important (1) to extremely important (7).

Each participant's perception of hazing was measured using an eight-item, seven-point Likert scale revised from several scales from Campo, Poulos and Sipple's (2005) study on hazing among college students ($\alpha = 0.842$). Organizational identity was measured using a six-item, seven-point Likert scale revised from Mael and Tetrick's (1992) organizational identity scale ($\alpha = 0.86$). Organizational loyalty was measured using a four-item, seven-point Likert scale revised from Patchen's (1965) study on employee morale ($\alpha = 0.9$). The three scales all ranged from strongly disagree (1) to strongly agree (7).

Procedure and Analysis

The instrument contained five parts. Section A consisted of informed consent. Section B provided the scale measuring the participants' hazing perceptions. Section C asked the participants to provide the first organization that they would use to provide membership information and attitudes, and then provided the scales measuring frequency of hazing, importance of hazing, organizational identification and organizational loyalty. Section D prompted participants who are members of more than one organization to complete the previous section for a second organization. Section E provided a debriefing statement and demographic questions. The survey was pre-tested among 20 students, and pretest data were used to review individual questions and edit the questionnaire.

Results

Table 11.2 provides the mean scores for frequency of hazing and importance of hazing for the 11 types of organizations.

Table 11.2. Mean Scores of Frequency and Importance of Hazing by Organization.

Type of Organization	Frequency of Hazing	Importance of Hazing
Academic/Professional Organization	1.84	1.94
Intramural or Club Sports Team	1.78	1.58
Fine Art and Music Organization	2.55	1.95
Cultural and Ethnic Organization	2.30	2.80
Greek Organization	2.36	2.00
Honor Society	2.00	1.21
Political Organization	2.00	1.00
Religious Organization	1.38	1.24
Service Organization	1.67	1.46
College or University-Sponsored Organization	2.17	2.00
Special Interest Organization	1.50	1.68
Organization Not Reported	1.98	1.75

Scores are based on a scale from 1 (low) to 7 (high)

The first two hypotheses examined relationships related to an organization member's perception of hazing. Specifically, the first hypothesis examined the relationship between frequency of hazing and perception of hazing. Based on the analysis, there was a positive, significant correlation between frequency of hazing and perception of hazing ($r = 0.174$, $p = 0.001$); therefore, *hypothesis 1 was supported.* The second hypothesis examined the relationship between importance of hazing and perception of hazing. Based on the analysis, there was a positive, significant correlation between importance of hazing and perception of hazing ($r = 0.327$, $p < 0.001$); therefore, *hypothesis 2 was supported.*

The next two hypotheses examined relationships related to a member's identification with an organization. Specifically, the third hypothesis examined the relationship between frequency of hazing and organizational identification. Based on the analysis, there was not a significant correlation between these two variables ($r = 0.045$); therefore, *hypothesis 3 was not supported.* The fourth hypothesis examined the relationship between importance of hazing and organizational identification. Based on the analysis, there was not a significant correlation between these two variables ($r = -0.002$); therefore, *hypothesis 4 was not supported.*

The next two hypotheses examined relationships related to a member's loyalty to an organization. Specifically, the fifth hypothesis examined the relationship between frequency of hazing and organizational loyalty. Based on the analysis, there was not a significant correlation between these two variables ($r = -0.065$); therefore, *hypothesis 5 was not supported.* The sixth hypothesis examined the relationship between importance of hazing and organizational loyalty.

Based on the analysis, there was a significant correlation between these two variables; however the correlation was negative (r = -0.166, p = 0.002). Therefore, *hypothesis 6 was not supported.* Table 11.3 provides the correlations between the frequency of hazing and the organizational outcomes (perception of hazing, organizational loyalty, and organizational identification) for each organization type.

Table 11.3. Relationships between Frequency of Hazing and Organizational Outcomes.

Type of Organization	Perception of Hazing *r*	Organizational Loyalty *r*	Organizational Identification *r*
Academic/Professional Organization	0.242	0.310	0.300
Intramural or Club Sports Team	0.099	0.148	-0.088
Fine Art and Music Organization	0.306	-0.185	-0.279
Cultural and Ethnic Organization	0.292	0.262	0.323
Greek Organization	0.088	-0.172	-0.156
Honor Society	-0.167	0.307	0.112
Political Organization	0.134	-0.674	-0.638
Religious Organization	0.158	-0.083	0.053
Service Organization	0.107	-0.236	0.181
College or University-Sponsored Organization	-0.052	-0.189	0.164
Special Interest Organization	0.484*	-0.090	0.137

* *means* $p < 0.05$

Table 11.4 provides the correlations between the importance of hazing and the organizational outcomes for each organization type.

Table 11.4. Relationships between Importance of Hazing and Organizational Outcomes.

Type of Organization	Perception of Hazing *r*	Organizational Loyalty *r*	Organizational Identification *r*
Academic/Professional Organization	0.436*	0.284	0.274
Intramural or Club Sports Team	0.259	-0.086	0.095
Fine Art and Music Organization	0.552*	-0.267	-0.210
Cultural and Ethnic Organization	0.205	-0.271	-0.190
Greek Organization	0.272*	-0.173	-0.201
Honor Society	0.372	-0.597*	-0.236
Political Organization	n/a	n/a	n/a
Religious Organization	0.011	-0.166	-0.030
Service Organization	0.363	-0.443*	-0.018

Type of Organization	Perception of Hazing *r*	Organizational Loyalty *r*	Organizational Identification *r*
College or University-Sponsored Organization	0.234	-0.347	0.146
Special Interest Organization	0.519	-0.220	-0.064

* *means* $p < 0.05$

The sole research question examined differences in perceptions of hazing, organizational identification, and organizational loyalty among members of organizations with a history of hazing issues compared to members of other organizations. There were 143 self-reports (41.4%) of membership in a type of organization with a history of hazing issues. Independent samples t-tests determined mean differences in these three variables. Based on the analysis, members of organizations with a history of hazing issues have a more positive perception of hazing ($M = 2.9$, $SD = 1.15$) than members of other organizations ($M = 2.52$, $SD = 1$; $t\,(339) = 3.187$, $p = 0.002$). Members of organizations with a history of hazing issues also identified more with their organizations ($M = 5.93$, $SD = 0.92$) than members of other organizations ($M = 5.21$, $SD = 1.18$; $t\,(334) = 6.153$, $p < 0.001$). Finally, members of organizations with a history of hazing issues also were more loyal to their organizations ($M = 6.27$, $SD = 1.05$) than members of other organizations ($M = 5.88$, $SD = 1.07$; $t\,(337) = 3.373$, $p = 0.001$).

Discussion

This study was designed to uncover the impact of hazing on organizational members' perceptions of hazing, identity with the organization, and loyalty to the organization. Specifically, the researchers looked at the importance of hazing to the organization and the frequency of hazing during the initiation, recruitment/membership process, and their relationships with the previous organizational outcomes. Results revealed that while members' perceptions of hazing are more positive when hazing is more frequent and more important, the importance of hazing yields a stronger correlation. Results also revealed that neither importance of hazing, nor frequency of hazing, had a relationship with organizational identification or organizational loyalty. In fact, the more important hazing was reported to be to an organization, the less loyal the members were to that organization.

Types of organizations with a history of hazing were compared to other types of organizations. Results indicate that members of organizations with a history of hazing had a more positive perception of hazing, identified more with the organization, and were more loyal to the organization.

Hazing Deemed Important

Based on our hypotheses and research questions, the answers yielded from study participants provide many implications to assist in examining hazing as a destructive organizational communication tool. Destructive organizational communication examines taboo organizational practices that are used in many organizations to maintain control and power (Lutgen-Sandvik & Sypher, 2009). Both authors find interest in exploring and critiquing organizational practices that impede supportive organizational growth and destroy the constructive practice of organizing. This research shows the correlation between hazing, a type of destructive organizational practice, in an organization and organization members' perceptions of hazing, and organization identity and loyalty. In the case of this study, organizational members that frequently engage in hazing and perceive the practice as important to the success of the organization also indicate that this organizational practice is positive. This information suggests that the more frequently an organization engages in hazing practices, the less likely members will see the practice as demeaning or damaging to the organization even though hazing is considered a destructive form of communication.

If an organization has a history of hazing, results indicate that not only is hazing more frequent, but organizational members deem this practice as important more so than members of organizations without a known history of hazing. Despite the demeaning rituals associated with organizational hazing, members seem to deny challenging historical organizational practices that may cause harm to members. Results from this study highlight that organizations with a history of hazing have a higher frequency of hazing, and members show greater enthusiasm for the practice of hazing. These results indicate a deep rooted problem associated with hazing that warrants a closer examination of how influential the practice has been on organizational culture.

Despite laws that prohibit hazing, this study indicates that members of organizations continue to participate in this rite of passage. Understanding that organizational members who participate in hazing consider it important and not harmful to the success of an organization indicates that more positive communicative messages must be cycled throughout organizations to stop hazing. Communication must be used as a tool to support hazing as a device that destroys the success of organizations and harms members associated with the organization. Study participants indicate that hazing is positive because the organization heavily engages in the practice of hazing, suggesting that members in an organization are more willing to participate in harmful activities based on the historical organizational acceptance of the action. As a destructive tool, hazing is more widely accepted in organizations with a history of participating in hazing and by members who know that hazing is part of organizational entry. Combating these practices and looking for alternative solutions for organizational rites of passages are issues organizations must address to stop the cycle of hazing. Organizational leaders must understand the dynamics associated with hazing in order to sustain healthy organizations and organizational members. Failing to address hazing as a destructive tool perpetuates the cycle of toleration for destructive organizational communities. These results, most importantly, display that history is a greater indicator of the sustainability of organizational practices.

Identification through Hazing

Although hazing frequency and organizational importance is a clear indicator of members' positive perception of hazing, these indicators have no bearing on members' organizational identity. Data from this study did not support hypotheses three and four. Even though hazing works as a practice to provide entry into an organization, this practice does not impact the identity of the members who participated in this study. Based on this study, hazing does not assist in defining members' identities. Past organizational identity work indicate that organizational members employ established organizational policies and use these policies to create a sense of organizational and personal connections (Chaput et al., 2011; Miller, 2002). Understanding this dynamic, the authors assumed that organizational members who were subjected to frequently being hazed would deem those practices as important to helping members shape their individual identity. However, this study indicates that there is no significant indicator that the more a person is hazed, as a result of membership into an organization, the more that member would designate hazing as an important practice. Based on these results hazing does not cultivate individual identities or assist in the establishment of member-related identities. This development from the research baffled the authors. Based on historical identity work (Tajfel & Turner, 1986), people participate in hazing to help create a sense of self. Our research indicates a shift in earlier hazing and identity work, suggesting that hazing frequency and importance to the organization does not enhance a member's identity. These results open doors for additional research to be conducted that examines the correlation between organizational identity and hazing. Participants indicated that they perceive hazing as important to the organization, but then indicate that hazing has no relation to the construction of their organizational identity. Even though hazing is deemed important, the authors wonder what specific organizational practices assist in creating member identification.

Organizational Loyalty

Based on the analysis, the frequency of hazing in an organization was not a significant predictor of members' loyalty to their respective organization. Study participants indicated that frequently participating in hazing activities had no relation to how loyal they remained to the organization. However, if the practice of hazing is important to an organization, the members are more likely to be loyal to the organization. These results support hypotheses six and strike down hypotheses five.

Many people join organizations because they are looking for a connection and have a desire to belong. If organizations have deemed hazing as an important practice, based on results from this study, members are willing to comply with the practice of hazing and follow the established organizational practices. Since hazing is a practice that occurs during the initial entry period into an organization, new members observing the importance of these practices equate hazing to a system that works for the organization and must continue. When organizational members participate in hazing, they display a level of commitment to the established policies and procedures of an organization. Even though the data deemed that the frequency of hazing is not an indicator of organizational loyalty, the data ultimately revealed that established practices are more important than the amount of time the organization spends focusing on the practice.

Practical Utility

Results provided in this study shed light on a few issues that scholars and practitioners can review. First, this study identifies hazing as a form of destructive organizational communication. Even though destructive organizational communication is relatively new to the field of organizational studies, past research has rarely identified hazing as a tenet of destructive practices. This study highlights that instances of hazing occur in many different organizations, and overall it is an accepted practice. Second, we learn from this study that when hazing is deemed important to the organization and a part of the organization's assimilation process, members deem the practice as important. This revelation highlights historical and cultural issues present in many organizations. If hazing is important to members, educators and practitioners can use this information to create educational programs that will highlight dangers associated with organizational hazing. Participants indicated that they perceive hazing as important to the organization, but then indicated that hazing has no relation to the construction of their organizational identity. As such, there are opportunities to investigate why hazing continues and ways to terminate it. Lastly, this study helps to identify the types of organizations that employ hazing practices and the impact those practices can have on member affiliation and future hazing instances.

Limitations, Future Research, and Conclusions

This study, while informative of the effects of hazing on several organization outcomes, has several limitations, some which can be examined through future research. This study looked at hazing by conducting a self-reported survey. An obvious limitation due to this methodology is hesitation by participants to disclose the true nature of hazing in their organizations, despite the researchers attempt at assuring participants confidentiality and anonymity of responses. Future research should attempt to strengthen this assurance.

Another oversight by the researchers was not including military organizations (e.g., ROTC). Future research should address this subset of organizations in order to examine their members' stances on hazing and the effects on organizational outcomes. Also, quantitative research, due to its closed-ended style of questioning, only provides a limited amount of information. Qualitative research could be used to dig deeper into the rationale behind hazing and its effects on certain organizational outcomes. Overall, this chapter defines hazing as destructive organizational communication and highlights the need to engage in more work that will secure ways to end the mentally and physically damaging practices associated with this dark side practice.

References

Allen, B. J. (2009). Racial harassment in the workplace. In P. Lutgen-Sandvik & B. D. Sypher (Eds.), *Destructive organizational communication: Processes, consequences, & constructive ways of organizing* (pp. 164–183). New York, NY: Routledge.

Barker, J. R. (1993). Tightening the iron cage: Concertive control in self-managing teams. *Administrative Science Quarterly, 38*(3), 408–437.

Burke, K. (1972). *Dramatism and development.* Barre, MA: Clark University Press.

Butt-Thompson, F. W. (1908/2003). *West African secret societies: Their organisation, officials and teaching.* Whitefish, MT: Kessinger.

Campo, S., Poulos, G., & Sipple, J. (2005). Prevalence and profiling: Hazing among college students and points of intervention. *American Journal of Health Behavior, 29,* 137–149.

Chaput, M., Brummans, B., & Cooren, F. (2011). The role of organizational identification in the communicative constitution of an organization: A study of consubstantialization in a young political party. *Management Communication Quarterly, 25,* 252–282.

Cheney, G. (1983). The rhetoric of identification and the study of organizational communication. *Quarterly Journal of Speech, 69*, 143–158.
Cheney, G., & Christensen, L. (2001). Organizational identity: Linkage between internal and external communication. In F. M. Jablin & L. L. Putnam (Eds.), *The new handbook of organizational communication: Advances in theory, research, and methods* (pp. 231–269). Newbury Park, CA: Sage.
Crow, B., & Rosner, S. (2005). Hazing and sport and the law. In H. Nuwer, (Ed.), *Hazing reader* (pp. 200–223). Bloomington, IN: Indiana University Press.
Dougherty, D. S. (2009). Sexual harassment as destructive organizational process. In P. Lutgen-Sandvik & B. D. Sypher (Eds.), *Destructive organizational communication: Processes, consequences, & constructive ways of organizing* (pp. 203–226). New York, NY: Routledge.
Dutton, J. E., & Dukerich, J. M. (1991). Keeping an eye on the mirror: Image and identity in organizational adaptation. *Academy of Management Journal, 34*, 517–554.
Eisenberg, E., & Riley, P. (2001). Organizational culture. (2001). In F. M. Jablin & L. L. Putnam (Eds.), *The new handbook of organizational communication: Advances in theory, research, and methods* (pp. 291–322). Newbury Park, CA: Sage.
Finkel, M. A. (2002). Traumatic injuries caused by hazing practices. *The American Journal of Emergency Medicine, 20*(3), 228–233.
Gill, M. J., & Sypher, B. D. (2009). Workplace incivility and organizational trust. In P. Lutgen-Sandvik & B. D. Sypher (Eds.), *Destructive organizational communication: Processes, consequences, & constructive ways of organizing* (pp. 53–74). New York, NY: Routledge.
Hoover, N. C. (1999). *National survey: Initiation rites and athletics for NCAA sports teams.* Retrieved from http://www.alfred.edu/sports_hazing/docs/hazing.pdf
Hoover, N. C., & Pollard, N. J. (2000). *Initiation rites in American high schools: A national survey.* Retrieved from http://www.alfred.edu/hs_hazing/docs/hazing__study.pdf
Howard, L. A., & Geist, P. L. (1995). Ideological positioning in organizational change: The dialectic of control in a merging organization. *Communication Monographs, 62*, 110–131.
Jablin, F. M., & Putnam, L. L. (2001). *The new handbook of organizational communication: Advances in theory, research, and methods.* Newbury Park, CA: Sage.
Jablin, F. M. (2001). Organizational entry, assimilation, and disengagement/exit. In F. M. Jablin & L. L. Putnam (Eds.), *The new handbook of organizational communication: Advances in theory, research, and methods* (pp. 732–818). Newbury Park, CA: Sage.
Jeong, E. (2003). The status of hazing in South Korean university soccer programs. *Dissertation Abstracts International: Section A. Humanities and Social Sciences, 64*, 441.
Keashly, L., & Neuman, J. H. (2009) Building a constructive communication climate: The workplace stress and aggression project. In P. Lutgen-Sandvik & B. D. Sypher (Eds.), *Destructive organizational communication: Processes, consequences, & constructive ways of organizing* (pp. 339–362). New York, NY: Routledge.
Leemon, T. A. (1972). The rites of passage in a student culture: A study of the dynamics of transition. New York, NY: Teachers College Press, Columbia University.
Lewis, A. P. (2009). Destructive organizational communication and LGBT workers' experiences. In P. Lutgen-Sandvik & B. D. Sypher (Eds.), *Destructive organizational communication: Processes, consequences, & constructive ways of organizing* (pp. 184–202). New York, NY: Routledge.
Lewis, D. M. H. (1992). Corporate and industrial hazing: Barbarism and the law. *Labor Law Journal, 43*, 71–83.
Lutgen-Sandvik, P., Namie, G., & Namie, R. (2009). Workplace bullying: Causes, consequences, and corrections. In P. Lutgen-Sandvik & B. D. Sypher (Eds.), *Destructive organizational communication: Processes, consequences, & constructive ways of organizing* (pp. 27-52). New York, NY: Routledge.
Mael, F., & Tetrick, L. (1992). Identifying organizational identification. *Educational and Psychological Measurement, 52,* 813–824.
McCarl, R. S. Jr. (1976). Smokejumper initiation: Ritualized communication in a modern occupation. *The Journal of American Folklore, 89*, 49–66.
Miller, K. (2002). The experience of emotion in the workplace: Professing in the midst of tragedy. *Management Communication Quarterly, 15*(4), 571–600.
Montague, D. R., Zohra, I. T., Love, S. L., McGee, D. K., & Tsamis, V. J. (2008). Hazing typologies: Those who criminally haze and those who receive criminal hazing. *Victims and Offender: An International Journal of Evidence-based Research, Policy, and Practice, 3*, 258–274.
Mumby, D. K. (1996). Feminism, postmodernism, and organizational communication: A critical reading. *Management Communication Quarterly, 9*, 259–295.
Mumby, D. K. (2001). Power and Politics. In F. M. Jablin & L. L. Putnam (Eds.), *The new handbook of organizational communication: Advances in theory, research, and methods* (pp. 732–818). Newbury Park, CA: Sage.

Nuwer, H. (1999). *Wrongs of passage.* Bloomington, IN: Indiana University Press.

Parks, G., & Brown, T. L. (2005). "In the fell clutch of circumstance": Pledging and the black Greek experience. In T. L. Brown, G., Parks, & C. M. Phillips (Eds), *African American fraternities and sororities: The legacy and the vision,* (pp. 437–464). Lexington, KY: University Press of Kentucky.

Patchen, M. (1965). Some questionnaire measures of employee morale: A report on their reliability and validity. *Institute for social research monograph no. 41.* Ann Arbor, MI: Institute for Social Research.

Rogers, C. (2007). *Informal coalitions: Mastering the hidden dynamics of organizational change.* New York, NY: Palgrave Macmillan.

Schlegel, A., & Barry, H. (1979). Adolescent initiation ceremonies: A cross-cultural code. *Ethnology, 18,* 199–210.

Seibold, D. R., Kang, P., Gailliard, B. M., & Jahn, J. (2009). Communication that damages teamwork: The dark side of teams. In P. Lutgen-Sandvik & B. D. Sypher (Eds.), *Destructive organizational communication: Processes, consequences, & constructive ways of organizing* (pp. 267–290). New York, NY: Routledge.

Shaw, D. L. (1992). A national study of sorority hazing incidents in selected land-grant institutions of higher learning. *Dissertation Abstracts International: Section A. Humanities and Social Sciences, 53,* 1077.

Sirhal, M. (2000). Fraternities on the rocks. *Policy Review, 99,* 1–14.

Sweet, S. (2004). Understanding fraternity hazing. In H. Nuwer, (Ed.), *Hazing reader* (pp. 1–13). Bloomington, IN: Indiana University Press.

Tajfel, H., & Turner, J. C. (1986). The social identity of intergroup behaviour. In S. Worchel & W. G. Austin (Eds.), *Psychology of Intergroup Relations* (2nd ed., pp. 7-24). Chicago: Nelson-Hall.

Van Maanen, J., & Schein, E. H. (1979). Toward a theory of organizational socialization. In B. M. Staw, (Ed.), Research in organizational behavior (pp. 209–264). Greenwich: JAI Press.

Waldron, V. (2009). Emotional tyranny at work: Suppressing the moral emotions. In P. Lutgen-Sandvik & B. D. Sypher (Eds.), *Destructive organizational communication: Processes, consequences, & constructive ways of organizing* (pp. 9–26). New York, NY: Routledge.

Appendix

Types of Organizations

Participants based answers to the questionnaire on their affiliation with one or two organizations. These are the categories provided for the initial question "Please indicate the first [second] organization whose membership experience you will complete this section about."

- Academic/Professional Organization (This organization provides academic support and/or promotes the attainment of academic and professional excellence.)
- Intramural or Club Sports Team (NOT including intercollegiate, school-sponsored sports, i.e. NCAA athletics)
- Fine Art and Music Organization (This organization promotes the enjoyment and/or performance of art and/or music.)
- Cultural and Ethnic Organization (This organization provides support of various ethnic and minority groups during their academic careers and promotes cultural understanding. Does NOT include fraternities and sororities.)
- Greek Organizations (fraternities and sororities)
- Honor Society
- Political Organization (This organization provides a means for students to support political issues, political parties, and candidates seeking office. Does NOT include Student Government Association and its corresponding organizations)
- Religious Organizations

- Service Organizations (This organization provides students volunteer opportunities for on-campus and off-campus service related projects. Does NOT include fraternities and sororities.)
- College or University-Sponsored Organization (This organization is funded by the college or university (i.e., Student Government, intercollegiate (NCAA, NJCCA) athletic team, Spirit Squads (cheerleaders, dance teams).
- Special Interest Organization (This organization does not fit any of the previous descriptions.)

Context Three

The Dark Side of Health Communication

TWELVE

Exploring the Dark Side of Social Support among African Americans with Prostate Cancer

Jason Thompson, Rockell Brown-Burton, and Devlon Jackson

Statistics concerning men diagnosed with prostate cancer are alarming. According to latest data from the U.S. Cancer Statistic Working Group and reported by the Centers for Disease Control and Prevention (2014), more than 209,292 men in the United States were diagnosed with prostate cancer in 2011. During that same year, 27,970 men died from the disease. The National Cancer Institute (n.d.) estimated that 233,000 new cases of prostate cancer would be diagnosed in 2014 and that more than 2,000,000 U.S. men would live with the diagnosis. Also, the American Cancer Society (2015) estimated that roughly 220,800 new cases of prostate cancer will occur in 2015 and that 27,540 men will likely die of the disease that year.

For the African American community, the picture looks considerably bleaker than the general trends portray. More African Americans are diagnosed and subsequently die from prostate cancer than men of other ethnic groups (National Cancer Institute, n.d.). Biological factors (e.g., androgen receptor function) along with exposure to "higher levels of testosterone because of diet-related hormonal influences" (McIntosh, 1997, p. 188) are attributed to higher incidence of prostate cancer cases among African Americans (Farrell, Petrovics, McLeod, & Srivastava, 2013). African Americans also tend to present with more advanced prostate cancer, making curative treatment less likely to benefit the recipient (Brooks, 2013). The greater number of advanced-phase incidents of prostate cancer among African Americans may be caused by a lower utilization rate of screenings that detect the disease in early stages (Merrill & Lyon, 2000). Historically, African Americans have been very suspicious of the medical community, and therefore, many have distanced themselves from preventative health care measures (Dula, 1994).

Researchers have discovered that prostate cancer causes a range of physical and psychosocial issues that decrease the quality of life for those with the diagnosis (Templeton & Coates, 2003). For example, individuals with prostate cancer, including those undergoing treatment, may experience erectile dysfunction, overactive bladder, bowel dysfunction, and urinary incontinence (Held-Warmkessel, 2006; Weber & Sherwill-Navarro, 2005). Those physical problems can lead to embarrassment, low self-esteem, poor self-concept, restricted or limited social activities, and an overall inability to lead a normal life (Fowler et al., 1995; Harris, 1997; Shavers & Brown, 2002). Furthermore, emotional instability, depression, and tremendous angst may affect those dealing with acute physical problems, as does an increased awareness of one's own mortality (Held-Warmkessel, 2006). Also, because an inordinate number of African American men live in underserved communities, they may not have access to the medical assistance necessary to help them deal with the disease (Bennett et al., 1998; Brooks, 2013). In addition to these issues, African American men may also endure distressing uncertainty as it pertains to prostate cancer screening, specifically as it relates to prostate-specific antigen (PSA) testing. A test that at one point was highly recommended for prostate cancer screening now has mixed opinions in the medical community, especially when caring for African American men (U.S. Preventative Services Task Force, 2012).

The disproportionate impact of prostate cancer on African Americans means that families and medical professionals need a richer understanding of the experiences of these men after diagnosis. Not surprisingly, family social support is often provided when a member faces a difficult health care situation. However, we give attention to the dark side of communication by exploring the descriptions and effects of social support not given to African Americans with prostate cancer.[1]

Social Support

While facing the challenges encountered along the path of life, individuals commonly seek assistance from members of their primary social network, such as family and friends, especially those with whom the person has established a significant personal bond (Bachman & Bippus, 2005). Family members primarily provide *social support*, defined as that which "[helps] the recipient see realistic alternatives to a stressful situation, gain skills, and recognize that help and resources are available from others" (Robbins & Rosenfeld, 2001, p. 279), or assistance that improves the welfare of the person who receives it (Schumaker & Brownell, 1984). Cutrona (1996) defined social support as the "responsiveness to another's needs, and more specifically, as acts that communicate caring; that validate the other's worth, feelings or actions; or that facilitate adaptive coping with problems through the provision of information, assistance, or tangible resources" (p. 10).

Researchers within a variety of disciplines have declared the beneficial effects of social support in several contexts. Perhaps the most beneficial aspect of social support is that it enables people to withstand challenges and obstacles; i.e., social support is communicated to help people as they deal with stressful life events (Gottlieb, 1983). Because the period after a prostate cancer diagnosis reflects a particularly troubling time, men who do not receive support may not effectively cope with the diagnosis, maintain essential psychological balance while facing the symptoms, or secure the practical means for seeking treatment.

Social Support among African Americans with Cancer

The research most relevant to the current study examined social support among African Americans diagnosed with different forms of cancer. For instance, Henderson, Gore, Davis, and Condon (2003) discovered that, while using personal coping strategies such as prayer and positive thinking, African American women with breast cancer also benefited from members of their social networks who provided them with empathy, care, and listening. In another study, Ford, Tilley, and McDonald (1998) found that African Americans with diabetes relied heavily on their informal social networks to cope with the physical and psychosocial effects of the disease. Although the studies by Henderson et al. (2003) and Ford et al. (1998) examined social support among African Americans with cancer, they did not specifically address those diagnosed with prostate cancer, nor did they focus on social support, or the lack thereof, offered in the family unit. Our study fills these information gaps.

A few researchers have examined social support received by African Americans with prostate cancer. Hamilton and Sandelowski (2004) discovered that the men in their study identified different types of support (e.g., emotional and instrumental) they received from members of their social network. Emotional support, for example, included encouraging words and positive, distracting activities. Instrumental support included prayer, assistance in maintaining social roles (e.g., knowing that your job will be held for you as you get well), and assistance in continuing religious routines. In other research on African Americans with prostate cancer, Jones et al. (2008) proclaimed, "The men in this study recognized the significance of family and friend support during this chronic illness" (p. 218). Family members, in particular, supported the men by becoming actively involved in the treatment decisions for the cancer. Also, wives and female friends proved invaluable sources of support as they demonstrated loyalty during a time when their mate experienced this often debilitating illness.

Although Hamilton and Sandelowski (2004) offered a well-designed study that provided a window looking into the lives of African Americans diagnosed with prostate cancer, they did not delve into supportive communication in the family unit. Instead, they took a broader approach by examining social support received from all members of one's social network, which includes friends and coworkers as well as family members. Jones et al. (2008) also took the wide view by considering support from both family and friends. Furthermore, neither Hamilton and Sandelowski (2004) nor Jones et al. (2008) considered the lack of social support among African Americans diagnosed with prostate cancer. Specifically, they overlooked the types of supportive communication that their study participants failed to receive from their families. As a result, one might assume that African Americans with prostate cancer naturally, as a matter of course, receive social support due to the close, intimate, familiar connections considered conventional in the family unit. However, this ideal situation may not reflect the reality for all men.

The deficient scholarly understanding of family social support among African Americans with prostate cancer, combined with the literature explanations about the construct of social support, informed the following research question:

> RQ_1. What types of social support do African Americans diagnosed with prostate cancer report not receiving from family members?

Based on their firsthand experience, African Americans diagnosed with prostate cancer can offer unique and important perspectives on familial communication of social support. These findings could be used to educate families about the best ways to assist and encourage a member with a prostate cancer diagnosis, and as a result, turn a dark side of communication into more positive messaging. Therefore, the second research question asks:

> RQ_2. What advice do African Americans with prostate cancer give to African American families about the best social support to give a relative with prostate cancer?

Although researchers have studied people with cancer, including African Americans, much has been left unexplored. The present study thus contributes to the intellectual discussion and an expanded volume of literature on cancer in the African American community.

Method

Participants

African American men who either have prostate cancer or are experiencing remission were qualified for inclusion in the research. Forty-one African American men qualified and agreed to participate in the study. Their ages ranged from 39 to 85 (M = 61.9, SD = 14.9) years. Over half of the participants had earned a bachelor's degree. Also, more than half were married.

Procedure

Subsequent to obtaining consent from the City University of New York–Brooklyn College Institutional Review Board, we recruited participants by sending a pithy description of the study to members of our social network. Through snowball sampling, the recruited individuals continued to circulate the written study description and requested that others relay it to potential participants (as per Lindlof & Taylor, 2002). In addition, we used an online website to recruit participants.

Instrument

Braithwaite, Moore, and Abetz (2014) argued that "qualitative methods are well suited to study close relationship forms, processes, and meanings, as scholars seek to understand how individuals, relational partners, families, and others in close relationships perceive, understand, experience, enact, and negotiate their relational worlds" (pp. 491–492). Therefore, to understand experiences with social support from the participants' points of view (as per Kvale & Brinkmann, 2009), a self-administered, open-ended questionnaire was used to collect data from participants.

Through the questionnaire, participants were asked about types of social support that they did not receive from their family members while they underwent treatment for prostate cancer and their recollections of their emotions during this difficult period: "Can you please list and explain types of social support that you recall NOT receiving from your family? Also how did it make you feel when you did not receive the support?" They were also asked about types of social support that they received from family members and their reactions to this support: "Can you please list and explain types of social support that you recall receiving from your family? Also how did it make you feel to receive the support?" Additionally, participants were asked to formulate a detailed description of advice about ways to communicate social support to a family member diagnosed with prostate cancer: "Can you please take a moment to describe any advice you would give to families about offering social support to a family member with prostate cancer? How should families act toward these members?"

For the questionnaire, we applied Robbins and Rosenfeld's (2001) description of social support: "[helps] the recipient see realistic alternatives to a stressful situation, gain skills, and recognize that help and resources are available from others" (p. 279). Several examples of social support were offered as prompts: "affection" (e.g., hugs and kisses), "listening," and "humor." The survey gave two additional directives to participants: "Please do not in any way allow these examples to affect and or hinder your response on the questionnaire," and "We invite your response to go beyond the samples listed here. In other words, you are not confined to the listed examples."

Data Analysis

To answer RQ_1 and RQ_2, we developed categories to organize and classify the descriptive open-ended responses provided by respondents. First, we gained both familiarity and understanding of the textual data. Second, we established inducted coding schemes from the data (as per Bulmer, 1979) by utilizing both open and axial coding (as per Strauss & Corbin, 1998). Open coding enables researchers to discover new thematic categories, add them to existing categories, and continue refining them (Strauss & Corbin, 1998).

Axial coding entails searching for similarities in the open-coded data (Strauss & Corbin, 1998). As we identified connections, the data were collapsed and combined. From this effort, a final set of thematic categories emerged. The interpretive method of negative case analysis was then used to ensure that none of the data contradicted any of the themes established (per Erlandson, Harris, Skipper, & Allen, 1993), thus ensuring the validity of the data. Finally, a data conference was coordinated to further establish validity of the findings (Baxter & Babbie, 2004). To achieve this aim, two experts familiar with qualitative research methods and knowledgeable about social support research were asked to offer critical feedback relating to interpretation of the results.

Results

Research Question 1: Types of Social Support Not Received

In response to RQ_1, regarding social support that respondents reported not receiving from their family members, two major categories of social support types emerged from the data analysis: (1) nurturant and (2) action-facilitating support. Within each major category, several subcategories were found (see Table 12.1).

Table 12.1. Reports of Social Support That African Americans Fail to Receive From Family.

Social Support Main and Sub Category	Frequency Reported
Nurturant	*n* = 28
Emotional	*n* = 18
Affection	
Listening	
Prayer	

Social Support Main and Sub Category	Frequency Reported
Humor	
Esteem	n = 10
Encouragement	
Masculine Affirmation	
Action Facilitating	n = 6
Tangible Assistance	n = 4
Financial Aid	
Motor Vehicle	
Informational	n = 2
Advice	

Nurturant Support

The most cited type of social support respondents indicated they did not receive was nurturant support (n = 28). Two subcategories, emotional and esteem support, also emerged within this category.

Emotional support. Respondents noted various forms of emotional support such that this subcategory was comprised of the most frequently cited type of specific support (n = 18). By and large, participants who felt a lack of emotional support described instances when they did not feel cared for, loved, and secure. For instance, participants explained that they did not receive affection from their children. As one man reported:

> My kids won't ask how I'm doing (stuff like "How's it going Dad? How are you feeling?"). That makes me feel they don't care about my condition. They also don't do the little things like wrap their arms around me squeezing to make me feel like they love me in this time. Maybe it's because I'm a man and they don't think I need it.

The children described in this example may demonstrate a belief that men do not require emotional support. However, the father with prostate cancer clearly articulated the need for specific forms of affection.

Participants also suggested that they received insufficient affection from their wives. These participants attributed the diminished warmth from their spouse to the penile symptoms (e.g., impotence) caused by prostate cancer. The lack of tenderness from their wives added to the distress of living with prostate cancer. As one participant reported:

> My wife does not give the type of affection she used to before I had prostate problems. She would kiss and hug me often. But now I am limited in how far I can go sexually because of impotence. I noticed that she has cooled off a lot.-

Similarly, another participant shared, "I want more love from my wife as I go through this, but I'm not getting it. I think she's as frustrated as I am." The participants in this study regard affection from family members as an important source of comfort, and those without this support clearly expressed discontent.

Listening, prayer, and humor represented several other forms of emotional support participants mentioned as missing from family members. Listening involves allowing the relative with prostate cancer to vent his emotions. For instance, one man explained, "Listening is something I want my family to do more. I know they have busy lives, but I am the sick one. Maybe they think I don't need to talk. Sometimes I need to let it out." Also, participants revealed a desire for prayers that God would heal the prostate cancer. One survey respondent explained, "I am a praying man. My wife does not pray like me and I want her to pray a lot for my condition. By her not praying I am less secure. Matt 18:19." Finally, regarding humor, one participant expressed disappointment that his siblings failed to help keep him in a lighter mood: "The lack of support from my siblings is shocking. I figured they would call to make me laugh…and everything. But they did not."

According to study participants, listening, prayer, and humor provide integral forms of support. When people demonstrate support by listening, they attend fully to the other person, thus symbolizing care and concern for the person. Prayer provides psychological reassurance as one may find comfort in knowing that a greater authority is working on their behalf so that bodily healing might manifest. Humor as a source of distraction can relieve pressure during any stressful period in life.

Esteem support. The second-most frequently reported form of nurturant support that participants lacked from family members involved efforts to maintain esteem ($n = 10$), which reinforces a person's self-confidence. Esteem support helps a person feel better about himself, but participants explained that family members did not provide that type of encouragement. They wanted more reassurance from family members that they had the fortitude to overcome the disease. As one man explained,

> [My cousins, siblings, and children] are not really there for me for the stuff like reassuring me that I will get better. It would be nice to hear them say stuff like "you will pull through" or "hang in there." Maybe they know this but I want to hear it more.

Not surprisingly, some patients diagnosed with prostate cancer may doubt their ability to defeat the disease even when medical evidence suggests otherwise. As this example shows, affirmation from family members affects the patient's state of hope.

Because of the serious physiological problems associated with prostate cancer diagnosis and especially without encouragement from meaningful others, a man can profoundly struggle with his identity and lose confidence. However, participants explained that family members do not always provide masculine affirmation to help reconstruct a fragmented identity: "This disease has challenged my manhood. I'm used to confidence. Now I'm always embarrassed. My wife definitely doesn't say enough things to lift my spirits." A father explained, "More could be done by my children. They know my disease has destroyed my confidence because I told them about that. I want them to give me a boost from time to time, make me feel I'm still dad!" Some adjectives associated with manhood include *self-assured* and *strong*. Prostate cancer can leave a man feeling insecure and weak.

Although the description of nonexistent support informs the focus of this study, some participants mentioned receiving esteem support from family members, but it was not always helpful; i.e., the interactions with relatives, although intended as support, did not contribute to an improved sense of self-esteem or ameliorate other negative feelings. For example, one man offered these details:

> My wife is always doing stuff to try to make me feel better about myself because she knows about my disease. To be honest, when she makes the effort to do this I feel worse off because her actions reinforce that I am the weak one in the relationship. I know she doesn't mean it that way, but I don't like being reminded of that.

Another participant revealed a similar sentiment as he explained the intended esteem support communicated to him from his parents:

> My parents call me quite often now that I've been diagnosed. They call with the intent to make me feel better about who I am but it doesn't help. I'm still pretty young [and] compared to them, I should be strong. So what they do to lift me up and help actually does the opposite.

This finding, that not all well-intentioned support creates the desired confidence boost, underscores the dark side of social support; i.e., the noblest efforts, executed and assumed to exert only a positive impact, do not necessarily translate to beneficial outcomes.

Action-Facilitating Support

The second-most frequently occurring kind of social support not received from family members involved action facilitation ($n = 6$). Participants' responses within this category revolved around tangible assistance and informational support.

Tangible assistance. Participants ($n = 4$) clearly stressed that they wanted their family members to provide useful concrete aid or physical resources to help make dealing with the prostate cancer more manageable. For example, one participant wanted financial aid:

> Going through prostate cancer has been a tough thing for me and my wife and kids financially. It costs a lot of money for treatment, checkups, and everything else involved and insurance doesn't cover everything. It makes me feel bad that my brother and some of my relatives on my dad's side with a lot of money won't offer some to me to help.

Other participants described a need for a vehicle. For example, one respondent needed a means to travel to the prostate cancer treatment center: "I don't have a car. Without a car I am not always able to make it to the center for care. My sister has an extra car and she won't let me use it." Another participant expressed disappointment that his cousin would not help him get to meetings that address cancer-related issues. He explained, "Without a car I need a ride to attend support groups. I have a cousin who can help but won't."

Informational support. The second-most frequently reported action-facilitating support missing from family members was information ($n = 2$); i.e., some participants did not receive advice from family members. Advice entails information about steps men with prostate cancer might take to advocate for and precipitate their healing. One participant expanded on the complicated need: "The biggest [issue] is probably not being there for advice. I think that's because this disease is new territory for my family. You can't give advice for something you don't know a lot about." Advice provides information capital that can help people solve problems and improve their circumstances. Although cancer patients can receive information and direction from medical professionals, the men in this study would value advice from family members. Perhaps they would appreciate the personalized input of a person who knows them well in addition to the clinical advice of physicians.

Research Question 2: Advice to Family Members about Social Support

Through RQ_2, we obtained data about advice that study participants offer to family members about appropriate social support to other African Americans diagnosed with prostate cancer. Two main categories of responses emerged in the study: nurturant and action-facilitating support (see Table 12.2).

Table 12.2. Advice for Families from African Americans with Prostate Cancer.

Advice on Social Support by Main and Sub Category	Frequency Reported
Nurturant	*n* = 36
Emotional	*n* = 27
Maintain a Presence	
Pray	
Do Not Pity	
Show Affection	
Listen	
Esteem	*n* = 9
Communicate Reassurance	
Action Facilitating	*n* = 12
Informational	*n* = 8
Become Informed	
Tangible Assistance	*n* = 4
Financial Aid	

Nurturant Support

The most frequently cited category of advice that participants offered to family members involved nurturant support (*n* = 36). Specifically, participants suggested that other men with a prostate cancer diagnosis would appreciate emotional and esteem support.

Emotional support. Respondents indicated various forms of emotional support would help other men with prostate cancer (*n* = 27). Some recommended that family members should maintain a presence; i.e., they should stay connected to the one in the family with prostate cancer, providing care and comfort throughout the disease's duration. As one participant explained, "Embrace them, since they are feeling scared and vulnerable. Don't be standoffish because you feel like you have nothing to say." Two other men urged consistency and endurance, "Don't allow them to go through it alone. Some patients may want to withdraw and give up and they need to know that there is someone in their family who cares for them," and "There's a lot of tips to give. The main one is to be there every step of the way. The person should never feel alone in the fight."

Participants also advised family members to pray steadfastly for the man diagnosed with prostate cancer: "Keep God in everything. Pray with and for the family member with cancer. Remind him that healing is the children's bread." The men who cited prayer deemed it a spiritual change agent that can effectively improve the condition of the man for whom the prayer was intended. One participant articulated, "If you believe in God and Jesus, then the thing to do is pray for the family member. Pray with them while around them. There is a supernatural power that is found in prayer."

Participants also admonished families to refrain from expressing pity toward the diagnosed man. Specifically, they suggested that family members abstain from excessively commiserating with the patient because it can precipitate recurrence of unpleasant emotions, such as sadness and depression. As one respondent explained, "I hate that some of my family members have a pity party for me. I feel sad already with this condition and that kind of stuff doesn't help matters." Similarly, another man advised, "Take care to ask them how they would like to be treated, helped, or supported rather than supposing pity is best. By all means don't offer pity."

Esteem support. A prostate cancer diagnosis can cause a man to feel helpless and hopeless. The incidence rate of prostate cancer for African Americans remains disproportionately high, and more than twice as many African Americans, compared to all other ethnic or racial groups, will not see the 5-year survival benchmark that generally indicates survivability (National Cancer Institute, n.d.).

Commensurate with the dire prospects associated with the disease, those in this study (n = 9) explained the importance of communicating reassurance that buoys the spirit of the family member who has prostate cancer. As one man suggested, "Try to encourage them that they can get through it even when they feel like they can't." Another participant asserted, "I cannot emphasize enough how important encouragement is. The patient needs to feel like he can make it through. Sometimes his mind won't tell him that."

Action-Facilitating Support

The second-most frequently reported category of advice for family members involved action facilitation (n = 12). Two subcategories, informational support and tangible assistance, emerged within this category.

Informational support. Survey respondents (n = 8) suggested that family members should learn about prostate cancer so that they can relay pertinent, supportive information to their diagnosed relative. Such advice may help him heal or cope. One participant recommended: "Get an education about the cancer so that you know what the person needs in order to get better. Most of the folks in my family don't have the knowledge, so educate yourself."

Tangible assistance. Participants (n = 4) recommended that family members use their physical and financial resources to help those trying to cope with the disease. One participant revealed, "Financial aid never hurts no matter how old or how young the man with cancer is." Pecuniary resources matter to those receiving treatment and trying to support themselves and their families amidst a devastating cancer diagnosis.

Discussion

In answer to RQ1, African American men described two main types of social support they felt family members had not offered: nurturant and action facilitating. Participants wanted family members to exert more effort at comforting and consoling them as they lived with the challenges of prostate cancer. They also revealed that their needs for financial resources and information remained unmet by family. In answer to RQ2, the men suggested apposite advice including maintain a presence, pray, do not offer pity, show affection, listen effectively, communicate reassurance, become informed, and offer financial aid.

Findings of the current study expand existing literature that provides concrete evidence for the importance of social support for people coping with challenging life events, such as prostate cancer (Hobfoll & Parris-Stevens, 1990; Petrie & Stoever, 1997). It also highlights the social support that men with prostate cancer seem to lack, particularly within the context of family, calling attention to the dark side of family communication (Duck, 1994). Study participants acknowledged varying negative emotions (e.g., sadness, anger) when they did not receive social support from their family members during their time of tremendous challenge. They felt the impact of an unmet need for encouragement, affirmation, and resources. The findings, perhaps more than any specific outcomes, reinforce the importance of social support (Allen, Ciambrone, & Welch, 2000; Castillo & Hill, 2004). Context notwithstanding, social support fortifies people, enabling them to cope with devastating diagnoses and life-altering situations. Furthermore, the human need for support is perhaps revealed most dramatically when it is withheld from those in crisis.

Results of the present study do not contradict previous research that describes the support African Americans diagnosed with prostate cancer receive from family members (Hamilton & Sandelowski, 2004; Jones et al., 2008). However, they add another layer to the understanding of the men's experience by demonstrating that recipients of familial social support may perceive it as unhelpful. Specifically, our study calls attention to the dark side of social support, which receives relatively little scholarly attention, but affects lives and must be recognized. Specifically, one may intuitively conceptualize social support as a wholly positive construct that strengthens people during challenging life events. However, our study shows that even the best-intended social support can lead to unintended negative consequences as it perpetuates unpleasant thoughts and emotions experienced by those receiving it. Specifically, study participants described feeling sad that the social support communicated from their relatives psychologically reinforced their frail, weak, and mortal existence.

The results also seem to challenge a long-standing narrative that prescribes gender role expectations (Cross, 2008; McRobbie, 2009). To this end, some of the African Americans with prostate cancer expressed disappointment when family members did not communicate emotional support. Perhaps the view of men as undesirous of emotional support, as based on the prevailing adjectives (e.g., *strong, controlled, dominant, tough*) and life scripts most commonly associated with masculine norms or manhood (Kimmel, 2008), needs to be reevaluated. Preconceived notions concerning male behavior and temperament, as evidenced with the esteem associated with terms such as *unshakable* and *fortitude*, amount to stereotypes that limit the type of social support offered by the most well-meaning and intimate relations.

Furthermore, fixed notions about masculinity manifest in formulated templates for behavior that may obstruct family members' abilities to see a loved one's need and may culminate in unsupportive treatment of their relative with prostate cancer. The outcomes of little or misapplied emotional support for the men in this study, all of whom were experiencing or had faced challenging life events associated with prostate cancer, were expressed in statements of disappointment, emasculation, sadness, and frustration.

Future Research in Communication

Future researchers should investigate the reasons some family members do not offer certain types of support to members facing prostate cancer. This future exploration would offer insight into the impact of complex family interactions on social support communication, especially in context of crisis. Also, future researchers might consider examining social support needed by all family members affected by the diagnosis. Caring for a family member with prostate cancer requires time-consuming, expensive, and distressing responsibilities. Those offering support likely need encouragement to persist and remain positive. Although researchers have examined the social support offered to wives who care for their spouse with prostate cancer (see Gottlieb & Maitland, 2014), scholars can take a more comprehensive approach by investigating social support offered by and for close family members (e.g., parents, cousins, and children). In sum, we hope this research has inspired other scholars to take the next steps toward understanding complex family dynamics surrounding important health issues.

Notes

1 Support for this project was provided by a PSC-CUNY award jointly funded by The Professional Staff Congress and The City University of New York.

References

Allen, S. M., Ciambrone, D., & Welch, L. C. (2000). Stage of life course and social support as a mediator of mood state among persons with disability. *Journal of Aging and Health, 12*(3), 318–341.

American Cancer Society. (2015). *Prostate cancer statistics*. Retrieved from http://www.cancer.org/cancer/prostate cancer/detailedguide/prostate-cancer-key-statistics

Bachman, G., & Bippus, A. (2005). Evaluations of supportive messages provided by friends and romantic partners: An attachment theory approach. *Communication Reports, 18*(1), 85–94.

Baxter, L. A., & Babbie, E. (2004). *The basics of communication research*. Belmont, CA: Wadsworth.

Bennett, C., Ferreira, M., Davis, T., Kaplan, J., Weinberger, M., Kuzel, T., … Sartor, O. (1998). Relation between literacy, race, and stage of presentation among low-income patients with prostate cancer. *Journal of Clinical Oncology, 16*(9), 3101–3104.

Braithwaite, D. O., Moore, J., & Abetz, J. (2014). "I need numbers before I will buy it": Reading and writing qualitative scholarship on close relationships. *Journal of Social and Personal Relationships, 31*(4), 490–496.

Brooks, D. (2013). *Why are Black men negatively affected by prostate cancer more than White men?* Retrieved from http://www.cancer.org/cancer/news/expertvoices/post/2013/09/24/why-are-black-men-negatively-affected-by-prostate-cancer-more-than-white-men.aspx#continue

Bulmer, M. (1979). Concepts in the analysis of qualitative data. *Sociological Review, 27*(4), 651–677.

Castillo, L. G., & Hill, R. D. (2004). Predictors of distress in Chicana college students. *Journal of Multicultural Counseling & Development, 32*(4), 234–249.

Centers for Disease Control and Prevention. (2014). *Prostate cancer statistics*. Retrieved from http://www.cdc.gov/cancer/prostate/statistics/

Cross, G. (2008). *Men to boys: The making of modern immaturity*. New York, NY: Columbia University.

Cutrona, C. (1996). *Social support in couples*. Thousand Oaks, CA: Sage.

Duck, S. (1994). Strategems, spoils, and the serpent's tooth: On the delights and dilemmas of personal relationships. In B. Spitzberg & W. Cupach (Eds.), *The dark side of interpersonal communication* (pp. 3–24). Hillsdale, NJ: Erlbaum.

Dula, A. (1994). African American suspicion of the healthcare system is justified: What do we do about it? *Cambridge Quarterly of Healthcare Ethics, 3*(3), 347–357.

Erlandson, D., Harris, E., Skipper, B., & Allen, S. (1993). *Doing naturalistic inquiry: A guide to methods*. Newbury Park, CA: Sage.

Farrell, J., Petrovics, G., & McLeod, D., & Srivastava, S. (2013). Genetic and molecular differences in prostate carcinogenesis between African American and Caucasian American men. *International Journal of Molecular Sciences, 14*(8), 15510–15531.

Ford, M., Tilley, B., & McDonald, P. (1998). Social support among African-American adults with diabetes. *Journal of the National Medical Association, 90*(7), 425–432.

Fowler, F., Barry, M., Lu-Yao, G., Wasson, J., Roman, A., & Wennberg, J. (1995). Effect of radical prostatectomy for prostate cancer on patient quality of life: Results from a Medicare survey. *Urology, 45*(6), 1007–1015.

Gottlieb, B. H. (1983). *Social support strategies: Guidelines for mental health practice*. Beverly Hills, CA: Sage.

Hamilton, J., & Sandelowski, M. (2004). Types of social support in African Americans with cancer. *Oncology Nursing Forum, 31*(4), 792–800.

Harris, J. (1997). Treatment of post-prostatectomy urinary incontinence with behavioral methods. *Clinical Nurse Specialist, 11*(4), 159–166.

Held-Warmkessel, J. (2006). *Contemporary issues in prostate cancer: A nursing perspective* (2nd ed.). Sudbury, MA: Jones and Bartlett.

Henderson, P., Gore, S., Davis, B., & Condon, E. (2003). African American women coping with breast cancer: A qualitative analysis. *Oncology Nursing Forum, 30*(4), 641–700.

Hobfoll S., & Parris-Stevens, M. (1990). Social support during extreme stress: Consequences and intervention. In B. Sarason, I. Sarason, G. Pierce (eds.), *Social support: An interactional view* (pp. 454–481). New York, NY: Wiley.

Jones, R., Taylor, A., Bourguignon, C., Steeves, R., Fraser, G., Lippert, M. … Kilbridge, K. (2008). Family interactions among African American prostate cancer survivors. *Family Community Health, 31*(3), 213–220.

Kimmel, M. (2008). *Guyland: The perilous world where boys become men.* New York, NY: Macmillan.

Kvale, S., & Brinkmann, S. (2009). *Interviews: Learning the craft of qualitative research interviewing.* (2nd ed.). Thousand Oaks, CA: Sage.

Lindlof, T. R., & Taylor, B. C. (2002). *Qualitative communication research methods.* Thousand Oaks, CA: Sage.

McIntosh, H. (1997). Why do African American men suffer more prostate cancer? *Journal of the National Cancer Institute, 89*(3), 188–189.

McRobbie, A. (2009). *The aftermath of feminism: Gender, culture, and social change.* Thousand Oaks, CA: Sage.

Merrill, R. M., & Lyon J. L. (2000). Explaining difference in prostate cancer mortality rates between White and Black men in the United States. *Urology, 55*(5), 730–735.

National Cancer Institute. (n.d.). *SEER stat fact sheets: Prostate cancer.* Retrieved from http://seer.cancer.gov/statfacts/html/prost.html

Petrie, T. A., & Stoever, S. (1997). Academic and nonacademic predictors of female student–athletes' academic performance. *Journal of College Student Development, 38*(6), 599–608.

Robbins, J. E., & Rosenfeld, L. B. (2001). Athletes' perceptions of social support provided by their head coach, assistant coach, and athletic trainer, pre-injury and during rehabilitation. *Journal of Sport Behavior, 24*(3), 277–297.

Schumaker, S. A., & Brownell, A. (1984). Toward a theory of social support: Closing the gaps. *Journal of Social Issues, 40*(4), 11–36.

Shavers, V., & Brown, M. (2002). Racial and ethnic disparities in the receipt of cancer treatment. *Journal of National Cancer Institute, 94*(5), 334–357.

Strauss, A., & Corbin, J. (1998). *Basics of qualitative research: Techniques and procedures for developing grounded theory.* Thousand Oaks, CA: Sage.

Templeton, H., & Coates, V. (2003). Informational needs of men with prostate cancer on hormonal manipulation therapy. *Patient Education Counseling, 49*(3), 243–256.

U.S. Preventative Services Task Force. (2012). Final recommendation statement, prostate cancer: Screening, May 2012. Retrieved from http://www.uspreventiveservicestaskforce.org/Page/Document/RecommendationStatementFinal/prostate-cancer-screening.

Weber, B., & Sherwill-Navarro, P. (2005). Psychosocial consequences of prostate cancer: 30 years of research. *Geriatric Nursing, 26*(3), 166–175.

THIRTEEN

"My Doctor Ruined My Entire Birthing Experience": A Qualitative Analysis of Mexican-American Women's Birth Struggles with Health Care Providers

Leandra H. Hernandez

The Hispanic/Latino population is the fastest growing minority in the United States (Brown & Lopez, 2013). With this increase, efforts are being made within the health care sector to increase physicians' cultural competence and understanding of minority health values, beliefs, and practices with a goal to improve minority patients' health experiences. One health care context within which these efforts are increasing is reproductive health, particularly birth. Women's preferences for birthing method, physician involvement during the birthing process, and physician communication styles can vary dramatically according to a woman's race, ethnicity, and cultural values, to name a few; thus, it is important to explore the ways in which minority women experience birth processes in the United States. This knowledge will increase our understanding of the ways in which race, ethnicity, and gender intersect to shape minority women's perceptions of their health care and relationships with physicians.

This study explores Mexican-American women's birthing experiences in the United States, their perceptions of their physicians' communication, and their struggles with physicians over birthing methods and experiential knowledge.[1] Despite research that has rhetorically (Gutierrez, 2008) and ethnographically (Galvez, 2011) explored Mexican-origin and Mexican immigrant women's birth experiences, there is a paucity of research that has explored second- and third-generation Mexican-American women's birth experiences and the unique intersections between and among their ethnicity, class, gender, and generational status as they influence their birth decision making. Research suggests that for Mexican-American women, higher levels of acculturation are associated with more undesirable prenatal behaviors and risk factors; a distancing from traditional Mexican cultural norms; and acceptance of Western ideals of individuality and autonomy, particularly within health care encounters (Zambrana, Scrimshaw, Collins, & Dunkel-Schetter, 1997).

Thus, this study asks: (1) What are second- and third-generation Mexican-American women's birthing experiences in the U.S.? and (2) How does their generational status shape and affect their birthing experiences?

Although much literature discusses the relationship between acculturation and Mexican-American women's perceptions of prenatal care and care during pregnancy, there is a paucity of research that examines Mexican-American women's perceptions of their birth and physicians' communication during this crucial time in their lives. Thus, the exploration of previous literature reviews what patient-centered communication is, as well as Mexican and Mexican-American women's perceptions of prenatal care during birth.

Patient-Centered Communication

According to Epstein and colleagues (2005), patient-centered communication's goal is to assist practitioners with providing care that encourages patients to actively participate in their encounter by stating their values, beliefs, feelings, and preferences. It also strives to include patients as much as possible in the decision-making process regarding their treatment regimen. Epstein and Street (2007) define patient-centered communication according to four main points:

> Eliciting, understanding, and validating the patient's perspective; understanding the patient within his or her own psychological and social context; reaching a shared understanding of the patient's problem and its treatment; [and] helping a patient share power by offering him or her meaningful involvement in choices relating to his or her health. (p. 2)

When using patient-centered communication, physicians place the patient at the heart of the medical encounter. Instead of following the traditional, scientific biomedical approach in which the doctor is the expert, patient-centered communication is patient-centric. It acknowledges that patients are individuals, not just bodies that need a diagnosis and a cure, and it recognizes patients' life histories, needs, and the meanings underlying their illnesses (Epstein et al., 2005; Mead et al., 2002; Zandbelt, Smets, Oort, Godfried, & de Haes, 2007). Within a birth context, patient-centered communication includes validating patients' birth preferences and fears about the birth process, as well as helping the patient become actively involved in making meaningful choices about birth methods, medication use, and postpartum care. Although patient-centered communication has been a widely supported framework for quite some time, a review of the literature suggests that Mexican-American women during their births do not perceive their physicians and nurses to be patient-centered communicators.

Mexican-American Women's Perceptions of Birth and Postpartum Care

Research that explores Mexican-origin and Mexican-American women's birth and postpartum care experiences suggests that Mexican-origin and Mexican-American women have specific expectations of their physicians pertaining to information exchange, reassurance, and comfort, yet their expectations are often not met. Engle, Scrimshaw, Zambrana, and Dunkel-Schetter (1990) examined psychosocial factors related to Mexican women's postnatal anxiety. Whereas their participants preferred health care providers who were knowledgeable, sympathetic, and friendly and who provided good medical explanations, Engle and colleagues (1990) found that explanations of the birthing and medical procedures were not provided for many patients, even in cases when the patients spoke English fluently. This led to more anxiety for the patients when they were giving birth and less satisfaction with their overall health care.

In addition to a lack of information exchange about the procedures, Galvez (2011) found in her ethnographic research of Mexican-origin women's birth experiences in the United States that Mexican women's births were characterized by verbal insults and humiliations from hospital staff and physicians, a lack of translation services, a lack of guidance and information, and power struggles between patients and physicians over perceptions of whether or not women were in labor, epidurals, moving around during the labor process, fetal monitoring, and birthing methods. Galvez (2011) noted that patients were often scared, unsure of what to expect from their physicians, scolded, and even "schooled into what was expected of them" (p. 120). Moreover, Galvez (2011) "observed well-meaning prenatal care providers reinforce dominant stereotypes and derogatory assumptions about Mexican patients" (p. 154), and that these assumptions played out within medical encounters as physicians offered no time for their patients to share their cultural practices and cultural attitudes about pregnancy and birth. This led Galvez (2011) to conclude that "culture has no place in the clinic" (p. 155).

Mexican-American women's postpartum care is also characterized by some of these medical barriers. Pope (2005) found that when compared to White women presumed to be primary English speakers, low-income postpartum Latinas suffered from more postpartum issues such as anemia, decreased vitamin instruction and use, higher rates of urinary tract infections, and higher rates of depression and even domestic violence. Pope (2005) concluded that many of these postpartum care issues can be attributed to limited English-language proficiency (LEP). She argued that "Hispanic LEP women need careful instruction about the importance of a follow-up visit and confirmation of a provider who accepts their care with a specific appointment" (p. 514). Therefore, a lack of caring physicians with limited translation abilities could exacerbate health disparities and make language barriers and culturally competent care problematic clinical issues.

Research Questions

This review of the literature about Mexican women's birth experiences in the United States suggests that women of Mexican origin are confronted with many barriers and patient-centered communication flaws during their birth processes. These barriers include lower levels of satisfaction with physicians pertaining to information-exchange, trust, and respect; and in extreme cases, verbal insults and humiliation. Given these findings, this chapter asks: (1) What are second- and third-generation Mexican-American women's birthing experiences in the U.S.? and (2) How does their generational status shape and affect their birthing experiences?

Methods

This project utilized qualitative methods because they allow researchers to explore human understanding, lived experience, and the nuances and negotiations that people experience in their everyday lives as they navigate the health care system and make important decisions about their health (du Pré & Crandall, 2011). Moreover, they help researchers unpack the processes surrounding health communication and explore what "really" is going on (Britten, 2011, p. 388). Lastly, qualitative research is a useful tool for understanding societal issues that arise from cultural contexts (Covarrubias, 2002; Kreuter & McClure, 2004; Tracy, 2013). Ethnicity, gender, race, and sexual orientation can be understood, critiqued, and transformed through contextual studies that examine how these categories are negotiated, ever-changing, and communicatively constituted (Tracy, 2013).

After receiving IRB approval, semi-structured, in-depth interviews were conducted with 30 first-, second-, and third-generation Mexican-American women (15 from Houston and 15 from San Diego) between the ages of 30 and 45 who had at least one pregnancy. All were proficient English speakers, and two-thirds of the participants spoke both English and Spanish. Eight participants had bachelor's degrees, two had master's degrees, and 20 had taken some college courses. The number of participants' individual pregnancies ranged from one to four.

The data come from a larger research project that sought to explore Mexican-American women's prenatal testing and birth experiences, which dictated the age range inclusion criterion. Moreover, the larger study sought to explore regional constructions of what it means to be Hispanic/Mexican/Mexican-American and the relationships between these ethnic identity categories and health beliefs and decision making, which dictated the location inclusion criterion. This population was chosen because no research has explored second- and third-generation Mexican-American women's negotiations of reconciling and making sense of their birth practices, beliefs, and traditions influenced by culture and older generations, and the birth practices, beliefs, and customs they have experienced and perhaps adopted as they have lived in the U.S. Given that the Hispanic/Latino population is booming in the U.S. (Lopez, 2014), knowledge of Mexican-American women's experiences of making birth decisions with spouses and clinicians can help inform future cultural competence curricula, as well as improve future health care encounters that deal with this complex decision-making process.

In-depth interviews allowed for the ability to explore the participants' views of reality (Reinharz, 1992) and support or disconfirm pre-existing statistics and generalizations (Sexton, 1982) about Mexican-American women's birth experiences and their perceptions of their health care providers' communication during this process. Moreover, in-depth interviews allowed for the elicitation of language used by participants to describe their experiences and garner their stories and explanations (Lindlof & Taylor, 2011).[2] The snowball sample recruitment method was utilized to recruit participants, and four key informants (two in each city) assisted with recruitment by suggesting friends, family members, and coworkers. Interviews were conducted in each city within a one-month time span, lasted on average one hour, and were mostly conducted at local coffee shops.

Once interviews were completed, a thematic analysis was conducted to explore the themes, categories, and codes that emerged from the data. According to Boyatzis (1998), a thematic analysis is, in its most basic sense, a way of seeing. More specifically, Braun and Clarke (2006) note that it is a "method for identifying, analyzing, and reporting patterns (themes) within the data" (p. 79). Thus, categories and a coding scheme were created based on patterns, similarities, and notable exceptions in the data (Lindlof & Taylor, 2010). Categories were identified pertaining to participants' birth experiences and perceptions of their physicians' communication, which were then collapsed into themes and their corresponding categories. In addition to the thematic analysis, reflexivity journals were kept throughout the data collection and analysis processes. Erlandson, Harris, Skipper, and Allen (1993) note that reflexivity journals are important research tools that support the credibility of a scholar's arguments, in addition to supporting the dependability, transferability, and confirmability of a study. As a second-generation Mexican-American woman who has had her fair share of disastrous medical encounters, I journaled about my own subjectivities and how they influenced my data analysis. Additionally, after each interview, I journaled about initial themes emerging from interview conversations. An average of 5 pages of handwritten notes were written after each interview, resulting in over 150 pages of post-interview reflections and initial data analysis.

Results

During interviews, most of the participants recounted their birth experiences with anger and resentment. Although a few participants did indeed have positive birth experiences due to their kind physicians and nurses, a majority of the participants disliked their physicians. As suggested by the three themes that emerged from data analysis, most participants disliked their physicians because of: (1) perceived low levels of medical understanding and information exchange, (2) power struggles with physicians, and (3) high levels of uncertainty experienced both during birth and postpartum care.

Perceived Low Levels of Medical Understanding and Information Exchange

The first theme, low levels of medical understanding, focuses on how participants perceived their lack of understanding about certain terms and procedures. Despite the fact that over half of the participants have at least a bachelor's degree (while a few have master's degrees), many participants experienced difficulty understanding birth procedures such as epidural necessity, Pitocin use, and other factors.

Many participants described the epidural during their birth as a "scary," "unknown" procedure that "wasn't explained accurately" to them. Participants experienced difficulty describing what the epidural does, as well as its side effects, and this lack of knowledge negatively impacted their birth expectations and experiences. Noelia, a 37-year-old second-generation Mexican-American mother of two from Houston, described her birth with her first son as an "upsetting" experience:

> And sure enough by the time I got to the hospital, what they did, I will *tell* you, it messed me up. I was kind of upset. Actually, I was very upset because of what they did and what they didn't tell me! Once I got there, I was 4 centimeters dilated and the Lamaze classes just went out the window. I was screaming, "Oh my god, give me something!" I was in so much pain. I was pretty much crawling on the floor, begging them to get the baby out of me. They gave me Pitocin to speed up the labor without even telling me or explaining it to me, but then my son was coming too fast. The doctor wasn't there yet, so they had to give me something to slow it down. That really messed me up. Then they gave me the epidural and I was numb from the waist down! I was *so* numb down there that I couldn't even tell if I was pushing! I hated it. I didn't want the epidural because I wanted to have a natural birth, and I wasn't even quite sure how the epidural works. All I knew is that I didn't want it, but they forced me to get one anyway.

In this birthing experience critical incident, Noelia described the interrelated factors that contributed to her lack of understanding about the epidural more specifically and about her birthing process more broadly. Factors such as time, physician unavailability, and lack of knowledge about the epidural contributed to her anxieties surrounding her son's birth and created an unsatisfactory experience. As the conversation progressed, Noelia noted that her physicians "never quite took the time" to explain to her what the epidural did or how it worked; she stated that she might have been more willing to accept the epidural procedure if she was more educated about what it entailed, yet a few years after her birth she is still unsure about how the epidural works and side effects associated with the procedure.

Similarly, Nayara had a problematic pregnancy with her fourth child. She recounted that they put her on Pitocin because her contractions weren't getting stronger, and the Pitocin led to the epidural, which led to her dissatisfaction with the process:

> I wasn't even quite sure what it was. I know it was supposed to speed up the contractions, but how? In what way? What sorts of effects was that going to have on my birth and on my baby? I didn't want it because I didn't want it to affect my baby, but at that point, I felt like I had no choice. It was the same with the epidural. They didn't even tell me how it was going to work or what. Nothing. It was the worst experience of my life.

A majority of the participants struggled with a lack of information during their pregnancies. This lack of information or understanding ranged from prenatal screening and testing to the use of drugs and procedures such as Pitocin and epidurals. In addition to struggling with lack of information, participants also struggled with their health care providers during their births.

Power Struggles with Physicians

Participants' issues with their health care providers during their births were twofold: participants' struggles to convince their health care providers that they were in labor, and participants' preferences for birth methods versus their physicians' preferences for birth methods. The first struggle that occurred in the participants' birth experiences occurred when they tried to integrate their experiential knowledge into their birth processes. According to Caron-Flinterman, Broerse, and Bunders (2005), experiential knowledge can be defined as "the ultimate source of patient-specific knowledge—the often implicit, lived experiences of individual patients with their bodies and their illnesses as well as with care and cure" (p. 2576). When patients integrate their experiential knowledge into their health care encounters they can experience better health outcomes because they are actively participating in their health care and discussing their personal expertise with physicians (Caron-Flinterman et al., 2005). However, within this specific context, participants noted that their experiential knowledge about their bodies' physical abilities during pregnancy (based upon previous pregnancy experiences and their bodies, which they "just know") was not welcomed when they were going into labor. Lourdes, a 32-year-old second-generation Mexican-American mother of four from Houston, noted that there was "absolutely no communication" during her births. She described her doctors and nurses as "just terrible" and recounted how her nurse injected her IV with unknown medication and then did not believe her when she expressed her concerns:

> "When I finally got in, they hooked me up to an IV and it immediately started burning. My arm felt like it was literally in flames, and so I started crying. The nurse just walked out! I was in so much pain. The nurse kept telling me that nothing was wrong, that maybe my pregnancy was getting to me. I kept telling her that I *knew* something was wrong!"

Nayara described her pregnancy with her third child as "horrible" because her doctors and nurses "just wouldn't listen":

> My water broke at home, and I went to the emergency room. Well, when I got there, my water broke some more and my pants were soaked! When I got there, the nurses and doctors kept telling me that my water didn't break, that sometimes pregnant women don't know when they're peeing on themselves—how ridiculous is that! They kept trying to send me home, and I kept refusing. I said, "No, you need to check again. I *know* my water broke." Then after a while of arguing with them, they finally realized my water broke. It was such a struggle! It was horrible. They wouldn't listen to me.

Similar to Nayara, Marita, a 34-year-old second-generation mother of two from San Diego, fought with her doctors and nurses about being in labor. She stated that when her water broke, she "went straight to the hospital" because she was anxious about her birth and "wanted to get it out of the way." However, when she arrived at the hospital, she struggled with her physicians: "I was ready, I expected it, and I knew I was in labor. They kept saying, 'No you're not. No you're not. You're not ready. Go back home and come back in a few days.' But I knew I was! It was so prolonged and so painful. The contractions were painful and unbearable. They finally admitted me after my husband and I just wouldn't budge." Lara, a 32-year-old first-generation mother of two from San Diego had a similar experience with her second pregnancy. She expressed anger about fighting to get admitted to the hospital and a triage nurse delivering her second child because her doctor was not at the hospital when she was in labor. She concluded: "I just wish they would've listened more to my needs as a patient, especially because when you're in labor, you are *not* a happy person."

In addition to struggling with physicians over their experiential knowledge about being in labor, participants expressed much dissatisfaction about their preferred birth method not being upheld and accepted when it was time to deliver their children. Most of the participants preferred to have vaginal births, yet certain circumstances mandated unexpected C-sections. This new requirement, which was spurred by health problems and other factors, was not welcomed by participants.

Yesenia, a 33-year-old second-generation Mexican-American mother of two from Houston, expressed concerns during her second pregnancy about her unexpected C-section:

> During her birth, I was pushing and pushing and then all of a sudden she went into distress. My doctor screamed, 'Stop pushing! We have to go in and cut her out!' I screamed, 'What?!' I did *not* want a C-section! It terrified me. I kept trying to ask my doctor if there was anything she could do, like more medication or something, and she kept saying no. I kept telling her I didn't want a C-section, but it didn't matter. Before I knew it, they were already wheeling me into the OR. She cut me open, pulled her out, sewed me up, and she was gone.

In Yesenia's situation, her baby's distress prompted the physician to mandate an emergency C-section so as to not exacerbate the baby during the birth. More information about what was occurring, according to Yesenia, would have assuaged her fears and anxieties during this process. Lastly, Anita, a 32-year-old second-generation Mexican-American mother of one from San Diego, described her birth with her son as a "highly emotional and empty" process because in a matter of hours her planned vaginal birth turned into a mandatory C-section. At the end of the process, she "felt like emptiness":

> To this day, I still don't understand the concept of what they were talking about. After the birth, they tried explaining it to me again, and I still feel clueless! They didn't do a good job of explaining anything. Anyway, they told me my baby would be born that day, and I straight up lost it. I kept telling them no, no, they had to be mistaken. Within a matter of minutes, I had an emergency C-section. It happened in the blink of an eye. I was a mess. I was crying so bad. It happened so fast. After it was over, I just felt like emptiness. Nothing happened the way I wanted it to, and I didn't have my baby. I know he was in NICU, but still. He wasn't with me, and I was so emotional.

Similar to Estrella and Yesenia's experiences, Anita grappled with her emergency mandatory C-section and resisted as much as she could until she had to undergo the C-section for her health and for her baby's health. As with Estrella and Yesenia, Anita noted that she might have felt a bit better about the situation if her physicians would have taken a few moments to discuss the emergency C-section with her and lessen her anxieties.

Thus, many participants experienced fear and anxiety when they went into labor about a variety of topics, including but not limited to dissatisfaction about mandatory emergency C-sections, lack of information about what was occurring, and uncertainty about the entire process. This uncertainty carried into their worries about their postpartum care and their newborns' care, as many of the participants mentioned that they "had no idea what was going on" in terms of their children's health. These high levels of uncertainty during birth and postpartum care are addressed in the ensuing section.

High Levels of Uncertainty during Birth and Postpartum Care

The third and final theme deals with uncertainties about birth and postpartum care that the participants experienced. One participant exemplified this theme when she noted, "It was so traumatizing. It was *very* scary not knowing what in the world was going on." Paula, a 33-year-old second-generation Mexican-American mother of three from Houston, had a traumatic birth with her twins, and the anxieties from this birth carried over to her postpartum care. Describing her physician's communication skills as "average" because she treated her "like any other patient," Paula noted that her birth plan throughout her entire pregnancy was to have a C-section because she was worried about delivering twins vaginally. Paula's physician agreed to this birth plan and, according to Paula, changed her mind at the last minute and suggested a vaginal birth. Naturally, this caused Paula a great deal of anxiety and fear because she had been set on a C-section for the duration of her pregnancy, and she mentioned that she did not receive enough information from her doctor:

> She basically just told me that whichever baby will come first, will come first. She basically just said that we'll figure it out when we get there, and that's what happened during the birth! I was terrified. I had no idea what was happening. We didn't really have a plan, which is why it went crazy. My child was about to die because he came down first and was on his cord, and it was all her fault.

Paula described her birth as a terrifying experience: "I was screaming bloody murder because I didn't have any pain medication, they were forcing me to have a vaginal birth, and my son sitting on the cord made them decide I needed an emergency C-section. This could have all been prevented!" After the birth, Paula noted that it "felt like forever" for her to figure out if her kids were fine: "I didn't see my children for hours, I didn't see my doctor, and I had no idea what was going on. I didn't know if my kids were okay, if my son was fine, and I didn't like that *at all*."

Similar to Paula, Mireia, a 44-year-old second-generation Mexican-American mother of two from Houston, had severe complications during her birth with her first child. After her emergency C-section, her stomach bloated tremendously. Once she returned home, the bloating evolved into a leaking, infected incision, but her physician and nurse told her to stay home one more night and return the next day if it did not heal. She returned to the hospital the next day to an unexpected surgery:

> Once I got back to the hospital, my doctor opened my incision to see inside, put her finger in my abdominal cavity, dropped what she was doing and immediately had me sent into pre-op. I was in surgery literally 30 minutes after that. They told me *the next day* that I was bleeding internally; at that moment, though, I didn't know what was wrong with me. She didn't explain anything! I didn't even know if my daughter was okay. I was so confused. To this day, it's still a blur.

Mireia and Paula's examples are but two of the uncertainty stories and experiences that were expressed by the participants during their interviews. Participants noted that they were often "left in the dark" about their pregnancy emergencies and about their children's well-being after their births. This uncertainty stemmed from a lack of information exchange between the participants and their physicians about the nature of their emergencies and the reasoning behind changing the birth plans.

Discussion

This study explored Mexican-American women's birthing experiences in the United States, their perceptions of their physicians' communication, and their struggles with physicians over birthing methods and experiential knowledge. Three main interrelated themes emerged from the data: (1) low levels of medical understanding, (2) power struggles with physicians, and (3) uncertainty about the birth and postpartum care processes. These findings support past research that has explored Mexican-origin and Mexican-American women's birth experiences in the United States. This smaller body of literature has two main conclusions: (1) Mexican-American women are typically more dissatisfied with their birth experiences than white women, and (2) Mexican-American women prefer friendly, warm physicians with expertise in jargon-free information exchange, yet often encounter physicians who provide little to no information about prenatal care, birth methods, and postpartum care (Bergman & Connaughton, 2013; Engle et al., 1990; Galvez, 2011). Participants repeatedly noted that their physicians provided little to no information about the birth and postpartum care processes, which often resulted in high levels of dissatisfaction.

Participants described their birth processes as somewhat terrifying experiences based upon their low levels of understanding about particular treatments, medications, and birth methods. Moreover, many participants fought with physicians about the onset of their labor, and under certain circumstances emergencies prompted physicians to radically change any birth plans that might have previously been agreed upon by the health care team and the participants. A majority of the participants described dissatisfied relationships with their health care providers, high levels of uncertainty regarding their birth plans, and an overall dissatisfaction with their birth experiences. These issues could have been mitigated with just a few moments of information exchange and efforts to foster the patient-provider relationship and provide reassurance—a few of the central tenets of patient-centered communication. Although in certain circumstances C-sections were mandatory for health reasons, participants' anxieties could have been mitigated with more explanation about the process and the reasoning behind the physicians' decisions to call for mandatory C-sections. The nature of the births—some involving slow labors, some very quick labors determined by outside factors such as time, health, emergencies—might have prompted physicians to engage in quick thinking to prevent major medical emergencies, yet the lack of information communicated in these contexts contributed to high levels of anxiety and uncertainty for the women involved.

This chapter concludes with two main findings. First, the participants' generational status certainly played a role in their birthing experiences, but not in a way that is supported by medical and communication literature. Research suggests that higher levels of acculturation could contribute to higher levels of medical literacy and lower levels of affiliation with cultural norms, but what I found in this study was quite the opposite. Participants had low levels of medical understanding pertaining to prenatal testing, birthing methods, and birthing medication, even though over half of them have college degrees and described themselves as well educated. This suggests that acculturation and education levels might not be one of the most significant predictors of patients' ability to understand, process, and apply medical information. This lack of medical understanding could also stem from physicians' lack of information provision and information exchange. Many of the participants noted that their physicians did not take the time to explain important health-related information to them, which could stem from time constraints limiting the medical encounter.

Secondly, pertaining to physicians' communication skills, participants discussed how they preferred physicians with a *personalismo* communication style, meaning they preferred physicians who were warm, empathetic, caring, and who sought to establish a relationship with them past the base level of information exchange. This supports past research on Hispanic patients' preferences for physicians' communication styles (Caballero, 2011; Galanti, 2003) and further supports how both a physician's communication style and a patient's perceptions of this communication style can profoundly shape a health care encounter and future memories of that encounter. When physicians did not take the time to establish a relationship with them, participants were offended and dissatisfied with their care. Overall, a majority of the participants were highly dissatisfied with their physicians' communication styles, noting that their physicians "wrecked their births" and "messed up the entire process" for them. This has tremendous implications for the role of communication during birth encounters.

Concluding Thoughts

This chapter is but one glimpse into what happens when the "dark side of communication" prevails in medical contexts, specifically Mexican-American women's birth experiences. Characterized by low levels of medical understanding, experiential knowledge struggles with physicians over the onset of labor and birth methods, and uncertainty during the birth and postpartum care processes, participants described their births as "traumatic," "scary," and "overwhelming." One main factor that could have improved the participants' experiences drastically is information exchange. If physicians would have taken a few extra moments to discuss birth methods, reasons for changing the birth plan, and care provisions for newborn infants, the participants would have experienced less anxiety and been more satisfied with their care. Information exchange, one of the main tenets of patient-centered communication, is not solely an exchange of health-related information; it also strengthens the patient-provider relationship, conveys care and respect, and has the potential to assuage any anxieties and fears that a patient might have about his or her health (Epstein & Street, 2007). In this case, a little bit of information exchange would have improved participants' birth experiences and perceptions of their relationships with their health care providers.

Notes

1 Parts of this chapter were presented at the 2014 National Communication Association Conference in Chicago, IL.
2 In order to protect participant confidentiality, all participants were given pseudonyms.

References

Bergman, A. A., & Connaughton, S. L. (2013). What is patient-centered care really? Voices of Hispanic prenatal patients. *Health Communication, 28*(8), 789–799.

Boyatzis, R. E. (1998). Transforming qualitative information: Thematic analysis and code development. Thousand Oaks, CA: Sage.

Braun, V., & Clarke, V. (2006). Using thematic analysis in psychology. *Qualitative Research in Psychology, 3*(2), 77–101.

Britten, N. (2011). Qualitative research on health communication: What can it contribute? *Patient Education and Counseling, 82*(3), 384–388.

Brown, A., & Lopez, M. H. (2013). *Mapping the Latino population by state, county, and city*. Retrieved from http://www.pewhispanic.org/2013/08/29/mapping-the-latino-population-by-state-county-and-city/.

Caballero, A. E. (2011). Understanding the Hispanic/Latino patient. *The American Journal of Medicine, 124*(10), S10-S15.

Caron-Flinterman, J. F., Broerse, J. E. W., & Bunders, J. F. G. (2005). The experiential knowledge of patients: A new resource for biomedical research? *Social Science & Medicine, 60*(11), 2575–2584.

Covarrubias, P. O. (2002). *Culture, communication, and cooperation: Interpersonal relations and pronominal address in a Mexican organization*. Lanham, MD: Rowman & Littlefield.

du Pré, A., & Crandall, S. J. (2011). Qualitative methods: Bridging the gap between research and daily practice. In T. L. Thompson, R. Parrott, & J. F. Nussbaum (Eds.), *The Routledge handbook of health communication* (pp. 532–545). New York, NY: Routledge.

Engle, P. L., Scrimshaw, S. C. M., Zambrana, R. E., & Dunkel-Schetter, C. (1990). Prenatal and postnatal anxiety in Mexican women giving birth in Los Angeles. *Health Psychology, 9*(3), 285–299.

Epstein, R. M., Franks, P., Fiscella, K., Shields, C. G., Meldrum, S. C., Kravitz, R. L., & Duberstein, P. R. (2005). Measuring patient-centered communication in patient-physician consultations: Theoretical and practical issues. *Social Science & Medicine, 61*(7), 1516–1528.

Epstein R. M., & Street, R. L., Jr. (2007) *Patient-centered communication in cancer care: Promoting healing and reducing suffering* (NIH Publication No. 07–6225). Bethesda, MD: National Cancer Institute.

Erlandson, D. A., Harris, E. L., Skipper, B. L., & Allen, S. D. (1993). *Doing naturalistic inquiry: A guide to methods*. Newbury Park, CA: Sage Publications, Inc.

Galanti, G. A. (2003). The Hispanic family and male-female relationships: An overview. *Journal of Transcultural Nursing, 14*(3), 180–185.

Galvez, A. (2011). *Patient citizens, immigrant mothers: Mexican women, public prenatal care, and the birth-weight paradox*. Piscataway, NJ: Rutgers University Press.

Gutierrez, E. R. (2008). *Fertile matters: The politics of Mexican-origin women's reproduction*. Austin, TX: University of Texas Press.

Kreuter, M. W., & McClure, S. M. (2004). The role of culture in health communication. *Annual Review of Public Health, 25*, 439–455.

Lindlof, T. R., & Taylor, B. C. (2011). *Qualitative communication research methods*. Thousand Oaks, CA: Sage Publications, Inc.

Lopez, M. H. (2014). *In 2014, Latinos will surpass whites as largest racial/ethnic group in California.* Retrieved from http://www.pewresearch.org/fact-tank/2014/01/24/in-2014-latinos-will-surpass-whites-as-largest-racialethnic-group-in-california/.

Mead, N., & Bower, P. (2002). Patient-centered consultations and outcomes in primary care: A review of the literature. *Patient Education and Counseling, 48*, 51–61.

Pope, C. (2005). Addressing limited English proficiency and disparities for Hispanic postpartum women. *Journal of Obstetric, Gynecologic, & Neonatal Nursing, 34*(4), 512–520.

Reinharz, S. (1992). *Feminist methods in social research.* New York, NY: Oxford University Press.

Sexton, P. C. (1982). *The new nightingales: Hospital workers, unions, new women's issues.* New York, NY: Enquiry Press.

Tracy, S. J. (2013) *Qualitative research methods: Collecting evidence, crafting analysis, communicating impact.* West Sussex, England: Wiley-Blackwell.

Zambrana, R. E., Scrimshaw, S. C. M., Collins, N., & Dunkel-Schetter, C. (1997). Prenatal health behaviors and psychosocial risk factors in pregnant women of Mexican origin: The role of acculturation. *American Journal of Public Health, 87*(6), 1022–1026.

Zandbelt, L. C., Smets, E. M. A., Oort, F. J., Godfried, M. H., & de Haes, H. C. J. M. (2007). Medical specialists' patient-centered communication and patient-reported outcomes. *Medical Care, 45*(4), 330–339.

FOURTEEN

Body Politics—Strategies for Inclusiveness: A Case Study of the National Breast Cancer Coalition

Annette Madlock Gatison

> *We don't need another pink product. We need a vaccine. Given the attention and resources dedicated to breast cancer, the public understandably believes we've made significant progress. The fact is it's quite the opposite. Billions of dollars have been raised, yet there's been no noticeable change in mortality rate. Let's change the conversation. Let's find a vaccine. Let's end breast cancer before it starts. (Fran Visco, Executive Director National Breast Cancer Coalition, 1998)*

In regards to fundraising for breast cancer by purchasing pink ribbons, pink products, organizing and participating in events such as the Susan G. Koman Race for a Cure, or other types of activities that bring awareness to breast cancer in the Black community, Black women are visible. We stand up to be counted and with good reason. Black women have a higher mortality rate from breast cancer than White women and other women of color, and we present at younger ages with more aggressive forms of the disease (Clarke et al., 2012; Kurian, Fish, Shema, & Clarke, 2010; Ma et al., 2013; SEER, 2014; Stead et al., 2009). As we stand up to be counted so are the dollars we spend on pink ribbon products. Although there is no official record of the total dollars spent on pink products by race or ethnicity, the collective amount raised in the name of breast cancer is in the billions. This estimate is based on the fact that the Susan G. Komen Race for the Cure alone raised over $208 million in 2013 and one three-day race garnered close to three million dollars (Susan G Komen for the Cure, 2013). The estimate does not include the thousands of other corporations and organizations that raise money for breast cancer awareness and research. It should also be noted that Komen has donated millions to research (Susan G. Komen for the Cure, 2012, p. 2).

Black women are prime commercial targets and represent a lucrative market for pink ribbon products. The Nielsen Company (2013a, 2013b) reports that in general Blacks in the U.S. have a current buying power of $1 trillion dollars, which is expected to reach $1.3 trillion by 2017. The same reports specifically identify Black women as the primary decision maker when it comes to where and how money is spent. This spending includes contributions and financial support to various charitable organizations and social causes of which breast cancer and pink ribbon campaigns are included.

Pink ribbon marketing is the dark side of health communication when it comes to breast cancer, as it emphasizes awareness and detection through product marketing. What started as a way to bring attention to a major health problem facing women has resulted in a perfect storm of pink ribbon marketing, commercialization, and the commodification of illness and women's bodies. For Black women's bodies, commodification is not new; it has just taken on a 21st-century face through cause marketing. *The Immortal Life of Henrietta Lacks,* a recent bestselling nonfiction book by Rebecca Skloot (2010), tells the story of Henrietta Lacks, a woman whose genetic material was stolen and used for research and commercial gain. Henrietta's story is one that has been recently told and is currently one of the most publically visible instances of a female Black body that has been used for medical and commercial profit. Harriet Washington, in her book *Medical apartheid: The dark history of medical experimentation on Black Americans from colonial times to the present* (2006), provides a historical account of the relationship between Black bodies and the medical establishment. Dorothy Roberts, in her seminal work *Killing the Black Body* (1998), also gives us cause to think politically about our bodies. Granted, this dark side of medical history has resulted in gains and changes to policy, procedure, and protocol that have benefitted human kind (Mukherjee, 2010); however, at the same time, the road to these advancements serves as a cautionary tale for Black women to be aware and vigilant when it comes to advocating for and supporting health care causes. As Black women, we need to be aware of legislation and public policy at all levels as it impacts access to quality and affordable health care, types of treatments that are available, and the choices being made for us about our bodies. Black women's health care is one of the many places where systems of inequality come together. I contend that Black women must have a stronger political presence (Harris-Lacewell, 2001; Harris-Perry, 2013) at the grassroots level when lobbying for legislation that specifically addresses their health issues. However, to do this we must cautiously work collectively and collaboratively with other organizations that have a similar agenda and welcome the presence of Black women. One such organization is the National Breast Cancer Coalition (NBCC).

Agenda Setting and the National Breast Cancer Coalition

The commercialization and commodification that is prevalent in the media's agenda setting of medicine and health care leaves out the political and legislative significance of advocating for one's health beyond purchasing products (Madlock Gatison, in press). I found this troublesome and decided to look for alternative ways to support breast cancer survivorship. In addition to personally supporting fundraisers through luncheons, dinners, marches, runs, walks, or buying pink products, I searched for something different and found the NBCC (2010), which was one of a few organizations that provided an alternative, political view of fighting breast cancer. The following four public policy priorities from NBCC (2014d) are what caught my attention:

- Priority #1. The *"Accelerating the End of Breast Cancer Act,"* S. 865 and H.R. 1830, defines the role that federal government must play to end breast cancer once and for all.
- Priority #2. $150 Million for the Department of Defense (DOD) Breast Cancer Research Program (BCRP) for FY2015.

Since its inception the NBCC has requested money every fiscal year for research.

- Priority #3 – Guaranteed Access to Quality Care for All. NBCC works to identify, advocate for, and support the implementation of laws such as the *"Patient Protection and Affordable Care Act"*.
- Priority #4 – Ensuring the Participation of Educated Patient Advocates in all Levels of Health Care Decision Making.

The priorities of the NBCC provide an alternative strategy for breast cancer survivors or others concerned to actively engage in finding a solution to eradicate breast cancer. For Black women this disease is a burden that affects contemporary daily life and future generations. To end breast cancer in the Black community we must identify points of connection with other groups that further the goal of ending breast cancer (Collins, 2000, p. 37). The priorities of the NBCC can be a place of coalition building and collaboration concerned with ending breast cancer in Black women.

What Is the National Breast Cancer Coalition?

Breast cancer activism and advocacy of the 1970s and 1980s focused attention on awareness, education, and support (Baird, Davis, & Christensen, 2009). In the early 1990s some of these organizations changed their mission and began to focus on the politics of breast cancer. The Women's Community Cancer Project in Massachusetts, Women's Cancer Resource Center in California, Breast Cancer Action in the San Francisco Bay area, and the Mary Helen Mautner Project in Washington, D.C., are representative of this organizational shift. Early on, each of these organizations had a decidedly political agenda aimed at confronting the oppression and silence surrounding the causes and prevention of breast cancer (Baird et al., 2009, p. 12–28).

In May of 1991, several different advocacy groups met in Washington, D.C., to form the NBCC. It promoted itself as a progressive grassroots political organization and emphasized three key areas: (1) increase research funding to end breast cancer; (2) increase of local, national, and global access to necessary information and lifesaving interventions for all women; and (3) to influence leaders regarding strategies to end breast cancer by assisting in the coordination of scientific research. The NBCC's mission is to disseminate this persuasive politically framed health information message to individuals who can best use this knowledge to reduce health risks associated with breast cancer, increase access to and effectiveness of health care, increase research funding, and assist in the coordination of scientific research (NBCC, 2014b; Visco, 1998).

Learning from the political activism of AIDS activists, the NBCC made certain that lobbying would be a critical component in meeting the organization's mission. To help facilitate this, the NBCC hired a professional lobbyist to represent it on Capitol Hill. One of NBCC's first efforts was collecting 175,000 signatures to deliver to Congress demanding more money for breast cancer research. The NBCC was overwhelmingly successful and delivered 600,000 signatures to Capitol Hill in the fall of 1991 (NBCC, 2013a, 2014a).

Framing breast cancer not only as a health issue but also a political issue that can be impacted through public policy and pressure, the NBCC held hearings in 1992 to ask scientists how much money it would take to eradicate the disease. Based on the hearings and other considerations, the coalition determined that the federal government should spend a minimum of $300 million more than the funds already allocated for breast cancer research.

In 1993, a second petition to Congress for similar reasons resulted in the delivery of more than 2.6 million signatures to then President Clinton (NBCC, 2010), resulting in a National Action Plan on Breast Cancer, an effort that involved policymakers, scientists, providers, and breast cancer survivors sitting at the same table and working together toward the same goal, to end breast cancer. To prepare advocates, the NBCC created Project LEAD® (Leadership, Education, and Advocacy Development), an intensive training program designed to prepare breast cancer survivors and others who shared the same concern for the important role they would have as advocates having input into what research projects received funding, how the projects were designed, and how the public was informed of the results (NBCC, 2014c, 2013b).

The Message

The NBCC has two prominent communication taglines: (1) to *change the conversation*, meaning move from awareness about breast cancer to political action and advocacy that coordinates the prevention and end of breast cancer, and (2) *Breast Cancer Deadline 2020®*, which is a specific call to action for researchers, policy makers, breast cancer advocates, and all other interested parties to know how to prevent and end breast cancer by January 1, 2020. As a grassroots organization, the NBCC has developed into a coalition of hundreds of organizations and tens of thousands of individuals across the United States to support and help spread its message. This information dissemination and persuasion process is often referred to as health education (Rubinson & Alles, 1988). But, how is this particular political message shared with Black women breast cancer survivors and other concerned members of the Black community? Are there any strategies used by the NBCC to include women who are underrepresented based on race or class? Is this inclusion validating our experience and way of knowing? I believe answering these questions is important, as Black women's experiences regarding their health and quality of life have often been silenced and marginalized.

Alternative Epistemology

Black women have developed distinctive interpretations of Black women's oppression, but have done so by using alternative ways of producing and validating knowledge. In the book *Black Feminist Thought,* Patricia Hill Collins (2000) gives four characteristics of alternative epistemologies or ways of knowing and validating knowledge that challenge the status quo. The first characteristic is that alternative epistemologies are built upon lived experience, not upon objectified position. Black feminist epistemology, then, begins with "connected knowers," those who know from personal experience. As a breast cancer survivor living in a Black body I am a "connected knower" (p. 259).

The second characteristic of Collins's (2000) alternative epistemology is the use of dialogue rather than adversarial debate. In the communication discipline and other social sciences, knowledge claims are assessed through adversarial debate. Using dialogue to evaluate implies the presence of at least two subjects, which sees knowledge as inclusive rather than having an objective existence outside the lived experience. In black feminist epistemology, the story is told and preserved in narrative form and not "torn apart in analysis" (Collins, 2000, p. 258). As an author, breast cancer survivor, and advocate, I am present; it is my narrative that is central to question the strategies for inclusiveness used by the NBCC.

The third characteristic of Collins' (2000) alternative epistemology is the ethic of caring. The use of dialogue and centering lived experiences, according to Collins, implies that knowledge is built around ethics of caring. She argues that all knowledge is intrinsically value laden and should therefore be tested by the presence of empathy and compassion. Valuing my own lived experience and caring about what happens to others who might share my experience was one of the driving forces behind this analysis. As I live and reflect on my experience I have been exposed to survivors who are also looking for ways to live beyond just surviving. Empathy and compassion are evident as we share our stories.

The fourth characteristic of Collins's (2000) alternative epistemology requires personal accountability. Sharing one's lived experience as a source of knowledge places one's character, values, and ethics under scrutiny. This perspective implies that knowledge is based upon beliefs and what one assumes to be true, and also implies personal responsibility. Sharing my lived experience about breast cancer survivorship and activism while acquiring information that builds knowledge through autoethnography is based on my beliefs and personal truth.

Ethnography and Autoethnography

To find answers, autoethnography and case study methods were used to collect data, and community organization theory (COT) (Rothman & Tropman, 1987) was the analytical tool. Ethnography alone is the art and science of describing a group or culture. It is a method and a research perspective that privileges the lived experiences of others as sources of knowledge worthy of documentation, translation, and interpretation. Ethnographers try to capture as fully as possible, and from the research participant's perspective, the ways people use symbols within specific contexts (Fetterman, 1998; Wrench, Thomas-Maddox, Richmond, McCrosky, 2008). Using personal experience (auto) creates the autoethnographic research method that allows the self to be studied within that cultural context. For me that is breast cancer survivor culture, as seen through a black feminist lens.

My focus on the NBCC for this case study evolved from The Pink and The Black Project*, a broader study that looked at the intersection of breast cancer culture, spirituality, and the idea of being a "Strong Black Woman" (Collins, 2000) on the quality of life of Black women who had survived breast cancer. Using case study as part of my research methodology places the focus on one organization. Case study also allows for data to be collected in various formats that include, autoethnography, texts of key documents, interviews, and social media in order to examine the communication strategies of one specific organization. This essay relies heavily on the data collected from autoethnography and organizational documents.

According to the National Cancer Institute (NCI) (2004), health communication is "the study and use of communication strategies to inform and influence individual and community decisions that enhance health" (p. 2). Health communicators use a wide range of methods to design programs to fit specific circumstances. The NBCC's communication strategies include multiple methods of influence such as media literacy, media advocacy, public relations and advertising, individual and group education, and partnership development (NCI, 2005).

The NBCC's strategy for media literacy is to show those who identify with their message and agenda how to deconstruct media messages so they can identify the advertiser or sponsor's motives. They also teach those concerned how to compose messages for their intended audience's point of view. An example of this would be messages that encourage an audience to think before they purchase pink products and question where their dollars are going.

The NBCC's media advocacy strategy seeks to set the agenda and change the social and political environment in which decisions that affect breast cancer health policy and research funding are made by influencing the mass media's selection of topics and by shaping the debate about those topics and moving from breast cancer awareness to political and legislative action. Media advocacy is the strategic use of mass media as a resource for advancing a social or public policy initiative. The core components of media advocacy are developing an understanding of how an issue relates to prevailing public opinions and values and designing messages that frame the issues so as to maximize their impact and attract powerful and broad public support. The NBCC uses the following core components to reach its audience (NBCC, 2014e, 2013a, 2014a):

- **Public relations and advertising** promote the inclusion of messages about a health issue from a political point of view in the mass media. The NBCC has done this through the use of paid or public service messages in the media or in public spaces to increase awareness of and support for their *Deadline 2020*® campaign.
- **Individual and group instruction,** an educational strategy, which influences, counsels, and provides skills to support breast cancer advocates. NBCC offers a variety of intensive training and programming for advocates of breast cancer to learn about legislative and public policies, the process of research and clinical trials, cancer biology, epidemiology, genetics, targeted therapies, branding and social media, just to name a few.
- **Partnership development** that increases support for a program or issue by harnessing the influence, credibility, and resources of profit, nonprofit, or governmental organizations. The NBCC has done this through their partnerships with organizations such as the National Institute of Health, NCI, Duke University Medical Center, and the Cochrane Institute.

Community Organization Theory

COT has its roots in theories of social networks and support. It emphasizes active participation, critical consciousness, and developing communities that can better evaluate and solve health and social problems. It has roots in several theoretical perspectives, including the ecological perspective, social systems perspective, social networks, and social support. It is also consistent with Social Learning Theory (SLT), which "asserts that people learn not only from their own experiences, but by observing the actions of others and the benefits of those actions" (NCI,

2004, p. 31). COT is the theoretical framework I chose to use for my analysis of NBCC, its message construction, and recruitment strategies. With connections to theories of social networks and support, COT emphasizes active participation and developing communities that can better evaluate and solve health and social problems (NCI, 2004).

According to the NCI (2004, p. 23–26), COT is composed of several alternative change models:

- **Locality development/**community development uses a broad cross-section of people in the community to identify and solve its own problems. It stresses consensus development, capacity building, and strong task orientation; outside practitioners help to coordinate and enable the community to successfully address its concerns. NBCC uses *Action Networks* composed of breast cancer advocates and community leaders in various states across the country. According to the NBCC (NBCC, 2014b, p. 4, 2014f, 2014g), Action Networks are powered by area activists and assisted by staff in the national office which provides various written, electronic, and social media resources. (NBCC, 2014g, para. 3).
- **Social planning** uses tasks and goals, and addresses substantive problem solving, with expert practitioners providing technical assistance to benefit community consumers. NBCC uses expertise from researchers, social workers, professors, politicians, nurses, and other leaders in the fields of medicine, science, technology, and politics.
- **Social action** aims to increase the problem-solving ability of the community and to achieve concrete changes to redress social injustice that is identified by a disadvantaged or oppressed group. Women's health advocates for the NBCC have used social action to pressure powerful institutions to address their problems; breast cancer now has global attention with a newfound focus on action and advocacy.

COT does not use one single concept for strategy development. The key concepts identified above are central to the various approaches that an organization uses for strategic planning, message construction, and community member empowerment. The culmination of the NBCC's use of these various change models has resulted in what it identifies as the *Blueprint for Breast Cancer Deadline 2020* (NBCC, 2014b), briefly:

> This blueprint describes how to harness the energy, resources and leadership around the world to achieve **Breast Cancer Deadline 2020**®. It is designed around three goals: research needed to end breast cancer; global access to the necessary information and lifesaving interventions; and the influence of leaders everywhere in the strategies to end breast cancer. NBCC will create and facilitate collaborations, formulate and implement plans of action, and identify and push for the policies needed (pp. 1–5).

Lived Experience

As a researcher, scholar, and breast cancer survivor, I participated in the social environment that I observed and systematically recorded and classified my observations (Rubin, Rubin, & Piele, 1996). I not only relied on my own observations, but also on information supplied from group members. Content analysis was the major qualitative research tool employed in this case study, as it allowed me to look at the characteristics of communication messages found in specific artifacts and create another avenue to learn about both the content of the message, those who produced the message, and the intended recipients of the message (Rubin et al., 1996).

Following the completion of a bilateral mastectomy in 2012, I decided to become actively involved with the NBCC. It was also during this time that I was asked by a board member of a local breast cancer support group if I would be interested in attending NBCC's The Project LEAD® Institute (Dickersin et al., 2001; NBCC, 2014c). I indicated that I would be interested and completed the application process, which included completing an essay, answering questions, and providing three letters of reference. I also applied for a travel and housing grant, but unfortunately, I was denied the opportunity to attend the 2012 institute. It was suggested by two of my references that I attend NBCC's Lobby Day in May of 2013 and perhaps I would have a better chance at being selected to attend The Project LEAD® Institute in 2013. Both of these events included heavy educational components (Madlock-Gatison, 2013), which are part of the NBCC's communication strategy—*instruction and education*.

In 2013 I raised the necessary $1,000 to attend the NBCC's Inaugural Advocate Leadership Summit held in Washington, D.C., May 4–6, 2013, which includes Lobby Day on Capitol Hill. Using the social media tools provided by the NBCC that consisted of a fundraising webpage and a generic, yet scripted solicitation letter I was able to personalize the solicitation letter, import my email contact list, and distribute the letter. The 2013 summit was touted as the first of its kind, as the emphasis was on building leadership at all levels by providing attendees with lifelong leadership skills (NBCC, 2013a). Participant/leaders also worked in small groups with noteworthy researchers and thought leaders of both scientific and political areas. In 2014 I again raised the necessary $1,000 to attend the May 2014 Summit and Lobby Day. The majority of the Black women who I met the previous year at the summit were not in attendance. This was disappointing, and White women participants also commented on the lack of Black women in attendance.

I was successful in my bid to participate in the 2013 Project LEAD® Institute. I felt that this particular educational experience was extremely worthwhile as it did just what it was set up to do—increase my scientific knowledge about breast cancer as a survivor advocate.

During this Project LEAD trip I also had the opportunity to meet a few other women who self-identified as Black or African American. Yet again, it was noticed that there were not very many Black women or women of color in general as participants. We were not the only ones to notice the lack of color as a White participant actually asked us why we were all sitting at the same table together. One of my newfound colleagues suggested we ask them why they are ALL sitting at the same table. However, we let it go; as professional women it is not the first time that we have found ourselves in the numerical minority. We were well aware that there is an air of privilege associated with participating with the NBCC at this level. This includes participating in the Advocate Leadership Summit in Washington, DC, which culminates in lobbying on Capitol Hill. Although a limited number of financial scholarships were offered to offset the cost of attendance at both programs, there is still a privileged position that one must have in order to have access to and complete the application process.

Reaching Black Women—NBCC Communication Strategies

The media is one mechanism for transmitting local grassroots concerns to the public and the NBCC has become well versed in its use of the media, particularly social media over the past several years. The NBCC's use of social media allows for a snowball effect that increases its web of contacts by using members' social networks and by old-fashioned word of mouth. The documents analyzed do not indicate any specific strategy for attracting Black women to its agenda.

NBCC use of specific communication strategies have found success as evidenced by its political influence to appropriate research funds and to assist with the coordination of research, researchers, and participants for significant clinical trials. This success is also evident when it comes to recruiting advocates, as NBCC uses true grassroots methods to recruit all women, as there is no specific emphasis on Black women beyond the core message. The strategy is to simply reach out to organizations that focus on Black women's health and other concerned organizations using social media and traditional word of mouth by advocates and concerned others. The limitation for this particular aspect of the case study discussed in this chapter is not having access to the exact numbers of participants by race and ethnicity. That information was not available in the text under review.

"To show the country and the world the breadth and diversity of the breast cancer advocacy community that supports Breast Cancer Deadline 2020" (NBCC Summit Documents 2014a, 2013a, n.p.) organizational endorsements have been received from groups such as The African American Breast Cancer Alliance, Inc., African-American Community Health Group of the Central Coast, African-American Women in Touch, Afro American Community Broadcasting INC/KBBG FM, BRECAN-Breast Cancer Association of Nigeria, Beta Alpha Chapter of Zeta Phi Beta Sorority, Inc., The Black Women's Health Imperative, The Cancer Support Network of Zambia, Sister's Journey, Sister's Network, and Triple Step for a Cure.

I asked Sharon Ford Watkins, Field Coordinator for the NBCC, the following questions: What are the strategies used by the NBCC for recruiting more Black women into the policy aspect of breast cancer and women's health? What strategies are needed for women who might need additional motivation to take the next step to work on policy issues, starting with those women who are "aware" and active when it comes to "awareness?" She answered: "I actually believe a straightforward approach is the best. You know your audience and how to frame the message but it's essentially that awareness, all well and good but it doesn't change things—action and activism is what brings about change (pause) we have to make our case for change" (personal communication, September 10, 2014). Ms. Ford Watkins never really answered the questions, but directed me to recent video comments and blog commentaries by Fran Visco, both of which contend that as a society we all have to move beyond the complacency that comes with awareness (Visco, 2014, n.p.). The video and blog do not address any strategies for inclusiveness specific to Black women and breast cancer policy; instead, the messages are designed for all women and those concerned with ending breast cancer. This speaks to the commonality of the issue for coalition building (Braithwaite, Taylor, & Austin, 2000).

A Charge to Black Women Confronted with Breast Cancer

In the words of Audrey Lorde (1988) "Caring for myself is not self-indulgence, it is self-preservation and that is an act of political warfare" (p. 131). Black women cannot afford to be complacent consumers of pink products and settle for just being aware. We are dying and must remember that most of the stressors that cause illness in the bodies of Black women are not the same for our non-Black counterparts. We have to be politically active when it comes to our health. There are other organizations that have a positive political agenda when it comes to Black women's health and sometimes we have to look a little harder for that message, as it is co-mingled with so many other messages that try to address a multitude of health disparities that burden the Black community. Some of which are the very health disparities linked to breast cancer (Vona-Davis & Rose, 2009).

The 21st century is still an age where race, class, gender, and sexuality influence media messages. Black women not only have to stand up and be counted as consumers of pink products, but we have to be counted politically. Our voice, our vote, and our lived experiences are an important component to the mission of ending breast cancer. Every woman has a militant responsibility to involve herself actively with her own health (Lorde, 1980/1997, p. 75), and we have to be diligent about ensuring that the interests of Black women are being served. This means we must have increased participation in the political process before the ballot box. We should know what policies and legislation are being proposed that impact our black bodies, from the food we eat to the air we breathe. We must call, write, email, or show up at the offices of our local and state officials. We must be visible and heard. Being counted should result in a decline in the mortality rates of Black women from breast cancer and the prevention of breast cancer in current and future generations.

References

Baird, K. L., Davis, D. & Christensen, K. (2009). *Beyond reproduction: Women's health, activism, and public policy.* Teaneck, NJ: Fairleigh Dickinson University Press.

Braithwaite, R. L., Taylor, S. E., & Austin, J. N., (2000). *Building health coalitions in the Black community.* Thousand Oaks, CA: Sage Publications.

Clarke, C. A., Keegan T. H., Yang, J., Press, D. J., Kurian, A. W., Patel, A. H., & Lacey, J. V. (2012). Age-specific incidence of breast cancer subtypes: Understanding the black-white crossover. *Journal of the National Cancer Institute, 104*(14), 1094–101. doi: 10.193/jnci/djs264.

Collins, P. H. (2000). *Black feminist thought: Knowledge, consciousness, and the politics of empowerment (2nd ed.).* New York, NY: Routledge.

Dickersin, K., Braun, L., Mead, M., Millikan, R., Wu, A. M., Pietenpol, J., Troyan, S., Anderson, B., & Visco, F. (2001). Development and implementation of a science training course for breast cancer activists: Project LEAD (leadership, education and advocacy development). *Health Expectations 4*(4), 213–220.

Fetterman, D. M. (1998). *Ethnography: Step by step* (2nd ed). Applied social research methods series. Volume 17. Thousand Oaks, CA: Sage Publications.

Harris-Lacewell, M. (2001). No place to rest: African American political attitudes and the myth of Black women's strength. *Women and Politics, 23(3)*, 1–33.

Harris-Perry, M. V. (2013) *Sister citizen: Shame, stereotypes, and black women in America.* New Haven, CT: Yale University Press.

Kurian, A. W., Fish, K., Shema, S. J., & Clarke, C. A. (2010). Lifetime risks of specific breast cancer subtypes among women in four racial/ethnic groups. *Breast Cancer Research, 12*(6), R99. doi: 10.1186/bcr2780.

Lorde, A. (1988). *A burst of light.* Ithica, NY: Firebrand Books.

Lorde, A. (1997). *The cancer journals: Special edition.* San Francisco, CA: Aunt Lute Books. (Original work published 1980)

Ma, H., Lu, Y., Malone, K. E., Marchbanks, P. A., Deapen, D. M., Spirtas, R., … & Bernstein, L. (2013). Mortality risk of black women and white women with invasive breast cancer by hormone receptors, HER2, and p53 status. *BioMed Central Cancer.* May 4; 13:225. doi: 10.1186/1471–2407-13–225.

Madlock Gatison, A. (In Press). *Embracing the pink identity: Breast cancer culture faith talk and the myth of the strong Black woman.* New York, NY: Rowman and Littlefield.

Madlock Gatison, A. (2013). BCC member attends advocacy, leadership and scientific training with the National Breast Cancer Coalition. In *Breast Cancer Consortium* online. Retrieved from http://breastcancerconsortium.net/bcc-member-attends-advocacy-leadership-scientific-training-national-breast-cancer-coalition/

Mukherjee, S. (2010). *The emperor of all maladies: A history of cancer.* New York, NY: Scribner Simon and Schuster.

National Breast Cancer Coalition. (2014a). 2014 Advocate Leadership Summit documents.

National Breast Cancer Coalition. (2014b). A blueprint for Breast Cancer Deadline 2020. Retrieved from http://www.breastcancerdeadline2020.org/assets/pdfs/breast-cancer-deadline-2020.pdf.

National Breast Cancer Coalition. (2014c). Center for Advocacy Training: The Project LEAD® Institute. Retrieved from http://www.breastcancerdeadline2020.org/get-involved/training/project-lead/project-lead-institute.html

National Breast Cancer Coalition. (2014d). Public policy. Retrieved from http://www.breastcancerdeadline2020.org/get-involved/public-policy/public-policy.html.

National Breast Cancer Coalition. (2014e). Using social media. Retrieved from http://www.breastcancerdeadline2020.org/get-involved/tools-and-resources/toolkit/using-social-media.html.

National Breast Caner Coalition. (2014f). Action networks. Retrieved from http://www.breastcancerdeadline2020.org/get-involved/take-action/join-with-us-to-end-breast-cancer-by-2020/.

National Breast Cancer Coalition. (2014g). Get involved take action. Retrieved from http://www.breastcancerdeadline2020.org/get-involved/take-action/join-with-us-to-end-breast-cancer-by-2020/Action-networks.html

National Breast Cancer Coalition. (2013a). 2013 Advocate Leadership Summit documents.

National Breast Cancer Coalition. (2013b). 2013 The Project LEAD® Course Documents.

National Breast Cancer Coalition. (2010). Breast cancer deadline: Why now? http://www.breastcancerdeadline2020.org/about-the-deadline/dl-2020-whitepaper.pdf

National Cancer Institute. (2004). *Making health communication programs work: A planner's guide* (NIH Publication No. 04–5145). Retrieved from http://www.cancer.gov/publications/health-communication/pink-book.pdf

National Cancer Institute. (2005). *Theory at a glance: A guide for health promotion practice* (2nd ed.; NIH Publication No. 97–3896). Bethesda, MD: National Cancer Institute.

Nielsen Company, The. (2013a). African American consumers are more relevant than ever: The African American consumer 2013 report. Retrieved from http://www.nielsen.com/us/en/insights/news/2013/african-american-consumers-are-more-relevant-than-ever.html

Nielsen Company, The. (2013b). Resilient, receptive, and relevant. Retrieved from http://www.nielsen.com/content/dam/corporate/us/en/reports-downloads/2013%20Reports/Nielsen-African-American-Consumer-Report-Sept-2013.pdf

Roberts, D. E. (1998). *Killing the black body: Race, reproduction, and the meaning of liberty.* New York, NY: Vintage Books.

Rothman, J., & Tropman, J. E. (1987). Models of community organization and macro practice: Their mixing and phasing. In F. M. Cox, J. L. Ehrlich, J. Rothman, & J. E. Tropman (Eds.), *Strategies of community organization* (4th ed.). Itasca, IL: Peacock.

Rubin, R. B., Rubin, A. M., & Piele, L. J. (1996). *Communication research: Strategies and sources* (4th ed.) Belmont, CA: Wadsworth.

Rubinson, L., & Ales, W. F. (1988). *Health education: Foundations for the future.* Prospect Heights, IL: Waveland Press.

Skloot, R. (2010). *The immortal life of Henrietta Lacks.* New York, NY: Random House.

Stead, L. A., Lash, T. L., Sobieraj, J. E., Chi, D. D., Westrup, J. L., Charlot, M., …& Rosenberg, C. L. (2009). Triple-negative breast cancers are increased in black women regardless of age or body mass index. *Breast Cancer Research. 11*(2), 1–10. doi:10.1186/bcr2242.

Surveillance, Epidemiology, and End Results Program. (2014). SEER state fact sheets: Breast cancer. Retrieved from http://seer.cancer.gov/statfacts/html/breast.html.

Susan G. Komen for the Cure. (2012). *Guided by hope: 2011–2012 annual report.* Retrieved from http://ww5.komen.org/AboutUs/FinancialInformation.html

Susan G. Komen for the Cure. (2013). *Consolidated financial statements and supplementary information.* Retrieved from http://ww5.komen.org/AboutUs/FinancialInformation.html

Visco, F. (1998). The National Breast Cancer Coalition (NBCC). *Breast Disease, 10(5),* 15–21.

Visco, F. (2014). The complacency of awareness. *Huffington Post.* Retrieved from http://www.huffingtonpost.com/fran-visco/the-complacency-of-awaren_b_5548046.html.

Vona-Davis, L., & Rose, D. P. (2009). The influence of socioeconomic disparities on breast cancer tumor biology and prognosis: A review. *Journal of Women's Health, 18*(6), 883–93.

Washington, H. (2006). *Medical apartheid: The dark history of medical experimentation on Black Americans from colonial times to the present.* New York, NY: Double Day.

Wrench, J. S., Thomas-Maddox, C., Richmond, V. P., & McCrosky, J. C. (2008). *Quantitative research methods for communication: A hands on approach.* New York, NY: Oxford University Press.

FIFTEEN

Being Detained: Time, Space, and Intersubjectivity in Long-Term Solitary Confinement

Michael P. Vicaro

In the decades following the so-called "discursive turn" in the social sciences and humanities, academics from a wide array of disciplines have argued persuasively that individual identity is largely shaped by communicative practices. A diverse range of contemporary political, critical, and empirical projects share the basic assumption that language is intimately involved in the formation of experience, rather than something added to an otherwise wordless "reality." Undermining the conceptual division between subject/object and mind/body, theorists have shown that persons are neither rational individual agents nor the passive recipients of raw sensory data. Subjective "feelings" and "ideas" as well as objective "things" are seen as outcomes or accomplishments of power-laden discursive practices rather than their points of origin. Many scholars, including phenomenologists, critical humanists, post-structuralists, feminists, post-colonial scholars, queer theorists, and others have argued that some recognition of the social and communicative construction of experience is a necessary element of any emancipatory project.

This chapter examines one "darker" implication of this interdisciplinary acknowledgment of the intersubjective locus of experience. It focuses on a set of new military and criminal justice technologies designed specifically to distort or dismantle individuals' basic communicative capacities. Specifically, I argue that the development and proliferation of "supermax" prison cells used to induce long-term solitary confinement allow practitioners to produce conditions tantamount to torture while typically evading the legal and moral prohibitions associated with that term. Rather than attacking the body or mind, long-term solitary confinement and related carceral techniques do violence by disrupting a detainee's capacity to maintain the discursive relationships that undergird the self/subjectivity. The supermax cell, therefore, should be understood not as "place" but as weapon able to exploit the "dark side" of human beings' inescapable reliance on communicative interaction for the formation and maintenance of identity and experience.

The Architecture of Control: Supermax Detention in U.S. Domestic Prisons

The supermax cell emerged in U.S. domestic prisons in the 1980s as a technological innovation in the domestic struggle for a state monopolization on violence amidst the "wars" on drugs and crime (Hartnett, 2003; King, 1999; Rhodes, 2004). Overcrowding, inmate organization ("gangs"), riots, escapes, and other acts of intra-penitentiary violence led to the development of a new architecture of confinement and control. Harrington (1997) has argued that the "modern history" of extreme isolation cells "began on Saturday, October 22, 1983, when Thomas Silverstein, an inmate at the federal maximum-security penitentiary in Marion, Illinois, stabbed a correctional officer forty times, precipitating a total lockdown of the prison" (p. 16). What happened at Marion was, at the time, a radical transformation in detention policy. While most U.S. maximum-security facilities have segregation cells for holding "exceptional" disciplinary violators (often referred to as "the hole"), the lock-down at Marion penitentiary was unique in that all the cells were turned into solitary segregation cells (Richards, 2008). The "total lockdown" model quickly spread and designers began experimenting with new architectural forms specifically designed for long-term solitary confinement, automation, and minute environmental control. Currently, the majority of U.S. state prison systems have supermax facilities (also known as Special Housing units or Administrative Segregation facilities), either as stand-alone units or as part of lower security prisons (Mears, 2006). The supermax cell has also been used to detain undocumented immigrants (Urbina & Rentz, 2013), as well as those deemed to be "unlawful enemy combatants" in the War on Terror (Toobin, 2008). The policy of incarcerating inmates in long-term solitary confinement cells has recently spread globally to Canada, Mexico, Brazil, Colombia, the United Kingdom, Denmark, Australia, New Zealand, Thailand, Russia, Jordan, and beyond (Ross, 2013). Deep isolation and communicative deprivation has become a prolific and acute form of state power.

Solitary confinement in penitentiaries is not a new idea. Isolation and communicative deprivation were essential parts of early nineteenth-century American prison design and were considered by many progressive architects and planners to be an effective tool for reform and rehabilitation. Critics, however, countered that long-term solitary confinement was destructive to the spirit and often led to irreparable social and spiritual damage, and in 1822, the governor of New York closed one of the world's first solitary confinement penitentiaries, concluding that the "health and constitutions of these surviving convicts had become alarmingly impaired" (as cited in Toch, 2003, p. 22). Those debating the merits of solitary confinement in the nineteenth-century agreed that discipline, reform, and rehabilitation were the desired ends of incarceration. The main question was whether solitary confinement would aid or hamper the achievement of these ends. Twenty-first century supermax detention policy, however, differs in that the explicit end is not rehabilitation and reform, but rather the achievement of total control of presumed-to-be dangerous populations (Rhodes, 2004). Prison officials are often quite willing to pursue practices that do damage to the mind and spirit of the inmate if they believe such techniques will help achieve the end of control. No longer is the inmate a creature to be reformed. Instead, he or she is treated as an inert (if threatening) object to be contained (Simon, 2007). From this perspective, supermax detention is different from the older "disciplinary" imprisonment; it amounts to an instrument for dismantling human integration.

Rather than an austere form of dwelling, the supermax cell becomes an instrument of torture that can destroy the capacity to dwell at all.

In addition to their role in the advancement of a radically new approach to the ends of incarceration, supermax prisons differ architecturally and materially from their nineteenth- and twentieth-century predecessors. In the older "radial" or "big-house" designs common to the mid-twentieth century, inmates share common spaces for "mess," "yard," and other daily activities. In contrast, supermax facilities totally eliminate common space in order to facilitate complete social isolation. Most domestic supermax inmates spend twenty-three hours per day in 8' x 12' cells (a bit larger than a standard elevator car). Inmates often receive as little as a single hour of solitary exercise in slightly larger security cells (often compared to animal cages) that are sometimes equipped with a pull-up bar but are otherwise empty (Rhodes, 2004). Cells are constantly illuminated with artificial light, which can only be dimmed or brightened from an external control panel. Communicative interactions between inmates can be completely eliminated, and staff/inmate interactions are minimized through the automation of most activities. Inmates are typically allowed no reading, writing, or art materials, and participate in no work, education, or rehabilitation programs. Many inmates have remained in these conditions for years-unto-decades. Many, after having served their time, are released directly onto the streets (Harrington, 1997).

Within the supermax cell, every aspect of the inmate's sensory experience can be controlled and, ultimately, degraded. The visual landscape consists of bare steel walls, a steel door with a small slot for food, an immovable toilet/sink unit, a small window (which can provide natural light when not purposefully blacked out by guards), a concrete or steel bed with a foam mattress, and sometimes a remotely controlled in-cell shower. Food distribution can be managed silently and automatically to further minimize the need for human presence and possibility of communicative interaction. Cells are often equipped with speakers used to broadcast commands from guards (in some cases, the voice is a computer simulation), soft music or white noise, or, potentially, louder and more disruptive sounds, which can be used for disciplinary purposes (Cusick, 2008). Social, aural, visual, and tactile stimulation are thus systematically controlled. Olfactory stimuli are likewise manipulated by harsh cleaning products and irritants such as pepper spray and tear gas (Rhodes, 2004). Taste is deprived by a predictably unpalatable cuisine that guards occasionally replace with the more extreme "Nutraloaf," a punitive food ration that provides a nutritionally complete meal but is carefully formulated to be utterly bland in taste and texture (Greenwood, 2008). And so, sensory experience in the supermax prison cell can be constantly micro-managed and, potentially, dialed down toward absolute zero. In such sterile vacuity, inmates struggle to maintain their perceptual and experiential capacities (Grassian, 2006).

The supermax facility represents a radical transformation in the ideology of detention. Rather than forcing inmates to conform to a highly regulated social world, it produces a worldless vacuity devoid of social relationships. Rather than disciplining bodies to the tyranny of the clock, it produces a timelessness without feature or dimension. The supermax cell does, therefore, seem both to complete and at the same time overturn Foucault's (1979) notion of the penitentiary as a disciplinary institution designed to produce normalized, docile subjects. It seeks to achieve maximum control over individual bodies, but rather than demanding disciplined and productive conduct, it seeks to render the detainee static and inert. Rather than reform, it abandons. Indeed, the supermax cell should be understood as part of a turn from what

Foucault called "disciplinary societies" to his late-formulated notion of "control societies" (Deleuze, 1992). In the former, prisons resembled other institutions like the factory, military unit, hospital, and school, all of which employed technologies of corporeal discipline and internalized self-surveillance in order to produce "docile bodies" useful for the productive demands of capital (Foucault, 1979). Control societies, however, are oriented toward consumption and circulation rather than production and linearity. As alleged by Deleuze (1992), "Indeed, just as the corporation replaces the factory, perpetual training tends to replace the school, and continuous control to replace the examination" (p. 5). New "campus" prisons can thus be seen as yet another break from the organizational model of the factory. Rather than massive, disciplined, and linear, organized by the clock and monitored though self-surveillance, these new prisons seek to technologically control with minute precision each and every aspect of inmates' sensorial experience. The supermax facility is thus "more akin to war than to the factories on which nineteenth-century penitentiary prisons were based" insofar as they "imagine that crime can never be normalized but only endlessly fought. (Simon, 2001, p. 113) This connection between war and supermax detention is not simply metaphorical. As the next section shows, the supermax detention cell has become an important technology of contemporary counter-insurgency warfare.

The Supermax Cell in Counter-Insurgency Operations

The previous section argued that the supermax cell should be seen as an architectural component of a broader change in the incarceration ideology. The current section explains how this new detention model, originally developed as a response to domestic intra-penitentiary violence, has become an important part of contemporary U.S. counter-insurgency military operations. In order to understand the significance of long-term solitary confinement in the Global War on Terrorism, one must first understand some of the unique challenges of contemporary "asymmetrical" military conflicts (Terriff, Karp, & Karp, 2008; Ucko, 2009). Modernist war theory presumes that war takes place between legally recognized sovereign nation-states and ends when the parties agree to the terms of a formal surrender (Scarry, 1990). In this model, an army is assumed to be a legally constituted entity, hierarchically organized and beholden to a centralized command structure. When the commander-in-chief issues a declaration of surrender, each citizen-solider, as a representative of the state, must find himself or herself addressed by this declaration and thus dutifully lay down arms. This modernist paradigm, which "sees warfare as essentially a contest between the armed forces of nation states" with "unity of command" (Hammes, 2008, p. 201), has failed in the face of contemporary CIW conflicts involving enemies who are "transnational in scope, nonhierarchical in structure, clandestine in approach and who operate outside the context of nation-states" (McFate, 2005, p. 48).

Recognizing the inadequacy of this nationalist/modernist model, the Department of Defense has initiated a "reorientation toward counterinsurgency and stability operations" (Ucko, 2009, p. 5). In counterinsurgency operations (COIN), the "Clausewitzian objective of defeating the enemy, has to be replaced by a focus on people, i.e., separating the population base from the insurgents worldwide" (Richards, Wilcox, & Wilson, 2008). As then-Secretary of State Condoleezza Rice explains in her 2005 opening remarks to the Senate Foreign Relations Committee, "our political-military strategy has to be to clear, hold, and build: to clear areas from insurgent control, to hold them securely, and to build durable, national...institutions"

(Rice, 2005). Rice's comments suggest that in the COIN/CIW paradigm, the traditional military goal—compelling the enemy to formally surrender—is replaced by a "managed" approach to the long-term control of populations.

Long-term solitary confinement plays an important role in this new paradigm for several reasons. First, the supermax cell helps to isolate hostile individuals and separate them from the broader civilian populations. As is well known, the U.S. has developed a policy in the Global War on Terrorism (GWOT) of preventative detention whereby so-called pre-combative individuals are captured and indefinitely detained without official charges or due process of law (Cole, 2003). The stated goal is to "physically and psychologically separate the insurgents or opponents from both external and internal sources of support…to 'cauterize' around the insurgency to keep it from spreading or acquiring support" (Hoffman, 2008, p. 191). To this end, a massive leaflet-dropping operation promised Afghan civilians financial rewards for information leading to the detention of suspected terrorists. Thousands were arrested and classified as "unlawful enemy combatants" on the basis of unsubstantiated allegations. The policy of large-scale detention was a central component in the Iraq war as well. In his 2004 report on detention policy at Abu Ghraib, Major General George Fay stated that, "as the pace of operations picked up in late November–early December 2003, it became a common practice for maneuver elements to round up large quantities of Iraqi personnel in the general vicinity of a specified target as a cordon and capture technique" (as cited in Greenberg & Dratel, 2005, p. 1042). In this and other similar asymmetrical military conflicts, long-term solitary confinement cells facilitate this goal of "physically and psychologically" separating potentially hostile parties.

Second, supermax detention can help control populations in situations in which the use of overwhelming military force is deemed to be unfeasible or inappropriate. Complex Irregular Warfare tends to have a hybrid identity, blending traditional military operations with police activity and border-control procedures. In these conflicts, detention, interrogation, and intelligence gathering, become more important than raw firepower. These hybrid conflicts are motivated by a so-called "low intensity" military doctrine concerned with "the establishment and maintenance of social control over targeted civilian populations" (Dunn, 1996, p. 4). In such situations, "military forces take on police functions, while police forces take on military characteristics" (Dunn, 1996, p. 4). The supermax cell, as an architectural form, embodies this logic of police/military hybridism. In CIW and other "low-intensity operations," such as border-security, riot-control, so-called "peace keeping" missions, etc., in which success is defined by the long-term monopolization of violence (rather than a declared surrender), the detention facility has become a powerful instrument of war.

Third, the supermax cell helps to "constitute" the enemy as such. In traditional warfare between nation-states, the existence of the enemy is typically taken for granted. By contrast, in the GWOT, the ability to wage war at all is dependent upon the prior creation of a localizable, recognizable enemy. Maintaining popular support for complex irregular conflicts can be difficult in cases in which the enemy may seem elusive or even illusory. Such operations thus depend upon the prior political achievement of establishing the tangible presence of a coordinated hostile force. The detention facility becomes a "territorial trap for a mobile enemy," helping to consolidate and identify an otherwise unlocalizable entity (Comaroff, 2007, p. 397). In conflicts that break with the modernist/nationalist paradigm, widespread detention serves an important political function: the display of detained bodies helps to maintain a vision of the "enemy" and to justify the use of military force.

This section has argued that supermax detention cells help fulfill several related counter-insurgency objectives: They separate suspected insurgents from the broader population, they help to monopolize violence in situations in which unconditional surrender is unrealistic, and they help to constitute the presence of the enemy as such. The next section argues that these cells can produce forms of deprivation and violence tantamount to torture. It argues that the supermax cell is not a traditional architectural form, not an austere form of dwelling, and not a place where violence takes place. Rather, the supermax cell is a weapon.

Being Detained: Time, Space, and Intersubjectivity as Concealed Weapons

The sense that a supermax cell can be an instrument of torture may be obscured by the seemingly innocuous term "detention." Evoking Kenneth Burke (1969), we may say that the supermax cell complicates the typical "grammar of motives" regarding torture. It is typically assumed that an act of torture must be undertaken by some actors, equipped with some tools ("agency," in Burke's terms), in some material place (a "scene"), for some purpose. Acts of torture typically take place in interrogation rooms or other special facilities; they are undertaken by military interrogators or other agents of the state. These interrogators have access to a set of tested and approved techniques, and they use these techniques for the purpose of procuring "intelligence." In this typical view, a detention cell is merely a "scenic feature" of the interrogation procedure—one is held in a specific space for a span of time. The cell, in this vocabulary, appears to be a place where torture may take place.

This section contests this description of "detention." Informed by the notion that human experience is an outcome of communicative praxis, it argues that a contemporary solitary confinement detention cell is not a "place" at all—it is a weapon. Communication theorists from a wide array of disciplines have shown that human beings are primarily "altercentric"—personal experience arises as an outcome of timely and "emplaced" engagement with others and with the material world (Braten, 2007; Schrag, 1999). This section develops this altercentricity thesis to show how solitary confinement degrades detainees' relationships to space, time, and other people, thereby disrupting their capacity for being-in-the-world (Guenther, 2013).

Space. In a contemporary supermax detention facility, space is not a neutral "scenic" feature or a container in which things take place. Merleau-Ponty (1962) refers to a "primitive spatiality" that "merges with the body's very being." The body, he claims, "is not primarily in space: it is of it" (p. 148). For this reason, the controlled, vacuous space of the supermax detention cell can itself become a weapon that does not require the presence of a torturer to do violence.

Because they are confined in a cell with few objects and have no opportunity to see the sky or the horizon, supermax detainees are subjected to a degraded relationship to space and place. In fact, one might say that a supermax detention cell is a space that is not a place. Casey (2001) defines space as "the encompassing volumetric void in which things (including human beings) are positioned" and defines place as "the immediate environment of my lived body—an arena of action that is at once physical and historical, social and cultural" (p. 683). Whereas the ancient Greeks distinguished *topos* (place) from *chora* (space), "one way of understanding modernity…is by its very neglect of this distinction" (Casey, 1997a, p. 270–271). In modern thinking, "to be is to be in space, where 'space' means something nonlocal and nonparticular…" (Casey, 1997a, p. 275). From this modernist standpoint a supermax detention cell is a space like any other—a certain three-dimensional volume of air that serves as an empty container that benignly accommodates things it contains.

In contrast with space, a place is not "something simply physical. A place is not a mere patch of ground, a bare stretch of earth, a sedentary set of stones" (Casey, 1997b, p. 26). One might say that to be "in" place means to be involved, rather than to be contained. In Heidegger's (1971) formulation, human beings are inevitably "ex-static"—always "emplaced" outside of themselves as they move through intentional "projects." This ability to occupy existential rather than simply physical space accounts for a certain resiliency among persons in conditions of confinement. Though one's body may be spatially constrained, one's projects can continue—letters, poetry, novels, political philosophy, and spiritual texts written in prisons attest to the fact that one's "dwelling" place is not equivalent to the volume of space in which the body is confined (Hauser, 2012).

However, the notion that human beings are essentially "projective" also entails certain vulnerabilities. The supermax detention paradigm relies on the fact that human beings are inevitably involved in "place," affected by and not simply contained within our surroundings. Donald Hebb, whose research on sensory deprivation has contributed to the development of the solitary confinement cell, has shown "how completely dependent the mind is on a close connection with the ordinary sensory environment," and how a person can fall into a state of psychosis if "cut off from that support" (as cited in McCoy, 2012, p. 68). Drawing on Hebb's insights, prison designers, along with military interrogators and other functionaries of state power, have come to use the detention cell as an instrument of violence that can create a space that has none of the attributes of placehood—an atopic "volumetric void" that is no longer an "arena for action."

Because human beings are inevitably engaged in and dependent on environmental emplacement, a supermax cell is not just a place where violent acts may take place. Indeed, from this perspective, a solitary confinement cell is not an architectural form at all. It is a weapon. However, the supermax cell serves as a concealed weapon—concealed in part by modern-instrumentalist vocabularies that collapse the distinction between space and place. Contemporary civilian and military prison officials thus exploit a dark side of the inextricable human communicative engagement with the emplaced material environment in order to produce a powerful technology of violence designed to look like a room.

Time. The supermax detention cell also does violence by disrupting detainees' normal experience of "lived" time. As with space/place, the ancient Greeks distinguished two notions of time. The term *kairos* refers to the "right moment" or "the opportune"—what Schrag describes as the element of "the good in the category of time" (Schrag, 1969, p. 78). The term *chronos*, by contrast, refers to abstract chronological time, removed from the lived experience of temporality. Modern thinking has generally tended to collapse the distinction between these two ancient Greek senses of time, viewing time exclusively from a "universal…cosmic standpoint" (Arendt, 1998, p. 288).

The vocabulary of physical, chronological time is full of spatial metaphors for temporal processes (Heidegger, 1985). In this vocabulary, time is assumed to move linearly, or else function as a container for things contained. The chronological/spatial notion of time suggests that detention is an event within a finite "time-box"—one must "do time," counting the hours and days until one may be released. However, the *experience* of time in long-term solitary confinement is not reducible to the chronological duration of imprisonment: sensory and communicative deprivation disrupts and potentially disables the lived experience of time. Stated otherwise, one may say that the contemporary solitary confinement cell does not just damage inmates "in time," but also *as* time. Human interdependence with temporality itself is turned toward violent ends.

The ancient Greek rhetorical tradition provides an alternative temporal vocabulary that better reveals the unique violence of indefinite detention. White (1987) notes that the ancient Greek term *kairos* grasps a notion of temporality, distinct from *chronos*, in which "the living present" serves as a "point of departure or inspiration for a purely circumstantial activity of invention" (p. 13). Poulakos (1995) argues that a "rhetor who operates mainly with the awareness of *kairos* responds spontaneously to the fleeting situation at hand, speaks on the spur of the moment, and addresses each occasion in its particularity, its singularity, its uniqueness" (p. 61). Rhetorical/kairotic temporality, by contrast with the empty and infinite progression of chronological clock-time, is shaped by both finitude and urgency. As Blumenberg (1987) has noted, the rhetorical orientation entails finitude insofar as we are limited by *perceptual* capacities that are easily overwhelmed by the complexity of the lived world; and the rhetorical orientation entails urgency because, despite *finitude* and *uncertainty*, rhetoric is tied to the imperative to act. In the moment of decision, no timeless ideals or eternal truths can replace the timely grasp of the possible. In sum, placing timeliness at the center of rhetorical praxis calls attention to the sense that human experience unfolds in relation to foresight, memory, and hope.

Given this alternative, rhetorical notion of time, the violence of long-term solitary confinement can be seen more clearly. There is a temporal dimension of indefinite detention that goes beyond the accumulation of hours, days, months, and years. This is because people are not "in" time like a beetle may be in a box. If experience takes place "in" time this is so in the sense that we say we are in love, or in touch, or in a mood. That is (as with "place"), the preposition would refer not to containment but to involvement. We are involved in time in a fundamental way—in anticipation and projection, in memory, and in the kairotic moment of situated action. Without memory and anticipation and, through these, the possibility of inventive action, life becomes mechanical and inhumane—producing what might be called "akairotic" time. In such a state, a person has "no anticipation of creative associations, and the past which he remembers is cut off from the present," resulting in a sense of "existential meaninglessness" (Schrag, 1969, p. 202). The violence of indefinite detention is, however, concealed by the more common vocabulary of time as mere chronological duration. Within this latter vocabulary, one might say that a detainee in total isolation remains "in" time: his or her heart would still beat and the clock would still tick. But he or she might no longer be involved in the lived time of experience. And so the inclination to describe the time of imprisonment in chronological terms may be obscuring the way solitary confinement can do damage *through* and *as* time.

Intersubjectivity. Theorists from a wide array of disciplines have argued that human beings are fundamentally *altercentric*—meaning that personality is in large measure an outcome of interpersonal interaction (Braten, 2007; Deetz, 1992; Ricouer, 1995). This in part explains why long-term solitary confinement can be so damaging. Grassian (2006) has shown how isolation and communicative deprivation can produce physical and psychological effects akin to the trauma associated with torture. He posits, "Solitary confinement...can cause severe psychiatric harm. It has indeed long been known that severe restriction of environmental and social stimulation has a profoundly deleterious effect on mental functioning" (p. 327). Grassian further argues that prolonged isolation can cause "a specific psychiatric syndrome" the features of which include hyperresponsivity to external stimuli; perceptual distortions, illusions, and hallucinations; panic attacks; difficulties with thinking, concentration, and memory; intrusive obsessional thoughts; overt paranoia; and problems with impulse control, including self-directed violence (p. 334–337). Haney (2003) reports that, "the absence of regular, normal interpersonal contact...creates a feeling of unreality that pervades one's existence" in supermax confinement; "in extreme cases prisoners may literally stop behaving" (p. 139).

Isolation and communicative deprivation are not typically recognized as forms of violence. This is in part because modernist theories of experience tend to view the individual as a locus of meaning and significance with "sovereignty over his or her own consciousness" (Peters, 1989, p. 393). To draw upon Burke's (1969) terminology again, this view suggests that other people are reducible to "scenic features" of one's primarily solitary experience. By contrast, Aristotle (trans. 1885) argues that the civic community of the *polis* is "prior to the individual," and anyone able to live without interpersonal interaction must be either "a beast or a god" (1253a). From this perspective, human experience emerges in the encounter with a world of others—a world that is always already meaningful, already compelling, already evoking obligation and response. It is this basic need for meaningful engagement with others that makes long-term solitary confinement so destructive. A world without otherness becomes a world without selfhood. Sensory and communicative deprivation creates an *apolitical* (i.e., without access to a *polis*) environment that is no longer a world in which an intersubjective being can dwell. A life of isolation in empty space and time, then, may cease to be a properly human life at all. This, then, is one of the "dark sides" of our need for human communicative engagement with otherness: pushed to extremes, a person in total isolation might remain corporeally alive but without the capacity for meaningful engagement in the world.

Conclusion

This chapter has examined the supermax detention cell as a new technology of violence designed to exploit the "dark side" of human beings' fundamental communicative altercentricity. Moreover, it has argued that the violent effects of long-term solitary confinement are in part obscured by vernacular and legal vocabularies that do not acknowledge this altercentricity and the vulnerabilities it entails. As a result of its ability to deploy the "concealed weapon" of communicative deprivation, the detention cell is able to produce conditions tantamount to torture while sustaining the corporeal "health" of the inmate and avoiding the moral and legal scrutiny brought on by the use of more traditional forms of violence. Theorists and critics attuned to the dark side of communication are thus well positioned to challenge the spread of this prolific instrument of state power and the narrow view of subjectivity and experience through which it appears to be justified.

References

Arendt, H. (1998). *The human condition.* Chicago, IL: University of Chicago Press.
Aristotle (1885). *Politics.* (B. Jowett, Trans.). Oxford: Clarendon Press.
Blumenberg, H. (1987). An anthropological approach to the contemporary significance of rhetoric. In K. Baynes, J. Bohman, and T. A. McCarthy (Eds.), *After philosophy: End or transformation?* (pp. 429–458). Cambridge: MIT Press, 1987.
Braten, S. (Ed.). (2007). *On being moved: From mirror neurons to empathy.* Amsterdam: John Benjamins Publishing Company.
Burke, K. (1969). *A grammar of motives.* Berkeley, CA: University of California Press.
Casey, E. S. (1997a). Smooth spaces and rough-edged places: The hidden history of place. *The Review of Metaphysics, 51,* 267–296.
Casey, E. S. (1997b). How to get from space to place in a fairly short stretch of time: Phenomenological prolegomena. In S. Feld & K. Basso (Eds.) *Senses of place* (pp. 13–52). Santa Fe, NM: School of American Research Press.
Casey, E. S. (2001). Between geography and philosophy: What does it mean to be in the place-world? *Annals of the Association of American Geographers, 91,* 683–693.
Cole, D. (2003). *Enemy aliens: Double standards and constitutional freedoms in the war on terrorism.* New York, NY: New Press.
Comaroff, J. (2007). Terror and territory: Guantanamo and the space of contradiction. *Public Culture 19*(2), 381–405.
Cusick, S. (2008). 'You are in a place that is out of the world': Music in the detention camps of the Global War on Terror. *Journal of the Society for American Music, 2*(1), 1–27.
Deetz, S. (1992). *Democracy in an age of corporate colonization: Developments in communication and the politics of everyday life.* Albany, NY: SUNY Press.
Deleuze, G. (1992). Postscript on societies of control. *October, 59,* 3–7.
Dunn, T. (1996). *The militarization of the U.S.-Mexico border: Low intensity doctrine comes home.* Austin, TX: University of Texas Press.
Foucault, M. (1979). *Discipline and punish: The birth of the prison.* New York, NY: Vintage Books.
Grassian, S. (2006). Prison reform: Commission on safety and abuse in America's prisons: Psychiatric effects of solitary confinement. *Journal of Law & Politics, 22,* 325–383.
Greenberg, K. J., & Dratel, J. L. (Eds.). (2005). *The torture papers: The road to Abu Ghraib.* New York, NY: Cambridge University Press.
Greenwood, A. (2008, June 24). Taste-testing Nutraloaf. *Slate.* Retrieved from http://www.slate.com/articles/news_and_politics/jurisprudence/2008/06/tastetesting_nutraloaf.single.html
Guenther, L. (2013). *Supermax: Social death and its afterlives.* Minneapolis, MN: University of Minnesota Press.
Hammes, T. X. (2008). Information operations in 4GW. In T. Terriff, A. Karp, and R. Karp (Eds.), *Global insurgency and the future of armed conflict: Debating fourth-generation warfare* (pp. 200–207). New York, NY: Routledge.
Haney, C. (2003). Mental health issues in long-term solitary and "supermax" confinement. *Crime & Delinquency, 49*(1), 124–156.
Harrington, S. (1997). Caging the crazy: 'Supermax' confinement under attack. *The Humanist, 57,* 14–19.
Hartnett, S. (2003). *Incarceration nation: Investigative prison poems of hope and terror.* Walnut Creek, CA: AltaMira/Rowman & Littlefield.
Hauser, G. A. (2012). *Prisoners of conscience: Moral vernaculars of political agency.* Columbia, SC: University of South Carolina Press.
Heidegger, M. (1971). *Poetry, language, thought.* New York: Harper Collins.
Heidegger, M. (1985). *History of the concept of time: Prolegomena.* Bloomington: Indiana University Press.
Hoffman, F. G. (2008). Combating fourth generation warfare. In T. Terriff, A. Karp, and R. Karp (Eds.), *Global insurgency and the future of armed conflict: Debating fourth-generation warfare* (pp. 177–199). New York, NY: Routledge.
King, R. (1999). The rise and rise of supermax: An American solution in search of a problem? *Punishment and Society, 1,* 163–186.
McCoy, A. (2012). *Torture and impunity: The U.S. doctrine of coercive interrogation.* Madison, WI: University of Wisconsin Press.
McFate, M. (2005). The military utility of understanding adversary culture. *Joint Force Quarterly, 38,* 42–48.
Mears, D. (2006). *Evaluating the effectiveness of supermax prisons.* Washington, DC: Urban Institute.
Merleau-Ponty, M. (1962). *The phenomenology of perception.* New York, NY: Routledge and Kegen Paul.
Peters, J. D. (1989). Locke, the individual, and the origin of communication. *Quarterly Journal of Speech, 75,* 387–399.

Poulakos, J. (1995). *Sophistical rhetoric in classical Greece.* Columbia, SC: University of South Carolina Press.
Rhodes, L. (2004). *Total confinement: Madness and reason in the maximum-security prison.* Berkeley: University of California Press.
Rice, C. (2005). Iraq and US policy. *Testimony Before the Senate Committee on Foreign Relations.* Washington, DC.
Richards, C., Wilcox, G., & Wilson, G. I. (2008). America in peril: Fourth generation warfare in the twenty-first century. In T. Terriff, A. Karp, and R. Karp (Eds.), *Global insurgency and the future of armed conflict: Debating fourth-generation warfare.* (pp. 115–131). New York, NY: Routledge.
Richards, S. (2008). USP Marion: The first federal supermax. *The Prison Journal, 88*(1), 6–21.
Ricouer, P. (1995). *Oneself as another.* Chicago, IL: University of Chicago Press.
Ross, J. (Ed.). (2013). *The globalization of supermax prisons.* New Brunswick, NJ: Rutgers University Press.
Scarry, E. (1990). Consent and the body: Injury, departure, and desire. *New Literary History, 21*, 867–896.
Schrag, C. (1969). *Experience and being: Prolegomenon to a future ontology.* Evanston, IL: Northwestern University Press.
Schrag, C. (1999). *The self after postmodernity.* New Haven, CT: Yale University Press.
Simon, J. (2001). Sacrificing Private Ryan: The military model and the new penology. In P. Kraska (Ed.), *Militarizing the American criminal justice system: The changing roles of the armed forces and the police.* (pp. 105–119). Boston, MA: Northeastern University Press.
Simon, J. (2007). *Governing through crime: How the war on crime transformed American democracy and created a culture of fear.* New York, NY: Oxford University Press.
Terriff, T., Karp, A. & Karp, R. (Eds.). (2008). *Global insurgency and the future of armed conflict: Debating fourth-generation warfare.* New York, NY: Routledge.
Toch, H. (2003). The contemporary relevance of early experiments with supermax reform. *The Prison Journal, 83*(2), 221–228.
Toobin, J. (2008, April 14). Camp justice. *The New Yorker.* Retrieved from http://www.newyorker.com/magazine/2008/04/14/camp-justice
Ucko, D. H. (2009). *The new counterinsurgency era: Transforming the U.S. military for modern wars.* Washington, DC: Georgetown University Press.
Urbina, I., & Rentz, C. (2013, March 24). Immigrants held in solitary cells, often for weeks. *New York Times*, p. A1.
White, E. C. (1987). *Karionomia: On the will-to-invent.* Ithaca, NY: Cornell University Press.

Context Four

The Dark Side of Computer-Mediated Communication

SIXTEEN

Catfished: Disenfranchised Grief for the Never-Existed

Jocelyn M. Degroot and Heather J. Carmack

Online dating has been increasing over the past several years; in fact, 1 in 10 Americans have used an online dating website (Smith & Duggan, 2013) and approximately more than one third of marriages in the United States initiated through online dating (Cacioppo, Cacioppo, Gonzaga, Ogburn, & VanderWeele, 2013). Numerous studies have also revealed the capability of computer-mediated communication (CMC), including the use of social media sites, texting, and instant messenger, to develop and sustain relationships (Parks & Floyd, 1996; Rabby & Walther, 2003; Tong & Walther, 2011; Walther, 1992). Moreover, Walther's (1996) hyperpersonal model posited that people who met and maintain a relationship online can actually be closer, or even more intimate, than their real-world couple counterparts. Rabby (2007) labeled couples who initiated and maintain their relationship online as *virtuals*. These people have never met in real life/face-to-face, but they communicate regularly and consider themselves to be in a relationship. The virtual relationship is maintained through communication and self-disclosure, primarily because they do not participate in other social activities as a couple (e.g., going out to eat).

Deception during online dating is a common concern for those utilizing online dating services (Donath, 1999; Ellison, Heino, & Gibbs, 2006) especially for virtuals (Rabby, 2007). Online dating profiles often contain some misrepresentations, including lying about one's physical features, career, or hobbies. Although deception can occur in an online context, Ellison et al. (2006) found that people are typically honest when engaging in online dating. Many online profiles were not completely accurate, but the researchers reported that online profiles were seldom entire misrepresentations of a person meant to deceive others. Rather, these inaccuracies were typically due to a user presenting an idealized self or unintentionally misrepresenting information due to a limited amount of self-knowledge.

In some cases, however, a person's dating profile is entirely fictitious. That is, someone uses another person's photos for the profile picture and invents that person's backstory. The profile's creator then establishes and maintains relationships with others using the faked profile as a front. This behavior is termed *catfishing*, based on a quotation given in a documentary of the same name (Jarecki, Joost, Joost, & Schulman, 2010). During the film, one individual discussed how catfish were kept in vats of cod that were shipped because the catfish would nip at the cod, keeping them agile and fresh. He likened certain people in the world (such as the movie's antagonist) to these catfish, which is where the term *catfish* originated. A recent social phenomenon, catfishing has changed the way individuals and researchers approach the study of online relationships

The framing of these relationships as "fake" and the emotions experienced by catfished individuals as inappropriate ultimately frames their loss as disenfranchised grief. Coined by Doka (1989), *disenfranchised grief* occurs when an individual is not afforded the "right to grieve" a loss (Doka, 2002b, p. 5). Societies create specific grieving rules that govern which relationships, losses, and emotional experiences are legitimate and appropriate (Kellehear, 2007). When an individual's loss is outside those rules, his or her grief is not recognized and acknowledged. Researchers have previously examined the disenfranchised grief of a number of groups including homosexual couples (Doka, 1989), individuals in extramarital affairs (Doka, 1989), coworkers (Bento, 1994; Eyetsemitan, 1998), ex-spouses (Doka, 1986, 2002a; Smith, 2006), survivors of suicide (Clark, 2001; Feigelman, Gorman, & Jordan, 2009), and parents who suffered miscarriages (Capitulo, 2005; Lang et al., 2011). These individuals are often denied the relational experiences associated with loss, such as validation, social support, and expressions of sympathy (Doka, 2002b).

With the Internet as a prominent, permanent fixture in our everyday lives, we need to revisit disenfranchised grief and examine potential additional instances of disenfranchised grief as examples of the dark side of communication. We argue that finding out that a significant other does not actually exist should be considered a "loss," because individuals are deprived of the meaningful relationship and must find some way to deal with that loss. One person invests time and emotions into the catfish, believing that the person is real. When the catfish is discovered, the other person in the relationship often feels the same grief that others feel when they experience a traditional breakup, but they have an added layer of emotional turmoil. Not only do these people lose the people they loved, they find out that this person *never* existed. We contend that this is a new, unstudied kind of disenfranchised grief.

In this chapter, we discuss disenfranchised grief for the never-existed, focusing on the aftermath of finding out one member of a romantic relationship did not exist (or did not exist as portrayed by the partner). This chapter first provides background information on grief following a breakup as well as traditional disenfranchised grief. It then describes the disenfranchised grief that exists as the result of catfishing. Finally, we use Manti Te'o's experience with catfishing as a case study to illustrate how disenfranchised grief for the never-existed extends our understanding of disenfranchised grief and the dark side of interpersonal communication.

Disenfranchised Grief of the Broken Hearted

LaGrand (1989) argued that the emotions experienced during and after a romantic breakup were those of loss and grief. Robak and Weitzman (1994–1995) found indirect evidence indicating several similarities between grief felt due to the loss of a life and the grief felt due to the loss of a relationship. Adolescents are especially vulnerable to this type of loss, as the losses are often dismissed as insignificant due to the parties' ages (Kaczmarek & Backlund, 1991; LaGrand, 1989; Robak & Weitzman, 1994–1995). LaGrand (1989) identified that "the root cause of disenfranchised grief following the breakup of a love relationship lies in the difference in perceptions of the event between support persons and the griever" (p. 178). Essentially, the support persons might try to minimize the griever's loss, which further intensifies the feelings of grief and loneliness. Robak and Weitzman (1994–1995) found that the grief following an ended romantic relationship during early adulthood is generally disenfranchised by family members (i.e., parents and siblings) but not by friends.

Dissolution of Romantic Relationships

Martin (2002) discussed the grief that often accompanies a failed relationship, or the breakup of a romantic relationship. He argued, "a non-death loss is more likely to affect a griever's self-esteem" (p. 237). This is due in part to the griever's belief that the loss could have been prevented (unlike death, which is inevitable) and observations of other couples successfully maintaining a romantic relationship. As evident, this can have far-reaching implications, affecting various aspects of the bereaved's life. Martin also discussed numerous factors that affect one's grief response to a breakup (originally identified by Sanders, 1993). These factors include the type of loss (i.e., a divorce vs. other romantic breakup), circumstances surrounding loss, significance of the lost relationship, age of the griever, and personal variables.

Breakups can lead to both emotional and physical distress. Morris and Reiber (2011) discovered that people who were broken up with had higher Physical Trauma Levels (PTL) and Emotional Trauma Levels (ETL) than people who initiated the breakup or when the breakup was mutual with no significant gender differences in expressions of Post-Relationship Grief. Physical responses to breakups include anxiety, insomnia, and appetite loss. Additionally, there is evidence indicating that a broken heart resulting from grief can cause death, a condition termed *bereavement mortality* (Parkes & Prigerson, 2010).

Often the magnitude of grief is abated by those in the bereaved's social network; however, when the loss is not viewed as legitimate, the bereaved does not get the social support that he or she needs. Weber (1998) argued that social support following a non-marital breakup is much less than the support following a divorce. One can surmise, then, that social support following a breakup with someone who never existed would be nearly completely absent. Secondary losses can also complicate grief. Rando (1993) described secondary losses as those that occur as the result of the primary loss. For example, one might lose half of his or her social network following a breakup, as that part of the network was "owned" primarily by the now-ex-partner.

Disenfranchised Grief

Experiencing the loss of a loved one is a common human experience, although how we deal with and communicate about the loss will be different. Bereavement, or the loss of something, is predicated on a valued relationship that has ended or significantly changed. The bereaved individual must identify ways to cope with that loss and find a way to live with the loss of the relationship (Corr, Nabe, & Corr, 2003). An important part of the study of loss is the emphasis on the ways through which others and relationships with them impact the grieving process. What happens when an individual is unable or not allowed to grieve the loss of another? What if friends, family, and society do not consider the loss valid? When this lack of validation happens, an individual experiences disenfranchised grief. Disenfranchised grief is defined as grief experienced when people "incur a loss that is not or cannot be openly acknowledged, publicly mourned, or socially supported" (Doka, 1989, p. 4). This type of grief might occur when the relationship, loss, or griever is not recognized by others as legitimate.

Disenfranchised grief "involves more than merely overlooking or forgetting to take note" of an individual's loss (Corr, 1998, p. 17). Originally associated with stigmatized relationships such as homosexual relationships and extramarital affairs (Doka, 1987), disenfranchised grief was the emotional experience of those in unsanctioned or societally inappropriate relationships. Doka referred to these relationships as "nontraditional" because they possessed, among other characteristics, limited public acceptance. The nontraditional framing of these relationships meant that the individuals in these relationships are unable to adopt traditional relationship roles; by extension, then, these individuals are denied the "transitional role" of a survivor that individuals are able to adopt after the loss of a loved one (Doka, 1987, p. 462). As a result, bereavement following a loss in a nontraditional relationship is often met with a lack of social support (Doka, 1987), a factor that is critical in facilitating healthy grief (Parkes, 1980, 1987). In order for grief to be effectively addressed, social support for the bereaved is needed, but this can only occur if the relationship and the emotions of the bereaved are acknowledged (Romanoff & Terenzio, 1998).

Disenfranchised grief is a multidimensional concept, with several key tenets (Doka, 1989, 2002b). First, and at its most basic, disenfranchised grief occurs when a relationship cannot be recognized as a legitimate or appropriate relationship. Relationships may be framed as "nontraditional" or unsanctioned, but can also include relationships that are not deemed to be close relationships (e.g., coworkers) or existed primarily in the past (e.g., ex-spouses or lovers, former friends). Important for this tenet is the fact that it is others who define this relationship, not the bereaved. This leads to the second tenet, which is that the loss is not acknowledged. The loss is typically not acknowledged because it is not considered significant. The relationship may be deemed insignificant because of the deceased (e.g., the loss of a pet) or because the relationship has not had enough time to be significant (e.g., miscarriages, abortions). Third, a person may experience disenfranchised grief because he or she is deemed incapable of understanding the loss or experiencing and expressing appropriate grief. Individuals with developmental disabilities, mental illness issues, extremely young children, and the elderly with dementia are often labeled as incapable of grieving. The way the deceased died may also contribute to disenfranchised grief. In this tenet, stigmatized illnesses or losses that are anxiety laden (e.g., suicides, deaths from AIDS) often result in disenfranchised grief because people do not know how to offer support to the bereaved. Finally, disenfranchised grief occurs because of the ways through which an individual grieves. Individuals who do not follow the appropriate grieving rules, including grieving too long and showing too much or not enough emotion, often find their grief turns into disenfranchised grief (Corr, 2002).

Understanding how grief becomes disenfranchised is important to appreciating the complexities of grief. First, an important element of disenfranchised grief is the temporal nature of loss. An individual's loss, once deemed acceptable and appropriate, can turn into disenfranchised grief if an individual grieves too long. There is an assumed endpoint to grieving; unfortunately, there is no identified timeframe for grief. This is difficult for the bereaved because family, friends, and coworkers can transform an individual's grief into disenfranchised grief if they believe a person needs to "move on," "stop dwelling on it," or "get over it." When this happens, it transforms the legitimacy of the loss. Second, disenfranchised grief occurs because of the high value that society places on "traditional" families (heterosexual relationships with children) in the United States. Families outside of this narrow definition are harder to define; disenfranchised grief is reinforced when individuals are unable to seek out support or are denied rights associated with loss (Kamerman, 1993).

Unfortunately, the bereaved's responses to disenfranchised grief are not usually positive. There is often less social support social support provided to bereaved individuals who experience disenfranchised grief and individuals are less likely to seek out social support if their grief is disenfranchised (Thorton, Roberston, & Miecko, 1991), meaning that bereaved individuals have few opportunities to communicate their loss (Doka, 2002b). Individuals who experience disenfranchised grief often find that, because they are unable to emotionally respond to the loss, they are likely to avoid expressing their emotions at all. This often leads to the individual's grief becoming intensified, resulting in anger towards others or isolation (Kauffman, 2002; Lenhardt, 1997).

For individuals who end a relationship (be it through dissolution or death), grieving the loss of that relationship is a common process to help the individual adjust and create a "new normal." But when individuals are marginalized, their relationships are not recognized as legitimate, or they do not follow societally prescribed grief rules, their grief can become disenfranchised. Individuals who are catfished might experience these same frustrations, but in different ways. They might become marginalized because of their naivety and their relationships, now deemed "fake," are no longer recognized as legitimate. Ultimately, this prevents those catfished individuals from being able to follow prescribed grief rules, potentially because of conflicting emotions, reactions from friends and family, and a need to justify their feelings.

Disenfranchised Grief and Catfishing

In 2010, the *Catfish* documentary was released, providing the public a label for this kind of online deception. The film followed Nev Schulman as he unknowingly engaged in an online relationship with a woman pretending to be someone else who did not exist. The woman used another person's photos online and created storylines and additional "characters" that existed in the fictional woman's life (Jarecki et al., 2010). Eventually, Schulman and his fellow filmmakers were able to reveal the truth, but the fact that the woman was fake did not negate Nev's emotions; he admitted in subsequent interviews that he had been in love with a person who had never existed. This movie gained a following, and a television show based on the premise of revealing online deceivers (the catfish) was developed for MTV and is currently in its third season (Warner, 2012). In addition, popular TV show *Glee* devoted a multi-episode storyline to the catfishing of one of the glee members.

The MTV show has documented several cases of catfishing, many of which continue to add to the definition of catfishing. Along with introducing a new catfishing story every week, the TV episodes explore the reasons why someone may engage in catfishing (Rothman, 2013). One of the primary reasons people catfish others is because of low self-esteem. Individuals may feel as if they are not attractive enough, are uninteresting to others, or lack the social skills required to engage in a face-to-face relationship. Some individuals engage in catfishing for revenge, focusing on a specific person who bullied them or to keep a person away from a shared love interest. A final reason catfishing may occur is because of frustration associated with sexual identity. Individuals uncomfortable with disclosing their sexual orientation or identity may create a fake persona to help them connect with a member of the same sex. This is often done because of the fear of people leaving or reacting negatively to their sexual identity. This was the reason behind the *Glee* catfish storyline, where Wade "Unique" Adams, a transwoman, catfished Ryder Lynn, creating an online persona "Katie" in order to communicate his feelings for Ryder (Goldberg, 2013). "Catfishing" has been used to describe instances of deception online since the film was released in 2010, but Manti Te'o's widely (and publicly) discussed relationship with a fictitious woman and subsequent fallout brought the term into the public vernacular.

The Case Study: The Catfishing of Manti Te'o

In 2013, Manti Te'o's experience with a non-existent girlfriend became the most explosive and public example of catfishing. Te'o was a football player for Notre Dame who gained sympathy and admiration for deciding to play a football game the day after both his grandmother and his girlfriend died. The story of his relationship was even covered during the game because his girlfriend, who died from leukemia, had been a topic of conversation during interviews; Te'o often talked about his girlfriend, their relationship, and the strength he found from watching her fight cancer (Goodman, 2013). Te'o's relationship became even more of a news story when an article on Deadspin.com revealed that his girlfriend not only did *not* die, but she never existed in the first place (Burk & Dickey, 2013). The Deadspin article revealed that Te'o's girlfriend was in fact Ronaiah Tuiasosopo, a man who used a former classmate's name, picture, and information to create an online profile. Tuiasosopo explained that he created the fake persona because he was in love with Te'o but was unable to tell him; this gave him the opportunity to develop a romantic relationship with Te'o. The pair never met, but did engage in online and phone conversations (Wolken & Myerberg, 2013). The story was "breaking news" on *ESPN*, and numerous outlets picked up the story after it first broke on Deadspin. In a statement issued after the story broke, Te'o revealed that he had developed an emotional relationship with a woman online. He continued, "To realize that I was the victim of what was apparently someone's sick joke and constant lies was, and is, painful and humiliating" ("Story of Manti Te'o," 2013).

Catfishing can turn grief into disenfranchised grief because blame is placed on both the person who created the fake profiles and the person who was catfished. People who were catfished are often confronted with addressing their naivety for developing relationships online and never meeting with their romantic partners. After the catfishing story broke, rather than focusing on Te'o's loss, reporters and the general public blamed him for "letting this happen," and in several instances, questioned if he was the catfisher or the catfished (Brady & George, 2013; Harris, 2013). Martin (2002) discussed how perceptions of the "preventability" of a breakup can negatively impact the intensity of grief, but in the case of being catfished, the only way to prevent the loss would be to have never met the catfish in the first place. Even if, after meeting the fictitious person online, the other party investigated the existence of the other person and discovered that he or she does not exist, a loss still results. In Te'o's case, his grief over the double loss of his girlfriend (her original "death" from cancer and her "death" from catfishing) was disenfranchised because he was partly to blame for what happened. Placing blame or responsibility on a bereaved individual acts as a deterrent from grieving by suggesting that a person has "no right" to grieve because the loss is, in some way, his or her fault (Doka, 1999).

Not only did Te'o need to cope with the loss of someone he loved (even though she did not exist), he also had to contend with a lack of social support for his loss. The response from the general public was extremely negative; people typically responded to the story with a "serves you right" attitude after finding out the Te'o had never met the woman in person. People also engaged in a fair amount of joking or making fun of Te'o's situation, including creating an Internet meme showcasing Te'o's "invisible girlfriend." People made fun of his situation rather than providing support during an emotional time as they likely would have provided in a situation that involved two "real" people. In many ways, a catfish relationship is very much akin to Martin's (2002) idea of a failed relationship. After people discover that one member in the relationship never existed, they typically view that relationship as fictional as well. Doka (1989) identified three reasons that mourners experience disenfranchised grief (i.e., an unrecognized relationship, loss, or mourner). Rando (1993) argued that a fourth reason for disenfranchisement exists: "whether the loss, once recognized, will be supported by the mourner's social group" (p. 498). Thus, secondary loss can be further intensified when the bereaved is aware that support is available but withheld (Rando, 1993).

It is likely that a social network's response to a member being catfished is similar to a social network's response to an adolescent's breakup, as described by LaGrand (1989). The "breakups" are minimized and the bereaved is told to "move on." Further, because the catfish never existed, those in social support groups might consider that the loss does not "count" as a loss and treat it accordingly. This lack of social support for those experiencing disenfranchised grief can also be problematic for the bereaved. Doka (1989) explained, "The very nature of disenfranchised grief exacerbates the problems of grief, but the usual sources of solace and social support may not be available or helpful" (p. 3). Romanoff and Terenzio (1998) argued that successful resolution of bereavement is influenced, among other factors, by the social support available to the griever; however, social support is not always readily available for losses not seen as legitimate by society. Pesek (2002) focused on the role of support groups for those experiencing disenfranchised grief. Pesek posited that those experiencing disenfranchised grief "need a supportive atmosphere because of the isolation, loneliness, shame, and guilt they may experience" (p. 133). Unfortunately, the type of grief following a catfishing experience is so marginalized that support groups do not even exist for this type of grief.

Because recognition of the grief and social support play a large role in both the grieving and ritual process, disenfranchised grievers need to enact a ritual in the presence of a supportive community. Doka (2002c) and Kollar (1989) discussed the role of rituals in disenfranchised grief and their benefits to the grieving process. The act of engaging in rituals when grieving (such as a funeral) can be therapeutic. However, after a catfishing experience, the grieving individual cannot even look back at his or her partner's photographs with good memories, as those photos are not of their ex; they are photos of a stranger who represented the ex. Weber (1998) also argued that we engage in an act of memorializing people who were important to us. Because a catfish relationship is primarily text-based online, the person usually retains a transcript of the relationship through chat messages and emails. Unfortunately, possessing these "souvenirs" might not be helpful, as they are reminders of a false relationship.

Additional concerns exist for bereaved catfishes. After a traditional breakup, people often follow a script (Lee, 1984). However, we do not have a script in place for a post-catfish relationship, which can result in further anxiety. Furthermore, Weber (1998) discussed the need for justification and closure following a breakup; however, when a catfish relationship ends, the catfish might not provide any explanation, leaving the bereaved without closure and a final termination of the relationship (as described by Davis, 1973). The grief associated with a relationship loss is further intensified in catfishing relationships because the catfish can involve the death of the fictitious person as a means for the catfish to end the relationship, which was the case of the catfishing of Manti Te'o. His initial grief for the "death" of his (completely fabricated) girlfriend became disenfranchised when he discovered she did not exist. His grief, which had been publically acknowledged, became a shameful act.

Te'o's response, as the person who was deceived, as well as the public's response to Te'o is not unique. In instances of being catfished, a person's emotional responses from his or her broken heart and the customary grief that follows is compounded with an added stigma of admitting to dating someone online for a substantial amount of time without ever meeting that person face-to-face. Broken hearts and the disenfranchised grief that follows is compounded with the added stigma when one admits to dating someone online for a substantial amount of time without ever meeting that person face-to-face. Parkes and Prigerson (2010) stated,

> It is obvious that these victims were broken-hearted. They also had to cope with the negative social stigma associated with online dating. [Stigma] causes other people to reject or ignore the needs of the stigmatized, on the other hand, it causes the stigmatized to give up trying to obtain support, having suffered rejection they may even imagine rejection where it does not exist. (p. 233)

In his seminal work, *Stigma: Notes on the Management of Spoiled Identity*, Goffman (1963) discussed the social role of stigma in ostracizing and discrediting certain individuals. Stigma, or the "socialized, simplified, standardized image of the disgrace of a particular social group" (Smith, 2011, p. 455), is a socially constructed phenomenon which socializes societal members into recognizing traits, characteristics, or physical attributes which identify someone as different. Going further, Goffman (1963) explained,

> Stigma is an illuminating excursion into the situation of persons who are unable to conform to standards that society calls normal. Disqualified from full social acceptable, they are stigmatized individuals. Physically deformed people, ex-mental patients, drug addicts, prostitutes, or those ostracized for other reason must constantly strive to adjust to their precarious social identities. (p. 154)

Stigma relies primarily on communication to highlight difference, devalue and segregate others, and reinforce what is considered acceptable in society (Smith, 2007). Stigma is an underlying element of disenfranchised grief (Doka, 2002d); it is difficult to disenfranchise an experience if there is not a stigma associated with the experience.

The emotional response to stigma is crucial to understanding the entire stigma experience, especially how individuals respond to stigma (Meisenbach, 2010). For stigmatized individuals, a key emotional response is shame. Shame, a reaction to criticism by other people (Shultz,

2000), is tied to embarrassment because unwanted attention is focused on the individual and is the product of being exposed, judged, and deemed different. The focus is on the individual who has been shamed, in this case stigmatized, and his or her emotional response is because the individual believes this is how the world sees him or her. The shame catfished individuals experience contributes to their disenfranchised grief. The catfished relationship is deemed inappropriate and others shame those who are catfished because of their inappropriate relationship.

Catfished individuals experience two different levels of stigma. On one level, they experience moral stigma, differing as a result of something that is regarded as sinful or associated with a dubious virtue (Ashforth & Kreiner, 1999). Catfished individuals experience this simply because they were not "clever" or vigilant about their online relationships. Although online relationships are becoming more commonplace, it is still understood that people need to be skeptical about what is online. One a second level, they experience social stigma (Page, 1984), which is separation as a result of association with someone stigmatized. Simply being friends with or associating with someone who is stigmatized would constitute social stigma. Individuals experiencing disenfranchised grief cannot seek out or receive social support to help cope with their loss. Likewise, individuals who are catfished are unable to communicate their grief or seek support because of the scandalized nature of catfishing. Moreover, they may see the catfished experience as shameful and be uncomfortable talking about it.

Implications and Recommendations

The Internet can be a wonderful place to connect with people and develop relationships, but it can also be used to deceive and manipulate (Lohmann, 2013). In this chapter, we focus on one form of deceptive relational communication—catfishing—and the ways through which grief and loss are disenfranchised when individuals find out their partner was not real. The Manti Te'o case serves as a contemporary and popular example of catfishing and provides an inside look into how a person who was catfished deals with the loss, the disenfranchisement of his or her grief, and the role of friends, family, and society in reinforcing that disenfranchisement.

The exploration of catfishing loss as an example of disenfranchised grief complicates theoretical understandings of grief and loss in relationships. As discussed in the chapter, disenfranchised grief is temporally bound, and grief can become disenfranchised if someone mourns a loss for too long. Moreover, disenfranchised grief is often associated with a specific type of socially "inappropriate" relationship. Individuals in these relationships are aware that these relationships are considered inappropriate and might anticipate disenfranchised grief (Doka, 1998). The disenfranchised grief associated with being catfished is also temporally bound, but in different ways. In this case, the relationship is not considered inappropriate until after the fact. It is only after individuals find out that the individual was fake and share it with others that their grief becomes disenfranchised. Up until then, the grief associated with the end of the relationship is considered appropriate because the relationship is appropriate.

The emphasis on "appropriate" relationships underscores another important implication of the study of catfishing and disenfranchised grief. A common assumption in the study of relationship dissolution is that there was a relationship to dissolve. A relationship was not only appropriate, but existed. Discussions of catfishing are quick to point out the "fakeness" of a relationship and to question the "realness" of the emotions associated with the relationship. This further disenfranchises those who have been catfished. By questioning the "realness" of the relationship, family, friends, and the general public further de-legitimize the catfished's experiences. Calling the relationship "fake" discourages individuals from communicating their feelings of loss (because they are not considered feelings of loss) and might prevent them from seeking out support to help recover from the loss. Throughout the chapter, we have been careful to not say that the relationship was fake, but that the individual who was catfishing created a fake persona. That is because the relationship was not fake for the person who was catfished. For those who are catfished, they began and developed a relationship, becoming emotionally invested.

Using the "realness" of a catfished relationship to disenfranchise the relationship calls into question societal acceptance of online relationships. The growing numbers of relationships which began online via dating match sites and social networks indicates that we, as a society, are accepting of online relationships, up to a point. Online relationships are considered appropriate as long as the couple eventually moves the relationship from online to a face-to-face relationship. One of the questions family, friends, and the general public have with catfishing is how an individual would "let" the catfishing happen; this was a question regularly asked of Te'o. For Te'o and many others who are catfished, the easiest answer is that they never actually met. Catfished relationships exist completely online (as virtuals; Rabby, 2007) and the individuals who are catfishing go to great lengths to keep from meeting (D'Costa, 2014). Interestingly, this unintentionally transfers some of the blame onto those who are catfished. Catfishing is partly their "fault"; if they had just met their partner in person and taken the relationship offline, this would not have happened. Placing the blame on the catfished individual further shames and disenfranchises their grief because they are blamed for this happening.

Legitimizing the relationships of those who are catfished will help to enfranchise their experiences and loss. Martin (2002), who equated the end of a failed relationship with the grief associated with individuals who died, recommended family, friends, and other support systems find ways to enfranchise the brokenhearted and rebuild an individual's self-esteem. This will help an individual deal with the loss of the relationship. This is also needed for those who have been catfished. Victims of a catfishing scam need to be able to grieve as they would a traditional breakup. They should express their emotions and talk to others about the situation (Weber, 1998).

Those in the catfished individual's social network should support them as they work through the breakup and death of the relationship. This can be accomplished by recognizing the grief, listening to an individual's stories and concerns, understanding that they are wrestling with not only the loss of the relationship but also the shame which comes from being catfished, and facilitating conversations with others to help communicate their feelings (Hocker, 1989).

The continued prevalence of catfishing serves as a call for the continued study of this phenomenon. First, there needs to be more research on cases of catfishing, specifically focusing on the ways individuals who have been catfished negotiate the impact of being catfished. Additionally, in order to understand catfishing, researchers need to focus on the narratives of those who were catfishes, examining their motives for catfishing and how they communicated with others after the catfishing was revealed. The Te'o case showed how family, friends, and even strangers might not understand how catfishing happened and the emotions experienced by the one in the relationship. More research is needed to examine the support systems individuals turn to when they are catfished. Catfishing represents one of the extreme forms of online deception, which impacts the catfished individual and his or her family, and friends.

Understanding the communication between catfishes and the catfished, communication about the catfishing, and the ways those catfished grieve their loss will help to shed light on this unfortunate consequence of looking online for love.

References

Ashforth, B. E., & Kreiner, G. E. (1999). "How can you do it?": Dirty work and the challenge of constructing a positive identity. *Academy of Management Review, 24*, 413–434.

Bento, R. F. (1994). When the show must go on: Disenfranchised grief in organizations. *Journal of Management Psychology, 9*, 35–44.

Brady, E., & George, R. (2013, January 18). Manti Te'o's catfish story is a common one. *USA Today*. Retrieved from usatoday.com

Burk, T., & Dickey, J. (2013, January 16). Manti Te'o's dead girlfriend, the most heartbreaking and inspirational story of the college football season, is a hoax. *Deadspin*. Retrieved from deadspin.com

Cacioppo, J. T., Caacioppo, S., Gonzaga, G. C., Ogburn, E. L., & VanderWheele, T. J. (2013). Marital satisfaction and break-ups differ across on-line and off-line meeting venues. *Proceedings of the National Academy of Sciences, 110*(25), 10135–10140. doi:www.pnas.org/cgi/doi/10.1073/pnas.12

Capitulo, K. L. (2005). Evidence of healing interventions with perinatal bereavement. *American Journal of Maternal Child Nursing, 30*, 389–396.

Clark, S. (2001). Bereavement after suicide—How far we have come and where do we go from here? *Crisis: The Journal of Crisis Intervention and Suicide Prevention, 22*, 102–108.

Corr, C. A. (1998). Enhancing the concept of disenfranchised grief. *Omega: Journal of Death and Dying, 38*, 1–20. doi:10.2190/LD26–42A6–1EAV-3MDN

Corr, C. A. (2002). Revisiting the concept of disenfranchised grief. In K. Doka (Ed.), *Disenfranchised grief: New directions, challenges, and strategies for practice* (pp. 39–60). Champaign, IL: Research Press.

Corr, C. A., Nabe, C. M., & Corr, D. M. (2003). *Death and dying: Life and Living* (4th ed). Belmont, CA. Wadsworth/Thomson Learning.

Davis, M. S. (1973). *Intimate relations*. New York, NY: The Free Press.

D'Costa, K. (2014, April 25). Catfishing: The truth about deception online. *Scientific American*. Retrieved from scientificamerican.com

Doka, K. J. (1986). Loss upon loss: The impact of death after divorce. *Death Studies, 10*, 441–449.

Doka, K. J. (1987). Silent sorrow: Grief and the loss of significant others. *Death Studies, 11*, 455–469. doi:10.1080/07481188708252210

Doka, K. J. (1989). Disenfranchised grief. In K. J. Doka (Ed.), *Disenfranchised grief: Recognizing hidden sorrow* (pp. 3–11). Lexington, MA: Lexington Books.

Doka, K. J. (1999). Disenfranchised grief. *Bereavement Care, 18*, 37–39. doi:10.1080/02682629908657467

Doka, K. J. (2002a). A later loss: The grief of ex-spouses. In K. Doka (Ed.), *Disenfranchised grief: New directions, challenges, and strategies for practice* (pp. 155–166). Champaign, IL: Research Press.

Doka, K. J. (2002b). Introduction. In K. J. Doka (Ed.), *Disenfranchised grief* (pp. 5–22). Champaign, IL: Research Press.

Doka, K. J. (2002c). The role of ritual in the treatment of disenfranchised grief. In K. J. Doka (Ed.), *Disenfranchised grief* (pp. 127–133). Champaign, IL: Research Press.

Doka, K. J. (2002d). How we die: Stigmatized death and disenfranchised grief. In K. J. Doka (Ed.), *Disenfranchised grief* (pp. 265–274). Champaign, IL: Research Press.

Donath, J. S. (1999). Identity and deception in the virtual community. In M. A. Smith & P. Kollock (Eds.), *Communities in cyberspace* (pp. 29–59). New York, NY: Routledge.

Ellison, N., Heino, R., & Gibbs, J. (2006). Managing impressions online: Self-presentation processes in the online dating environment. *Journal of Computer-Mediated Communication, 11*(2), 415–441. doi:10.1111/j.1083-6101.2006.00020.x

Eyetsemitan, F. (1998). Stifled grief in the workplace. *Death Studies, 22*, 469–479.

Feigelman, W., Gorman, B. S., & Jordan, J. R. (2009). Stigmatization and suicide bereavement. *Death Studies, 33*, 591–608.

Goffman, E. (1963). *Stigma: Notes on the management of spoiled identity*. Englewood Cliffs, NJ: Prentice-Hall.

Goldberg, L. (2013, May 9). *Glee* season 4 finale: The 10 biggest reveals. *The Hollywood Reporter*. Retrieved from hollywoodreporter.com

Goodman, T. (2013, January 24). Manti Te'o, "catfish," Katie Couric, Oprah and the sports world: Paging Dr. Phil! *The Hollywood Reporter*. Retrieved from hollywoodreporter.com

Harris, A. (2013, January 17). Manti Te'o: Catfish? Or catfish victim? *Slate*. Retrieved from slate.com

Hocker, W. V. (1989). Unsanctioned and unrecognized grief: A funeral director's perspective. In K. J. Doka (Ed.), *Disenfranchised grief: Recognizing hidden sorrow* (pp. 257–269). Lexington, MA: Lexington.

Jarecki, A., & Joost, H. (Producers), & Joost, H., & Schulman, A. (Directors). (2010). *Catfish* [Motion picture]. United States: Supermarché and Hit The Ground Running Films.

Kaczmarek, M., & Backlund, B. (1991). Disenfranchised grief: The loss of an adolescent romantic relationship. *Adolescence, 26*(102), 253–259.

Kamerman, J. (1993). Latent functions of enfranchising the disenfranchised griever. *Death Studies, 17*, 281–287.

Kauffman, J. (2002). The psychology of disenfranchised grief: Liberation, shame, and self-disenfranchisement. In K. Doka (Ed.), *Disenfranchised grief: New directions, challenges, and strategies for practice* (pp. 61–77). Champaign, IL: Research Press.

Kellehear, A. (2007). *A social history of dying*. Cambridge: Cambridge University Press.

Kollar, N. R. (1989). Rituals and the disenfranchised griever. In K. J. Doka (Ed.), *Disenfranchised grief: Recognizing hidden sorrow* (pp. 271–285). Lexington, MA: Lexington.

Lang, A., Fleiszer, A. R., Duhamel, F., Sword, W., Gilbert, K. R., & Corsini-Munt, S. (2011). Perinatal loss and parental grief: The challenge of ambiguity and disenfranchised grief. *Omega: Journal of Death and Dying, 63*, 183–196.

LaGrand, L. E. (1989). Youth and the disenfranchised breakup. In K. J. Doka (Ed.), *Disenfranchised grief: Recognizing hidden sorrow* (pp. 173–185). Lexington, MA: Lexington.

Lee, L. (1984). Sequences in separation: A framework for investigating endings of the personal (romantic) relationship. *Journal of Social and Personal Relationships, 1*, 99–173.

Lenhardt, A. M. C. (1997). Grieving disenfranchised losses: Background and strategies for counselors. *Journal of Humanistic Education and Development, 35*, 208–216.

Lohmann, R. C. (2013, April 30). The two-sided face of teen catfishing. *Psychology Today*. Retrieved from psychologytoday.com

Martin, T. L. (2002) Disenfranchising the brokenhearted. In K. J. Doka (Ed.). *Disenfranchised grief* (pp. 197–216). Champaign, IL: Research Press.

Meisenbach, R. J. (2010). Stigma management communication: A theory and agenda for applied research on how individuals manage moments of stigmatized identity. *Journal of Applied Communication Research, 38*, 268–292. doi:10.1080/00909882.2010.490841

Morris, C. E., & Reiber, C. (2011). Frequency, intensity and expression of post-relationship grief. *The Journal of the Evolutionary Studies Consortium, 3*(1), 1–11.

Page, R. M. (1984). *Stigma*. London: Routledge and Kegan Paul.

Parkes, C. M. (1980). Bereavement counseling: Does it work? *British Medical Journal, 281*, 3–6.

Parkes, C. M. (1987). *Bereavement: Studies of grief in adult life* (2nd ed.). Madison, CT: International Universities Press.

Parkes, C. M., & Prigerson, H. G. (2010). *Bereavement: Studies of grief in adult life* (4th ed.). New York, NY: Routledge.

Parkes, M. R., & Floyd, K. (1996). Making friends in cyberspace. *Journal of Communication, 46*, 80–97.

Pesek, E. M. (2002). The role of support groups in disenfranchised grief. In K. J. Doka (Ed.), *Disenfranchised grief* (pp. 127–133). Champaign, IL: Research Press.

Rabby, M. K. (2007). Relational maintenance and the influence of commitment in online and offline relationships. *Communication Studies, 58*, 315–337. doi:10.1080/10510970701518405

Rabby, M. K., & Walther, J. B. (2003). Maintaining on-line relationships. In D. J. Canary & M. Dainton (Eds.), *Maintaining relationships through communication: Relational, contextual, and cultural variations* (pp. 141–162). Mahwah, NJ: Lawrence Erlbaum.

Rando, T. A. (1993). *Treatment of complicated mourning*. Champaign, IL: Research Press.

Robak, R. W., & Weitzman, S. P. (1994–1995). Grieving the loss of romantic relationships in young adults: An empirical study of disenfranchised grief. *Omega: Journal of Death and Dying, 30*(4), 269–281.

Romanoff, B. D., & Terenzio, M. (1998). Rituals and the grieving process. *Death Studies, 22*, 697–711.

Rothman, L. (2013, January 23). The Manti Te'o Hoax: 5 reasons people create fake girlfriends (according to *Catfish*). *Time*. Retrieved from time.com.

Sanders, C. M. (1993). Risk factors in bereavement outcome. In M. S. Stroebe, W. Stroebe, & R. O. Hansson (Eds.), *Handbook of bereavement: Theory, research, and intervention* (pp. 255–267). New York, NY: Cambridge University Press.

Shultz, K. (2000). Every implanted child a star (and some other failures): Guilt and shame in the cochlear implant debates. *Quarterly Journal of Speech, 86*(3), 251–275. doi:10.1080/00335630009384296

Smith, A., & Duggan, M. (2013, October 21). Online dating and relationships. *Pew Research Center.* Retrieved from pewinternet.org/2013/10/21/online-dating-relationships/

Smith, H. I. (2006). Does my grief count? When ex-family grieves. *Illness, Crisis, & Loss, 14,* 355–372.

Smith, R. A. (2007). Media depictions of health topics: Challenge and stigma formats. *Journal of Health Communication, 12,* 233–249. doi:10.1080/10810730701266273

Smith, R. A. (2011). Stigma communication and health. In T. L. Thompson, R. Parrott, & J. Nussbaum (Eds.), *Handbook of health communication* (2nd ed., pp. 455–468). London: Taylor & Francis.

Story of Manti Te'o girlfriend a hoax. (2013, January 17). *ESPN.* Retrieved from ESPN.com

Thornton, G., Robertson, D. U., & Miecko, M. L. (1991). Disenfranchised grief and evaluations of social support by college students. *Death Studies, 15,* 355–362.

Tong, S. T., & Walther, J. B. (2011). Relational maintenance and CMC. In K. B. Wright & L. M. Webb (Eds.), *Computer-mediated communication in personal relationships* (pp. 98–118). New York, NY: Peter Lang.

Walther, J. B. (1992). Interpersonal effects in computer-mediated interaction: A relational perspective. *Communication Research, 19,* 52–90.

Walther, J. B. (1996). Computer-mediated communication: Impersonal, interpersonal, and hyperpersonal interaction. *Communication Research, 23*(1), 3–43. doi:10.1177/009365096023001001

Warner, K. (2012, July 5). "Catfish" MTV show brings online love stories to life. *MTV.* Retrieved from MTV.com

Weber, A. (1998). Losing, leaving, and letting go: Coping with nonmarital breakups. In B. H. Spitzberg & W. R. Cupach (Eds.), *The dark side of close relationships* (pp. 267–306). Mahwah, New Jersey: Lawrence Erlbaum Associates.

Wolken, D., & Myersberg, P. (2013, January 18). Manti Te'o's inspirational girlfriend story a hoax. *USA Today.* Retrieved from usatoday.com

SEVENTEEN

I Heard It through the Grapevine: How Organizational Rumors Impact Interpersonal Relationships via Traditional and CMC Channels

Stacie Wilson Mumpower and Megan Bassick[1]

Rumors are common in informal dialogue during social interaction among individuals in various relationships (Bordia, DiFonzo, & Schultz, 2000; Kniffin & Wilson, 2010). In organizations, rumors are considered informal communication and often perceived as latent dialogue present during investigation of other phenomenon (Kniffin & Wilson, 2010). Rumors emerge in organizations among employees, managers, and other stakeholders, particularly during instances of flux when information about the organization or its members is unclear or withheld (Bisel & Barge, 2011; DiFonzo & Bordia, 2000; Rosnow, 2001). When information is suppressed, people may create explanations to minimize uncertainty (Rosnow, 2001). These explanations often manifest as potentially false information—rumors.

Organizations provide an interesting environment to study rumors because rumors can be prevalent and associated with uncertainty, which can impact workplace relationships (DiFonzo & Bordia, 2000). For example, DiFonzo and Bordia (2000) suggested rumors can negatively affect trust between management and staff, decrease morale, create bad press, increase employee stress, and lead to public image issues. Informal relationships in workplaces constitute a relevant interpersonal context in which rumor manifestation can have consequential effects. Social capital may facilitate rumor flow, like gossip, because trusted resources offer an outlet to clarify information, particularly in highly connected social networks (Ellwardt, Steglich, & Wittek, 2012). Unverified information later falsified may have subsequent negative outcomes for these interpersonal relationships and, by extension, the organization.

Consistent with Allport and Postman's (1947) rumor definition, general assumptions presume rumors are spread primarily face-to-face (FtF); current literature lacks investigation of organizational rumor transmission via computer mediated channels. This chapter adds a technology dimension to the notion of typical rumor spreading and examines whether the rumor is perceived differently when communicated through technology.

Rumor and sensemaking theories direct our attention to an understudied phenomenon situated in interpersonal and computer-mediated contexts within organizations. Rumor theory provides insight into organizational rumors' effects, along with potential ramifications to interpersonal relationships (DiFonzo & Bordia, 2000; DiFonzo, Bordia, & Rosnow, 1994). Sensemaking (Weick, 1995) provides a framework for scrutinizing individuals' process of making sense out of unanticipated events that violate expectations about organizational events.

Organizational rumors' impact on interpersonal relationships and media choice for conveying rumors is a novel contribution to research investigating the dark sides of communication. In the following sections, we review relevant rumor literature, sensemaking, and rumor theory.

Literature Review

Organizational Rumors

Informal communication, like rumors, is ubiquitous in organizations and may develop trust and cohesiveness among organizational members (Charles, 2007; DiFonzo & Bordia, 2000; Luna & Chou, 2013). Anthropologists and sociologists view gossip as a necessary aspect of society and, from a researcher's perspective, gossip aids in understanding the organization's communicative fabric (Kniffin & Wilson, 2010). Conversely, informal talk, like gossip, can impact organizational culture negatively when perceptions are altered and mistrust develops (Burke & Wise, 2003; Michelson, van Iterson & Waddington, 2010). Workplace gossip is a neglected area in management research and often observed coincidentally while investigating other phenomena; consequently, gossip is conceptualized inconsistently in management research (Kniffin & Wilson, 2010). Though understudied in management research, psychology scholars have investigated gossip and rumor spreading in organizations related to perception, verbal aggression, bullying, victimization and resulting climates of psychological strain (see Farley, Timme, & Hart, 2010). Leader strategies for combating rumors have also been studied (see Bordia, DiFonzo, & Schulz, 2000). Communication research has investigated gossip and rumors when they surface in organizational change discourse (Bisel & Barge, 2011; Brown & Starkey, 2000).

DiFonzo and Bordia (2000) label gossip as one form of internal rumor, which stems from issues important to organizational members (e.g., job security and satisfaction or personnel changes). External rumors are associated with organizational reputation and service quality, and are important to stockholders, customers or the general public (DiFonzo & Bordia, 2000).

Previous research focused on understanding what impacts the believability of rumors, perceptions of individuals who deny rumors, how organizations manage rumors, and more generally, how and why rumors are formed (DiFonzo & Bordia, 2000; Rosnow, 2001). Little research has examined the impact organizational rumors have on interpersonal relationships. Informal relationships, like friendships, play an important role in organizations (Ellwardt et al., 2012). For example, employees are inclined to be more accommodating when formal relationships are complemented by informal ones (Sparrowe, Liden, Wayne, & Kraimer, 2001). Since informal, interpersonal relationships are intertwined in organizations, how rumors impact these relationships could provide a more comprehensive understanding of rumors' roles in organizations.

Rosnow (2001) states rumors can be threatening to relationships and aggravate already tense situations. Additionally, spreading false rumors can permanently damage a rumor carrier's reputation (DiFonzo et al., 1994). It is important researchers study the impact rumors might have on work relationships, not only between managers and staff, but also between same status coworkers. A rumor's truthfulness may also impact perceptions of the rumor spreader. Farley et al. (2010) found high gossipers who spread negative gossip at work were perceived as needing more control, wanting to control others, and being less emotionally close in interpersonal ability to express affection than low gossipers. This suggests rumor spreaders may be perceived differently based on how often they spread rumors. Rosnow, Yost, and Esposito, (1986) found a linear, non-causal relationship between belief that a rumor was true and the likelihood of transmitting the rumor; people are more likely to pass on a rumor they think is true than one they think is false. Likewise, rumors that are not believed are less likely to be spread further in organizations (DiFonzo & Bordia, 2000). Interestingly, Jaeger, Skleder, Rind, and Rosnow (1994) found high gossipers in a sorority were perceived by others as less likeable, but more socially connected. High gossipers were likely to be powerful transmitters of rumors, regardless of veracity, because they are central to the communication network (Farley et al., 2010; Jaeger et al., 1994). Further inquiry regarding rumor truth and subsequent interpersonal perceptions is warranted. This study considers organizational rumors as any unverified information associated with the organization that would impact the entity's integrity either internally or externally.

Computer-mediated communication (CMC). To our knowledge, no existing research specifically examines organizational rumors' impact on interpersonal relationships spread FtF or via CMC, despite widespread technology use in organizations. Organizational members use communication technologies for personal as well as task-related functions, such as sending personal emails (Garrett & Danziger, 2008). Since newcomers use both FtF and mediated channels (e.g., email and mobile chat functions) to acquire information that assists in uncertainty reduction during organizational assimilation (Waldeck, Seibold, & Flanagin, 2004), CMC may play a role in the circulation of organizational rumors because rumor transmission serves as a method for dealing with anxiety and uncertainty (Rosnow, 2001). It seems likely members are sharing rumors through communication technologies, perhaps for uncertainty diminution; therefore, we asked:

RQ1: What channels are most frequently used to spread organizational rumors?

RQ2: Is there a relationship between the channel used to spread the organizational rumor and perceived (a) rumor veracity, and (b) rumor valence?

RQ3: If the participant spread the rumor, is there a relationship between the channel used to spread the organizational rumor and perceived (a) rumor veracity, and (b) rumor valence?

Theoretical Foundation

Rumor theory. Rumor theory suggests organizational rumors impact interpersonal relationships. Rumors go through three stages: generation, evaluation, and dissemination (DiFonzo et al., 1994). In the generation stage, rumors begin to frame and structure ambiguous information and give meaning and predictability to uncertain events (DiFonzo et al., 1994). During the evaluation stage, individuals assess the rumor's truthfulness (DiFonzo et al., 1994). If an individual spreads false information, the circumstance could create undue anxiety for others, which in turn may impact the rumor spreader's reputation (DiFonzo et al., 1994). Rumors can disseminate quickly and cause damage to both individual and organization because of repeated circulation and mutation that mirrors what the rumor spreader believes true. The more the rumor circulates, the more plausible the proposition (DiFonzo et al., 1994).

Furthermore, holding a leadership position may influence interpersonal relationships. Leaders often need to manage rumor dissemination to prevent organizational damage and low morale (Bordia, DiFonzo, & Travers, 1998). Contrary to popular belief, top-level managers are not necessarily appropriate sources to quell rumors (Bordia et al., 1998). Bordia et al. (1998) found sources directly relevant to the rumor scope were most successful in stopping rumors than those occupying the highest positions. These findings suggest organizational leadership position holders of all levels, who spread false rumors, may experience a plummet in credibility because of their inherent believability, damaging interpersonal relationships. Therefore, we hypothesized:

> H1: Organizational rumor spreader's status will impact an organizational member's relationship with the rumor spreader.

Sensemaking. Like rumor theory, sensemaking suggests individuals attempt to make sense of rumor development and dispersal, particularly when it impacts interpersonal relationships in negative ways, when rumors are first spread or when they are proven untrue. Sensemaking occurs when a disruption in normal routine causes uncertainty about a particular state of affairs, leading to anxiety and prompting attention to a situation (Weick, 1995). Interruptions in perceived organizational homeostasis are what Weick (1995) calls "shock." The ambiguous stimulus elicits negative emotions that trigger sensemaking (Weick, 1995). Similarly, Allport and Postman (1947) stipulate that in the "law of a rumor," (p.33) the rumor itself must have importance to the individual and be obscured in ambiguity. Further, rumors function as an emotional component, conceptualized as anxiety, when individuals attempt to understand incidents of social change that violate mental models about how the world works Rosnow, 2001).

Sensemaking helps explain how organizational members engage in an ongoing process of navigating extracted cues from rumors, negotiating their identities in social context, and developing plausible conclusions about rumored events and rumor bearers (Weick, Sutcliffe, & Obstfeld, 2005). In short, individuals engage in sensemaking to determine what an event means to their identities and the organization with which they identify. Individuals then make choices about whether to change their perception of a rumor bearer, believe rumors, or spread rumors, as they interpret events in the ongoing state of their organizational experience. It is this action-interpretation interplay that helps people determine their next action (Weick et al., 2005). Interpretation of (1) the rumor in its respective context, (2) the rumor spreader's credibility, and (3) potential outcomes is primary in an individual's determination of an interpersonal relationship's state. Further, if rumors prove untrue, members will likely engage in additional sensemaking processes to evaluate the rumor bearer's credibility.

Rumors constitute a way for organizational members to regain some predictive control and make meaningful sense of ambiguous events (DiFonzo & Bordia, 2000). Sensemaking occurs to restore order to disrupted routine. Both rumor theory and sensemaking can be situated in the context of rumor spreading to describe, (1) a response to uncertainty during which organizational members attempt to frame events into more predictable outcomes, and (2) the impact of rumors on perception. Rumors constitute a way of knowing whom to trust as they navigate the organizational environment during instances such as change. When verifiable evidence becomes available, the veracity in circulated rumors can have damaging effects on the rumor bearer's credibility. We investigated the relationship between sensemaking about an untrue rumor and the rumor bearer's credibility, and hypothesized:

> H2: Organizational rumor veracity will moderate the relationship between sensemaking and credibility, such that an untrue rumor will increase an individual's sensemaking and negatively impact the relationship between them.

Methods

Participants and Procedure

A sample of 165 participants, from a large Southwestern university, were recruited via a university department subject pool and were compensated in course credit for a communication course. Respondents had to be current members of a Greek life organization (e.g., sorority, fraternity) for at least one month and have experienced rumors spread within their organization. Greek life organizations are rife with organizational change, which has been linked to decreased group identification and discussions about how to handle issues surrounding the change (Williams & Connaughton, 2008). Therefore, these organizations are an appropriate population to sample. Twenty-one surveys were incomplete or not eligible and eliminated from analysis.

Participants' average age was 19.35 (SD = 1.09). Forty-three (29.9%) were men, 100 (69.4%) were female, and one (0.7%) did not report gender. Most respondents identified as White (n = 126, 87.5%), followed by Hispanic or Latino (n = 7, 4.9%), Asian (n = 5, 3.5%), Native American (n = 4, 2.8%), Black or African American (n =1, 0.7%), and Pacific Islander (n = 1, 0.7%). This sample is representative of the university's population. Length of membership in a Greek organization ranged from one to 40 months with an average of 13.29 months (SD = 11.69).

Respondents completed electronic informed consent prior to receiving access to the online survey. The five-part survey consisted of multiple choice, Likert-type questions, and open-ended items. Respondents provided demographic and Greek life organization information, then proceeded to sensemaking and relational measures.

Instruments

Status within the organization. To determine whether organizational status influenced the participant–rumor spreader relationship, respondents were asked to indicate whether they held a leadership position in the organization with an option to enter position title. Participants also reported whether the rumor spreader held a leadership position and described that position.

Rumor items. In an open-ended item, participants were asked to describe the rumor event with as much detail as possible. They were then asked to rate how positive or negative they perceived the rumor to be on a 7-point Likert-type scale (1 = *very negative*, 7 = *very positive*). Respondents also rated their perceived rumor veracity on a 7-point semantic differential scale.

Lastly, to assess respondents' involvement with rumor dissemination, they were asked whether they spread the rumor with the following answer choices: "yes, I heard the rumor from someone, and told others," "yes, I started the rumor," and "no."

Channel choice. Two categorical items determined what channel was used most to spread organizational rumors, and the relationship of channel choice to rumor veracity and valence. Participants, who spread the rumor to others after hearing it, were asked to select what channel the rumor spreader used and what channel they used to spread the rumor *(face-to-face, text messages, email, phone call, video chat, social media, and other).* Those who reported not sharing the rumor only responded to the rumor spreader channel choice item. Finally, those who indicated they started the rumor only responded to the item identifying their channel choice.

Relational satisfaction. To measure participants' satisfaction with their relationship with the rumor spreader and their Greek organization, participants were asked to indicate agreement with five items adapted from Norton's (1983) Quality of Marriage Index (QMI; (e.g., *"My relationship with my sorority/fraternity is good"* instead of *"My marriage is good"*) on a Likert-type scale from 1 = strongly disagree to 7 = strongly agree. They responded to these items regarding their relationship with the organization and rumor spreader. For each relationship, satisfaction scores were calculated by averaging the sum of each item. QMI proved to be reliable in measuring relational quality with the Greek life organization (α = .95) and the rumor spreader (α = .96).

Credibility. To determine whether a rumor influenced Greek organization and rumor-spreader credibility, participants rated their agreement with six statements adapted from Wakefield and Whitten's (2006) semantic differential credibility scale, which was an adapted version of Ohanian's (1991) credibility scale (e.g., "*My sorority/fraternity is trustworthy*" and "*He/she is credible")* on a 7-point Likert-type scale (1 = *strongly disagree*, 7 = *strongly agree*). Both scale versions' reliability were adequate (Greek life, α = .85; rumor spreader, α = .90). To improve the Greek life version's reliability, the item, "*My sorority/fraternity is not credible,*" was deleted (α = .89). One item ("*He/she is not credible*") from the rumor spreader version was also removed for symmetry (α = .89).

Sensemaking. Participants responded to two open-ended items to record their retrospective accounts of (1) the "talk" that occurred about the rumor event, and (2) how they perceived the rumor disrupted their daily activities ("Describe how you and others in your sorority/fraternity talked about the rumor in your daily activities with the organization;" "Describe how the rumor affected your daily activities with your sorority/fraternity"). Respondents indicated agreement with statements adapted from Carrington and Tayles (2011) concerning how the organization responded to rumors (e.g., *"Discussion about the rumor in the sorority/fraternity is participative," "All members of the executive board participate in conversations about dealing with rumors on a regular basis,"*) on a 7-point Likert-type scale, 1 = strongly disagree to 7 = strongly agree, (α = .65). Participants also indicated agreement with seven statements, developed by the current researchers to measure anxiety, on a 7-point Likert-type scale (1 = *strongly disagree,* 7 = *strongly agree*; α = .82). The anxiety scale was used to determine anxiety levels associated with the rumor. Negative emotions like anxiety are often heightened when uncertain events trigger sensemaking and rumors are associated with higher levels of anxiety (Rosnow, 2001).

Intimacy. To measure intimacy level in the participant–rumor spreader relationship and the participant-organization relationship, an adapted version of Miller's (1982) Social Intimacy Scale (MSIS; items were reworded from questions to statements to better fit the organizational context [instead of *he/she, sorority/fraternity* was used] for intimacy with their Greek organizations) was employed. MSIS measures intimacy by relationship's contact (6-items) and intensity (11-items). Participants responded using a 7-point Likert-type scale (1 = *not much at all, 7 = a great deal*). Item three (*"How often do you show him/her affection?"*) was excluded in the measure of Greek organizational intimacy because the authors felt the question did not pertain to organizational relationships. To improve Cronbach's α from .84 to.89, item eight (*"How much damage is caused by a typical disagreement in your relationship with your fraternity/sorority?"*) was removed. For the rumor-spreader version, after alpha scores were calculated, item eight was removed from analysis to improve the reliability score from α = .94 to α = .96. For each relationship, intimacy scores were calculated by averaging the sum of each item.

Relational impact. Participants rated how positively/negatively the rumor impacted different aspects of their relationships with the rumor spreader and with the organization on a 7-point Likert-type scale (1 = *negatively*, 7 = *positively*). The relational impact scale consisted of items adopted from MSIS (three items), the credibility scale (two items), and the QMI (two items). The Greek relational impact scale and the rumor spreader relational impact scale reliability scores were acceptable (Greek life, α = .96; rumor spreader, α = .95).

Analysis and Results

RQ1 asked which channels were most frequently used to spread organizational rumors. To answer this question, a simple frequency count using listwise deletion was conducted for the rumor channel variable. Somewhat surprisingly, FtF was the most common channel used to spread organizational rumors, (N = 92, 62.6%), followed by text messages (N = 30, 20.4%), social networking sites (N = 16, 10.9%), and "other" (N = 3, 2%).

RQ2 asked whether there was a relationship between the channel used to spread the organizational rumor and perceived (a) rumor veracity findings revealed the leadership position held by the rumor spreader was significantly related to rumor outcomes (see results for H1), we chose to control for the rumor spreader's leadership position during this analysis. A two-way ANCOVA was conducted with channel choice (IV) and perceived (a) rumor veracity and (b) rumor valence (DVs,) while controlling the rumor spreader's leadership position. Neither test was significant (perceived rumor veracity F (6, 128) = .501, p = .481; perceived rumor valence, F (5, 129) = .962, p = .328); there was no relationship between channel used to spread the rumor and perceived rumor veracity or valence.

RQ3 asked whether there was a relationship between the channel used to spread the rumor and perceived (a) rumor veracity, and (b) rumor valence if the participant spread the rumor him/herself. The same procedure utilized for RQ2 was repeated, analyzing only those participants who indicated spreading the rumor. Again, neither test was significant (rumor veracity, F (6, 48) = .720, p = .635; rumor valence, F (5, 49) = .770, p = .576).

H1 predicted organizational rumor-spreader status would impact the participant's relationship with the rumor spreader. An independent samples t-test was conducted comparing mean relational impact of the rumor for rumor spreaders with an organizational leadership position to rumor spreaders with no position. The test was significant, t (132) = -2.445, p < .05, indicating when a rumor spreader held a leadership position, the rumor negatively impacted the participant's relationship with the rumor spreader more (M = 2.65) than rumor spreaders without a leadership position (M = 3.39). H1 was supported.

Finally, H2 predicted organizational rumor veracity would moderate the relationship between sensemaking and credibility, such that an untrue rumor would increase an individual's sensemaking process and negatively impact the relationship between sensemaking and perceived rumor spreader credibility. To test this hypothesis, a multiple regression was conducted with rumor veracity, sensemaking, and an interaction term of veracity and sensemaking as IVs, and perceived credibility as the DV. To guard against issues with multicollinearity, the variables were centered before creating the interaction term and conducting analysis. The overall regression was significant, $R = .365$, R^2 *adjusted* $= .110$, $F(3, 115) = 5.88$, $p < .01$. The main effect of rumor veracity was significant, ($B = .327$, $p < .001$); the more truthful the rumor, the more credible the rumor spreader was perceived. However, the main effect of sensemaking on credibility was not significant, ($B = .104$, $p = .511$), and the interaction was not significant, ($B = -.028$, $p = .722$). Although the hypothesis was not supported, we were able to provide support for the notion that rumor veracity impacts the perceived rumor spreader credibility.

Because the sensemaking scale had lower reliability than expected ($\alpha = .65$), we ran a post-hoc test substituting anxiety for the sensemaking scale. A disruption in organizational routine elicits anxiety, in turn, initiating an observable sensemaking process (Weick, 1995). Further, Rosnow (2001) suggests rumors contain an emotional component conceptualized as a form of anxiety when confronted with unforeseen occurrences; anxiety might be a sufficient alternative for the sensemaking scale in this case. We centered the variables, created an interaction term between rumor veracity and anxiety caused by the rumor, and entered the truthfulness measure, anxiety measure, and interaction term as IVs in a regression equation. The overall regression was significant $R = .359$, R^2 *adjusted* $= .11$, $F(3, 111) = 5.49$, $p < .01$. The main effect of truthfulness remained significant ($B = .292$, $p < .001$). The main effect of anxiety was not significant ($B = -.148$, $p = .200$). Therefore, H2 was partially supported. The interaction was also not statistically significant ($B = .006$, $p = .914$).

Discussion

To our knowledge, this was the first study to examine the impact organizational rumors have on interpersonal relationships. Further, this study explored the role of channel use in rumor dispersal. Consistent with Allport and Postman's (1947) rumor definition, results indicated most rumors were spread FtF. Originally, since rumors tend to be considered negative, and research shows individuals tend to spread negative information via communication technologies (Baruch, 2005), we assumed a mediated channel would be most popular. However, after further consideration, since sorority and fraternity members' identities tend to be tied to their Greek organization, it is possible spreading a rumor publicly would be considered too face-threatening, and so opted to disseminate potentially harmful information about their organization more privately. Additionally, Greek life members spend considerable amounts of time together; they attend classes, chapter meetings, meals, and events together and often live in a communal building. Rumor spreaders in Greek life organizations may have preferred spreading rumors FtF rather than via CMC because other Greek life members were co-present. Further, there are numerous components one might consider when selecting a channel (i.e., accessibility, ability to send a message to multiple people, avoiding a traceable, electronic record). Thus, one may opt for FtF message distribution to avoid a permanent record tying one to the rumor, especially if it turned out to be false.

Social affordances—how useful an individual believes a channel to be (Treem & Leonardi, 2012)—may explain a rumor spreader's channel choice. Persons may opt for a certain medium based on its perceived affordances (Treem & Leonardi, 2012). One's perception of affordances is based on an individual's ability to achieve his or her communicative goals with a medium, rather than technology features. Four types of affordances of social media can impact an organization's ability to share information, and its socialization and power progressions (Treem & Leonardi, 2012). Treem and Leonardi (2012) list four types of affordances: (a) visibility, or when a person's knowledge is visible and easy to locate; (b) persistence, when a communicative act, can be reviewed electronically by others in its original form; (c) editability, when a message can be crafted carefully before distribution; and (d) association, when it links social ties together in venues like blogs. Treem and Leonardi (2012) found employing assorted communication channels based on affordances effects organizational members' communicative behaviors. Consistent with Treem and Leonardi's (2012) typology, participants may have chosen a channel for rumor dispersal based on its social affordances. In the current case, some respondents avoided visibility, persistence, and association while others embraced the editability of text messages, and visibility, persistence, and association of social media networking sites.

Perceived rumor veracity and valence did not significantly influence channel selections for rumor spreading. Although non-significant results were unexpected, the findings may be attributed to small cell sizes for different channels used (only 30 participants used a cell phone, while a mere 16 used SNS). Cell size was even more problematic when answering RQ3; only 48 participants could be included in the media analysis. Therefore, larger sample sizes might yield more telling results.

We found rumor spreaders holding leadership positions had a negative effect on the rumor spreader–member relationship. Our findings support previous research that suggests leadership position influences whether others believe their involvement is appropriate (Bordia et al., 1998). Findings suggest the source of rumor distribution is consequential and organizational members held in high regard are likely to see negative effects on their interpersonal relationships when spreading rumors.

We additionally tested whether rumor veracity moderated the relationship between sensemaking and perceived rumor-spreader credibility. The interaction was not significant, but the relationship between perceived rumor veracity and rumor-spreader credibility was significant. Perceptions of rumor spreaders' credibility were more likely to be low for untrue rumors, suggesting spreading false rumors can hurt the rumor spreader's reputation (DiFonzo et al., 1994). Contrary to our expectations, sensemaking did not impact perceived rumor-spreader credibility. The likely culprit of non-significant results is the low reliability of the scale (α = .65). Future researchers should further investigate this potential relationship using more appropriate and reliable measures. For this study, we replaced the sensemaking scale with the anxiety scale and conducted the analyses again, since the literature suggests anxiety plays a role in the sensemaking process (Weick, 1995). The results were still not significant, but showed more promise than the sensemaking measure. Average anxiety was moderate, suggesting the rumor did not cause much anxiety or anxiety levels had changed since the incident ensued. Sensemaking is ongoing, occurring on latent levels in day-to-day activities (Weick, 1995). The sensemaking process is observable when an anxiety-producing event causes individuals to attend to a situation (Weick, 1995). Because of the retrospective nature of survey questions, participants' anxiety levels may not have been as heightened as at the beginning of an incident.

Perceptions about the event may not have been salient enough to yield maximum observation opportunity in survey responses.

Importantly, although quantitative results suggest moderate anxiety and little or no observable sensemaking process, narrative responses from participants suggest an untrue rumor will, in fact, increase sensemaking. A rumor that causes anxiety and ambiguity can trigger communicative processes in social interaction. Interpretations and meaning about events emerge when individuals attempt to make sense of a situation (Weick, 1995), often through descriptions of the event. Many participants indicated anxiety. Below is one examples of anxiety's presence:

> It made it uncomfortable being around the girl because so many girls knew about the rumor and made me believe that she may know about the rumors or that she will eventually hear the rumors and be upset at her sisters.

Individuals engage in sensemaking continuously to retrospectively rationalize what people are doing (Weick et al., 2005), particularly regarding how the event shapes their organizational identity (Weick, 1995). In other words, actors try to make sense of an event through discussion with others. This type of process was evident in one participant's response. She described a conversation about the rumor that illustrates how the sensemaking process is visible in her dialogue.

> It just got brought up in conversation—one, because it didn't make sense that someone would go and do that to our president because she's the sweetest. And two, it didn't make sense why that was even a rumor.

Based on this and similar responses, an analysis of participants' narrative descriptions may yield a better understanding of the sensemaking processes occurring during rumor events in this Greek organization. Past studies investigated sensemaking qualitatively (see Dougherty & Smythe, 2004), and narrative accounts more accurately offer a lens into organizational members' lived experiences (Lindlof & Taylor, 2011). Further examination of narrative accounts in this study would likely reveal sensemaking processes adjacent to the rumor event.

Future Directions, Limitations, and Conclusions

Our study supported DiFonzo et al. (1994) and Bordia et al.'s (1998) research concerning the role of rumors in interpersonal relationships, and extended their findings to an organizational context; however, Greek organizations tend to be closely tied to members' identities and, therefore, rumors might have more interpersonal impact in this context than in others. CMC might be a more typical way to spread rumors in a more conventional organization. Therefore, future researchers should study the interpersonal impact rumors have in a variety of organizations to better understand this process. Future researchers should continue to investigate the role sensemaking plays in the rumor process. Initially, our results did not seem to support sensemaking theory, but we believe this is an artifact of our measures, not the actual role sensemaking plays. Perhaps more reliable measures, or more in-depth qualitative data analysis, would add to our knowledge of this relationship. Although the convenience sample used in this study is representative of the institution from which we sampled, the sample is not representative of the overall population, limiting our findings' generalizability. However, since we were interested in the way rumors impact interpersonal relationships in Greek organizations, sampling from a college student population was appropriate. Finally, participants were asked to recall a past rumor within their Greek life organization, so retrospective accounts may have been degraded (e.g., a participant who is now a senior could be recalling a rumor from his/her first year).

This study's purpose was to investigate organizational rumors' impact on interpersonal relationships, channel choice, sensemaking processes, and organizational status's impact on perceptions of rumor spreaders. This chapter offers a more encompassing depiction of how rumors are spread within an organization and how rumors affect an individual's relationship with the rumor spreader and the organization—a significant contribution to organizational communication literature.

Notes

1 The authors would like to thank Eryn Bostwick for her assistance with data collection and analysis.

References

Allport, G. W., & Postman, L. (1947). The psychology of a rumor. *Journal of Clinical Psychology, 3*, 402–402.
Bisel, R. S., & Barge, J. K. (2011). Discursive positioning and planned change in organizations. *Human Relations, 64*, 257–283.
Baruch, Y. (2005). Bullying on the net: Adverse behavior on e-mail and its impact. *Information and Management*, 42, 361-371. doi: 10.1016/j.im.2004.02.001
Bordia, P., DiFonzo, N., & Schulz, C. A. (2000). Source characteristics in denying rumors of organizational closure: Honesty is the best policy. *Journal of Applied Social Psychology, 30*, 2309–2321.
Bordia, P., DiFonzo, N., & Travers, V. (1998). Denying rumors of organizational change: A higher source is not always better. *Communication Research Reports, 15*, 188–197.
Brown, A. D., & Starkey, K. (2000). Organizational identity and learning: A psychodynamic perspective. *Academy of Management Review, 25*, 102–120.
Burke, L. A., & Wise, J. M. (2003). The effective care, handling, and pruning of the office grapevine. *Business Horizons, 46*, 71–76.
Carrington, D., & Tayles, M. (2011). The mediating effects of sensemaking and measurement on the intellectual capital and performance linkage. *Electronic Journal of Knowledge and Management, 9*, 284–295.
Charles, M. (2007). Language matters in global communication. *Journal of Business Communication, 44*, 260–282.
DiFonzo, N., & Bordia, P. (2000). How top PR professionals handle hearsay: Corporate rumors, their effects, and strategies to manage them. *Public Relations Review, 26*, 173–190.
DiFonzo, N., Bordia, P., & Rosnow, R. L. (1994). Reining in rumors. *Organizational Dynamics, 23*, 47–62.
Dougherty, D. M., & Smythe, M. J. (2004). Sensemaking, organizational culture, and sexual harassment. *Journal of Applied Communication Research, 32*, 293–317.
Ellwardt, L., Steglich, C., & Wittek, R. (2012). The co-evolution of gossip and friendship in workplace social networks, *Social Networks, 34*, 623–633.
Farley, S. D., Timme, D. R., & Hart, J. W. (2012). On coffee talk and break-room chatter: Perceptions of women who gossip in the workplace. *The Journal of Social Psychology, 150*, 361–368.
Garrett, R. K., & Danziger, J. N. (2008), Disaffection or expected outcomes: Understanding personal Internet use during work. *Journal of Computer-Mediated Communication, 13*, 937–958.
Jaeger, M. E., Skleder, A. A., Rind, B., & Rosnow, R. L. (1994). Gossip, gossipers, gossipees. In R. F. Goodman & A. Ben-Ze'ev (Eds.), *Good gossip* (pp.154–168). Lawrence, KS: University Press of Kansas.
Kniffin, K. M., & Wilson, D. (2010). Evolutionary perspectives on workplace gossip: Why and how gossip can serve groups. *Group & Organization Management, 35*, 150–176.
Lindlof, T. R., & Taylor, B. C. (2011). *Qualitative communication research methods* (3rd ed.). Thousand Oaks, California: Sage.
Luna, A., & Chou, S. Y. (2013). Drivers for workplace gossip: An application of the theory of planned behavior. *Journal of Organizational Culture, Communications and Conflict, 17*, 115–129.
Michelson, G., van Iterson, A., & Waddington, K. (2010). Gossip in organizations: Contexts, consequences, and controversies. *Group & Organization Management, 35*, 371–390.
Miller, R., & Lefcourt, H. (1982). The assessment of social intimacy. *Journal of Personality Assessment, 46*, 514–518.

Norton, R. (1983). Measuring marital quality: A critical look at the dependent variable. *Journal of Marriage and Family, 45*(1), 141–151.

Ohanian, R. (1991). The impact of celebrity spokespersons' image on consumers' intention to purchase. *Journal of Advertising, 31*, 46–54.

Rosnow, R. L. (2001). Rumor and gossip in interpersonal interaction and beyond: A social exchange perspective. In R. M. Kowalski (Ed.). *Behaving badly: Aversive behaviors in interpersonal relationships* (pp. 203–232). Washington, DC: American Psychological Association.

Rosnow, R. L., Yost, J. H., & Esposito, J. L. (1986). Belief in rumor and likelihood of rumor transmission. *Language & Communication, 6*, 189–194.

Sparrowe, R. T., Liden, R. C., Wayne, S. J., & Kraimer, M. L. (2001). Social networks and the performance of individuals and groups. *Academy of Management Journal, 44*, 316–325.

Treem, J., & Leonardi, P. (2012). Social media use in organizations: Exploring the affordances of visibility, editability, persistence, and association. In C. T. Salmon (Ed.), *Communication yearbook 36* (pp. 143–189). Thousand Oaks, CA: Sage.

Wakefield, R. L., & Whitten, D. (2006). Examining user perceptions of third-party organization credibility and trust in an e-retailer. *Journal of Organizational and End User Computing, 18*, 1–19.

Waldeck, J. H., Seibold, D. R., & Flanagin, A. J. (2004). Organizational assimilation and technology use. *Communication Monographs, 71*, 161–183.

Weick, K. E. (1995). *Sensemaking in organizations.* Thousand Oaks, CA: Sage.

Weick, K. E., Sutcliffe, K. M., & Obstfeld, D. (2005). Organizing and the process of sensemaking. *Organization Science, 16*, 409–421.

Williams, E., & Connaughton, S. (2008). *Expressions of changing identifications: The nature of talk and resistance.* Paper presented at the annual meeting of 94th National Communication Association Conference, San Diego, CA.

Eighteen

"I Regret Nothing": Cyberbullying and Prosocial Influence in the First-Person Shooter Game

Kimberly L. Kulovitz and Edward A. Mabry

Bullying is aggressive, repeated, and intentional behavior toward an individual who cannot easily defend him- or herself, involves a power imbalance (Olweus, 2001), and is considered an "interpersonal activity that arises within the context of dyadic and group interaction" (Menesini, Melan, & Pignatti, 2000, p. 262). When bullying takes place in technologically mediated social contexts like online discussion groups and chat rooms it is typically referred to as *cyberbullying*; however, cyberbullying is far more than just bullying taking place online. In fact, recent studies (CDC, n.d.; Dooley, Pyzalski, & Cross, 2009; Kulovitz & Mabry, 2012; Spears, Slee, Owens, & Johnson, 2009) have found that cyberbullying has distinct (and potentially more damaging) characteristics that distinguish it from bullying that occurs face-to-face.

Face-to-face bullying is characterized by a power imbalance and is defined as aggressive, repeated, intentional behavior toward individuals who cannot easily defend themselves (Olweus, 2001) that can occur in dyadic and group interaction (Menesini et al., 2000). There are three main types of bullying, which include: physical (e.g., hitting, shoving), indirect or relational (e.g., third-party attacks, damage to reputation), and verbal (e.g., name-calling, teasing). Examples of bullying among middle school and high school students include verbal aggression, social exclusion/isolation, physical aggression, lies and false rumors, property damage, threats, and racial/sexual aggression (Menesini et al., 2000; Olweus, 2001), while the most common bullying behaviors in the workplace are social exclusion/isolation, rumor-mongering, and general domination (Hodson et al., 2006).

The Unique Nature of Cyberbullying

Only recently have studies begun to conceptually separate the act of bullying from the act of cyberbullying (see Dooley, Pyzalski, & Cross, 2009; Kulovitz & Mabry, 2012; Nocentini, Calmaestra, Schultze-Krumbholz, Scheithauer, Ortega, & Menesini, 2010; Pearce, Cross, Monks, Waters, & Falconer, 2011). Initial explorations into cyberbullying defined the phenomena narrowly as "bullying using an electronic medium" (Dooley et al., 2009, p. 182). The most comprehensive explanation of cyberbullying, and the one that is used as the conceptual framework for this research, is outlined by Nocentini et al. (2010), who identify cyberbullying as intentional, repetitive, power imbalanced, potentially anonymous, and public.

There is high potential for bullies to disguise or hide their identity online, something that is unique to information and communication technologies (ICT) or computer-mediated communication (CMC). ICT permits bullies to dispense aggression to large audiences (whether it is directed at one person or a group). For example, a bully that is targeting one individual can remain anonymous to the victim and bystanders by using a fake screen name and posting derogatory and false comments about the victim to a public forum, something that would be nearly impossible in face-to-face bullying.

Cybergroups and Cyberbullying

Bullying and cyberbullying implicates more than just the bully and the target, pressing researchers to look at bullying behavior as a group activity. Bystanders can involve themselves by intervening in the bullying, observing and doing nothing, or even being recruited by the bully to victimize (Dooley et al., 2009; Easton & Aberman, 2008; Kulovitz & Mabry, 2012); regardless, bystanders play an integral role in creating a group dynamic. Additionally, cyberbullying can also be present in already established virtual groups such as chatrooms, video games, or social media (see Kulovitz & Mabry, 2012). According to Easton & Aberman (2008), there is a need for self-preservation on the part of the bystander; regardless of whether they have empathy for the victim, they support the bully to save themselves. Kulovitz & Mabry (2012) also discovered that victims or targets of cyberbullying tended to be more sensitive to bullying when others were being bullied, but chose not to intervene.

Most applicable to understanding the context of video games as a group is that cyberbullying tends to be seen by players as punishment for poor performance. Shafter (2012) found increased hostility and enjoyment in games that position player against player (PvP, non-collaborative). These outcomes indicate that cyberbullying in online games may be used as a norming tactic to regulate the behavior of players in the game environment. If the players deviate from the standard pattern of interaction, they are punished, or coerced via cyberbullying, to adhere to the norm. The increase in aggression, and conversely enjoyment, may represent group cohesiveness in much the same way that hazing displays group cohesiveness when more sever forms of hazing are enacted; however, more research needs to be done on the potential connection between cyberbullying and group cohesiveness in online games.

Online Digital Video Games as Cybergroups

Historically, video games (meaning any digital form of gameplay) have received mostly negative attention in research and popular press, blamed for causing violence in adolescents, social isolation, and low empathy (Ivory, 2008). Conversely, contemporary research is finding that most video games have an "intrinsic social component" (Ivory, 2008, p. 363) that can be prosocial, interactive, responsive, and promote overall health and well-being (Jin, 2011; Maillot, Perrot, & Hartley, 2012). Video game players also become emotionally invested in the games they play as well as the online public spaces that surround the games; we enjoy the games, interact with other game players, and are challenged by the emotional impact of our experience.

Prosocial Gameplay

Prosocial video games (relatively nonaggressive collaboration-based games such as *Animal Farm* or *Super Mario Sunshine*) have been credited with helping people recover from stress, encouraging helping behavior, and increasing empathy (Greitemeyer & Oswald, 2010a; 2010b). Greitemeyer and Oswald (2010a) found that after playing a prosocial video game participants were more likely to help after a mishap and intervened when someone was being harassed, thus prosocial video games may mediate the relationship between cyberbullying and gameplay.

Video games also provide a space for shared identities, interests, and enjoyment, which have been found to promote overall health (Schott & Hodgetts, 2006). In a study examining the connection between exergames (games that promote physical movement such as those on the *Nintendo Wii* and *Xbox Connect*) and older adults, Maillot, Perrot, and Hartley (2012) discovered these types of games help improve player health. Older adults in the study increased their physical performance, and were emotionally and cognitively more productive.

Greitemeyer & Osswald (2010b) investigated whether prosocial games can reduce stress and lead to prosocial behavior. Greitemeyer & Osswald (2010b) conducted laboratory experiments where participants played video games and were subsequently tested with tasks and surveys. Prosocial video games affected the players' internal state and primes cognitive associative networks related to prosocial behavior. Essentially, playing prosocial video games triggers players to have more prosocial thoughts, which reduces stress. Players are also more likely to enact prosocial behavior after accessing their prosocial thoughts (Greitemeyer & Osswald, 2010b).

First-Person Shooters—*Left 4 Dead 2*

The first-person shooter (FPS) game is a type of video game that takes on a first-person perspective and often involves the use of weapons to support the player through the game (Hitchens, 2011). The FPS differs from other similar game types (such as first-person perspective role-playing games) because the focus of the FPS is on weaponry and tactical strategy and not on character or story development. In a comprehensive analysis of FPS games from 1991 to 2009, Hitchens (2011) determined that FPS avatars (the playable main character) across all consoles are mostly Caucasian males in the military (e.g., *Call of Duty: Black Ops*). While many FPS games are team-based and collaborative in nature, they are not considered prosocial video games. Unlike prosocial games, the FPS focuses on weaponry and fighting as the primary mode to advance the game.

Contrary to the prosocial genre, the FPS game provides players with realistic firearms and other weaponry that are used to develop often violent wartime narratives (see Hitchens, 2011). There are other video game types that contrast the prosocial genre (e.g., role-playing, multiplayer), but FPS games best capture the opposing elements while still maintaining some level of collaboration with other players. Prosocial games reduce player stress and emphasize well-being, whereas FPS games tend to foster rapid mental and physical responses, underscoring the anxiety of combat scenarios (Hitchens, 2011).

Left 4 Dead 2 (L4D2) both exemplifies the standard of the FPS genre and at the same time rejects the norm. Like most of the titles in the genre, L4D2 takes on a first-person perspective where the player sees the game environment through the eyes of the on-screen character and must use a range of weapons both melee (e.g., chainsaw, axe, strangulation) and projectile (e.g.,, shotgun, sniper rifle, spit) to make it through the levels. L4D2, takes place in Louisiana and Georgia during a zombie outbreak, it is rated "M" mature by the Entertainment Software Rating Board for blood and gore, intense violence, and language (ESRB, 2013), thus it does not fit into the prosocial type video game due to the aggressive content.

What sets L4D2 apart from other FPS games is the character environment and its "fiercely team-oriented style" (Onyett, 2009). The game takes place in the zombie infested cities and bayous of the deep South; there is no military support and the characters, just noncombatant civilians from diverse backgrounds. Unlike many of the FPS titles, L4D2 forces players to work cooperatively with other players to complete levels, tasks, and generally survive as a team. Many FPS are single-player games, where players compete individually against other players; however, L4D2 sets a new precedent for FPS games making team-based gameplay necessary if a player wants to complete the game segments. Exemplifying group collaborative play, "no other game emphasizes teamwork as strongly as this" (Onyett, 2009, para. 14); thus L4D2 provides an exceptional form of groups functioning in online video games.

Hypotheses

Game success is often measured by the satisfactory completion of a specific task within the game (e.g., effectively rescuing the princess or killing the end boss and completing the mission). In fact, there is a positive relationship between skillful game performance and the completion of game related tasks; such that game enjoyment is "massively threatened by insufficient performance" (Klimt, Hefner, Vorderer, & Roth, 2008, p. 10; see also Wirth, Fledberg, Schouten, van den Hoof, & Williams, 2012). Aggressive intra-group behaviors used to motivate better game performance is a possible consequence of players' game outcome expectations, especially since cyberbullying behavior within online games has been found to function as punishment for poor performance (Kulovitz & Mabry, 2012). The above implications of online team-based game play leads to the following hypotheses:

H1: Groups with lower amounts of cyberbullying will be more likely to experience game success.

H2: Groups with lower amounts of cyberbullying earlier in the game will be more likely to experience game success compared to groups with cyberbullying later in the game.

Methods

This experiment tested cyberbullying group behavior in the online video game *Left 4 Dead 2* (L4D2) focusing on game outcome, group cohesion, and leader influence. Using four participant observers (also referred to as confederates) to play and record L4D2 video game sessions and four coders to systematically analyze the recordings using a coding protocol, the cyberbullying behavior of L4D2 players was tested. An overview of the game L4D2, explanation of participant and confederate roles, and description of the measures, procedures, and experimental design follows.

The Game: *Left 4 Dead 2*

Left 4 Dead 2 was chosen for this research design since the mechanics of game programming help control for in-game variables that would otherwise be exceedingly difficult to account for. L4D2 is comprised of four-person teams, of which the characters cannot be leveled up or customized in anyway. Additionally, weapons cannot be upgraded and there is a set amount of time that players have to complete in-game campaigns, modes, or for specific tasks to be completed with a set end goal determining either task completion or task failure.

These L4D2 game attributes are important because they standardize the in-game resources that players have access to while also systematizing the style of game that players experience. In L4D2 player options are predetermined, thus each player experiences the same game flow. This allows the game outcome and experience to be determined by player involvement and skill rather than character role or game narrative.

Participants

Confederates/Participant Observers

Four confederates, also referred to as participant observers, were asked to play and record the gameplay as well as observe the behavior of the other three players in the four-person team. Using confederates was necessary because gameplay could only be recorded from a first-person perspective as the game was being played. Both ScreenFlow and FRAPS, the software used for recording (see procedure section for description), are only capable of capturing what is occurring on the player's screen, thus the confederates were used to play and record game sessions (see authors for additional confederate information). After finishing each game session, (approximately 40 minutes), confederates completed a 19-item survey (see Kulovitz, 2013). This procedure is consistent with participant observation research in online game communities (see Siitonen, 2011).

Non-Confederate Participants

Non-confederate participants were the three other players that had their gameplay recorded by each confederate. Non-confederate participants were not informed they were participating in research, did not fill out surveys and did not participate in the research process in any way other than having their gameplay recorded (waiver obtained from Institutional Review Board). Participation was random and anonymous based entirely on self-selection into the server and game that the participant observers occupied. After logging into L4D2 and selecting the game type (e.g., versus, campaign, etc.) players signal their readiness to play with human team members by clicking "play online" which then places them into the first available server/game.

Measures

Coders and Coding Protocol

Four undergraduate students were recruited to analyze the recorded L4D2 game sessions for which they received three credits in a Research Practicum course. The coding scheme used in this research (see Kulovitz, 2013) contained 18 codes in four categories (overt bullying, covert bullying, contextualizing categories, and contextualizing categories specific to L4D2). Only part of the total coding protocol was used for the present study.

A reliability analysis using Cohen's Kappa (Cohen, 1960) was conducted to determine inter-coder reliability. Approximately 20% of the 400 coded messages were randomly sampled and independently coded by the researchers. The inter-coder reliability was found to be Kappa = .917, indicating a strong agreement; however, code 2-threatening and code 9-slander were not represented in the random sample due to low incidences of these behaviors. As an alternative to rerunning Cohen's Kappa, all messages containing code 2-threatening and code 9-slander were deliberately sampled and coded by the researchers. The inter-coder reliability was found to be a 100% overlap.

Questionnaires

For each game session, confederates and coders were asked to fill out a survey that captured their observations. Confederates and coders were asked to identify who they perceived as group leader, if they perceived the group as cohesive overall, and asked how much influence the group leader (if any) had over the group. The perceived cohesion scale from Geidner (2012) was used (see Kulovitz, 2013) for both the confederates and the coders. This six-item scale used a 5-point Likert-type scale (1 = none; 5 = an extreme amount) to indicate the level of cohesion experienced and observed in the L4D2 game recordings. The perceive cohesion scale has an $\alpha = .94$ and includes questions such as "did you feel/observe that this group was the best of its type?" and "Did you feel/observe a sense of belonging to this group?"

Procedures

Recording Game Process

Data was collected using an online survey for confederates and coders (see Appendix A) and video capturing software, which was operated by the four participant observers. In-game participants playing *Left 4 Dead 2* had their gameplay recorded using FRAPS or ScreenFlow, video capturing software. FRAPS is a combination benchmarking and real-time video capture software designed specifically for online video game recording. FRAPS is a Windows-based application that can record games using DirectX or OpenGL graphics and captures both audio and video with custom frame rates (7680 x 4800 and 1 to 120 frames per second). The FRAPS software can be obtained for a nominal fee per license for unlimited video recording directly to .jpg, .png, and .tga formats, which then can be easily transferred to other software applications for data analysis and coding. ScreenFlow is the Mac equivalent of FRAPS produced by Telestream, also available for a fee-per-license basis and can record directly to Mpeg4 as well as .jpg, .png, and .tga formats.

Because FRAPS and ScreenFlow require the person recording the gameplay to be a participant in the game, four participant observers (also referred to as confederates) were recruited to play *Left 4 Dead 2* and use FRAPS or ScreenFlow to record their gameplay. Confederates were provided with the FRAPS or ScreenFlow software, the *Left 4 Dead 2* video game, a one-terabyte external hard drive, and a detailed set of instructions. The confederates were asked to return the external hard drive, but were welcome to keep the FRAPS or ScreenFlow software and the *Left 4 Dead 2* video game. Confederates were required to have played *Left 4 Dead 2* within the past year, have a PC or Mac that meets the minimum system requirements of *Left 4 Dead 2* (Windows 7/Vista/XP operating system; 4.3GHz CPU; 2GB RAM; ATIx800/nVidia 6600 graphics card; DirectX9c sound card), and had their skill level (i.e., in-game performance level) tested and cataloged, ensuring their skill level was average. Confederates also attended a training session on how to install *Left 4 Dead 2*, how to operate FRAPS or ScreenFlow, and how to operate the external hard drives.

Survey Procedures

The online survey for confederates and coders took approximately ten minutes and was completed for each of the 41 recorded/coded game sessions. Prior to completing the surveys confederates and coders read an informed consent disclaimer form/page. While participation was voluntary and confederates and coders could choose at any time not to fill out a survey, involvement was not anonymous. In order to track the completion scores, participant screen names, and other game-related data, it was necessary to identify the game sessions by the confederate and coder assigned to each.

Coder Training

Following Meyers and Seibold's (2012) process on coder training, a face-to-face training session took place with all coders and researchers present, which lasted approximately an hour. During the training the coders each received a 32 GB USB flash drive, the codebook (see Kulovitz, 2013), coding instructions, Excel file with codesheet for data recording and a test file of a previously recorded L4D2 game (not included in analysis). After reviewing the instructions, codebook, and Excel file, and answering any questions, all coders, and researchers independently coded the test file.

After independent coding of the test file by the coders and researchers another face-to-face meeting was held to discuss the reliability of the code choices. Similar to the experiences of Meyers and Seibold (2012) reliabilities were initially very low (approximately 35% agreement between code categories), thus changes were discussed between all coders and researchers and modifications were made to clarify codes and reduce redundancy. Coders and researchers independently coded a different test file using the modified coding scheme and gathered for another face-to-face meeting. Reliabilities using the altered coding scheme were high (approximately 95%), thus no additional changes were made. Each coder was then provided with the finalized coding scheme and a quarter of the recorded game sessions to begin the coding process, which took approximately six weeks.

Research Design

The hypotheses advanced for this study are based on game outcome and relative amounts of cyberbullying behavior observed among team members engaged in playing an online computer-based video game. Group-level independent variables for this study are game success (successful vs. unsuccessful) and relative level of observed in-game cyberbullying behavior (high vs. low).

Research Variables

Game Performance

Game performance was measured based on the successful completion of a task within the game. Advancement in *Left for Dead 2* is based on the successful completion of a series of campaigns or goals, which framed the task completion in the experimental design. Each game lasts approximately twenty to sixty minutes and is initiated when a group of four players indicate their readiness to play the game. The game ends either when the designated goal is reached (predetermined by the game) or when all four players are "dead" and the campaign has failed. Successful campaign completion will indicate task success, while unsuccessful completion of the campaign will indicate a failed task. A team outcome was indicated by the number of team members that survived to finish the game session. Five outcomes were possible: (1) no group members finished, indicating a failed game; (2) only one group member finished a game; (3) two group members finished a game; (4) three group members finished a game; and (5) all group members finished the game. Thus, game outcome could be analyzed either as an interval scale or five-category nominal scale.

Cyberbullying

Cyberbullying was measured using the coding scheme developed by Kulovitz and Mabry (2012) (see Kulovitz, 2013) described above. Cyberbullying was coded into target behavior and bully behavior and was then divided further into high amounts of cyberbullying behavior versus low amounts of cyberbullying behavior compared to all groups included for analysis in the study. Designation of bully and target roles was based on the relative amounts of bullying behavior initiated or received by each member during gameplay as observed by confederates and coders.

Results

Initial analyses assessing scale development, coding reliability and descriptive information about participants and game content are reported first. The remaining results are organized by hypotheses.

Descriptive Statistics

The descriptive statistics reported include general information on the number of groups and messages, message codes, game measures, and cohesion scale frequencies. The total number of groups recorded was 59 with 1,425 messages. After deleting coded messages that contained game statistics, player scores, and other game-generated communication, the effective sample size was comprised of 41 groups which contained a total of 415 messages. Out of the 41 groups 25 groups contained bullying behavior and 16 groups had no bullying behavior. Out of the 415 messages, 30 were coded as insults (7.2%), followed by exclusion at 28 (6.7%). The least used code was extortion, used 1 time (.2%). Table 18.1 shows the complete breakdown of code frequencies.

Table 18.1. Frequency Distribution of Message Codes.

Code	Frequency	Percentage
Harrassment (code 1)	8	1.9
Threatening (code 2)	5	1.2
Insults (code 3)	30	7.2
Initiating Conflict (code 4)	14	3.4
Disrupting Play (code 5)	6	1.4
Silence (Ignore) (code 6)	7	1.7
Extortion (code 7)	1	.2
Teasing (code 8)	13	3.1
Slander (code 9)	7	1.7
Exclusion (code 10)	28	6.7
Apologies (code 11)	7	1.7
Resisting Bullying (code 12)	17	4.1
Positive Task Reactions (code 13)	19	4.6
Positive Social-Emotional Reactions (code 14)	11	27
Helping Behavior (code 15)	241	58.1
Uncodable (code 18)	1	.2
Total	415	

Game outcome measured the relative team success or failure in the game. A team as indicated by the number of team members that survived to finish the game session. In 13 groups (31.7%) no group members finished, indicating a failed game. All group members finished in 13 groups (31.7%), indicating a successful game. Additionally, only one group member finished in 6 groups (14.6%), two group members finished in 4 groups (9.8%), and three group members finished in 5 groups (12.2%).

Hypothesis 1 and hypothesis 2 evaluated the amount of cyberbullying in each game session (high amounts vs. low amounts) and its effect on game success. The amount of cyberbullying in each game session was determined by calculating the distribution of cyberbullying messages across groups and then calculating a high-low cutoff point for only the groups that contained cyberbullying. Out of the 41 groups 25 groups had cyberbullying behavior present and 16 groups had prosocial behavior. Out of the 25 groups that contained cyberbullying behavior 13 groups (52%) had high levels of cyberbullying (3 or more occurrences) and 12 groups (48%) had low levels of cyberbullying (1 or 2 occurrences).

Hypothesis 1

Hypothesis 1 predicted that groups with lower amounts of cyberbullying will be more likely to experience game success. A chi-square test on game outcome results was performed. No relationship was found between high/low cyberbullying and game outcome, $\chi^2 = 7.37$, $df = 4$, $p = .117$; however, the Likelihood Ratio Chi-square test was significant $\chi^2 = 9.70$, $df = 4$, $p = .046$. A Likelihood Ratio test to report chi-square statistics is more effective when cell frequencies are below .5 (Ozdemir & Eyduran, 2005), which is the case for testing H1. Thus, these results suggest that there is a modestly strong relationship between cyberbullying and game outcome, such that high cyberbullying contributes to a successful game outcome.

Since H1 only tested groups that contained cyberbullying behavior and its effect on game outcome, *t*-tests were performed to test how groups with observed cyberbullying messages ($n = 25$) compared to groups with no observed cyberbullying messages ($n = 16$). There was a significant difference in game success, $t(39) = -2.25$, p = .030 for prosocial-only groups ($M = 2.69$, $SD = 1.81$) compared to cyberbullying-only groups ($M = 1.52$, $SD = 1.47$). Groups with prosocial messages were more successful than groups with cyberbullying messages.

Hypothesis 2

Hypothesis 2 predicted groups with lower amounts of cyberbullying earlier in the game will be more likely to experience game success compared to groups with more cyberbullying later in the game session. To determine early game versus late game midpoint, the total time of each game session was divided at the midpoint and the cyberbullying messages on each side of the midpoint split were tallied. Groups with no cyberbullying were labeled "prosocial" groups. Out of the 41 groups 14 (34.1%) had cyberbullying occurring early in the game, 11 (26.8%) had cyberbullying occurring late in the game, and 16 (39.1%) were prosocial groups.

There was no significant difference between groups, $t(23) = 1.61$, $p = .120$, with higher amounts of cyberbullying earlier in the game ($M = 1.93$, $SD = 1.54$) compared to groups with more cyberbullying later in the game ($M = 1.00$, $SD = 1.26$) on game success. Cyberbullying occurring either early or late in the game has no effect on game success although game outcomes are lower on the late side rather than the early side of play, suggesting that bullying earlier in the game leads to better team member survival.

Since H2 only tested groups that contained cyberbullying behavior and its effect on game outcome, a univariate analysis of variance (ANOVA) was performed to determine whether game outcome was affected by groups with cyberbullying occurring early in the game (n = 14) or groups occurring late in the game (n = 11) compared to groups characterized by only using prosocial behaviors (n = 16). The ANOVA analysis was significant, $F(2, 38) = 3.66$, $p = .035$. Subsequent analyses using the Scheffe post hoc criterion for significance indicated that prosocial groups ($M = 2.69$, $SD = 1.81$) were significantly different from groups where cyberbullying occurred late in the game ($M = 1.00$, $SD = 1.265$), but not significantly different from groups where cyberbullying occurred early in the game ($M = 1.93$, $SD = 1.54$).

Discussion

The most surprising finding overall is the presence of both cyberbullying and prosocial behavior within the same game sessions. While prosocial behavior by far has a larger impact on game outcome, cyberbullying was still found to be present. While further exploration is necessary, the type of game and intention of gameplay may be an explanation for the co-presence of both cyberbullying and prosocial behavior.

Since video games provide a space for shared interest and enjoyment (Schott & Hodgetts, 2006) and the FPS genre is marked by relative fast-paced combat gameplay (Hitchens, 2011), players may not want to ruin the gaming experiences with aggressive behavior such as cyberbullying. Additionally, the FPS game environment becomes extremely fast-paced and requires players to react and make decisions quickly. L4D2 is no exception to this aspect of the genre, compelling players to adapt and react rapidly to opponents in the game. The pace of L4D4 may have suppressed cyberbullying, leaving prosocial behavior as the easier and more efficient option to choose for motivating other players.

This study has a few limitations to consider. The choice of game genre, pace of the game, small sample sizes, and inability to survey the player participations require articulation. The investigation's limitations did not have a large impact on the results and analysis of this study, but should be taken into consideration when interpreting the hypotheses.

The FPS game chosen over other game types (e.g., massively multiplayer online role-playing games) provides the best opportunity to record the game sessions with very little ability for customization. While using L4D2 provided rich data, the fast-paced environment may not have provided optimal behavior episodes for studying cyberbullying. The speed of decision making and action within L4D2 may have truncated conversations that contained cyberbullying. Similarly, the lack of customization, while working in favor of controlling game experiences, did not allow for variety and may have inadvertently stifled interaction that may have contained cyberbullying.

Both larger sample sizes and participant perspectives would have provided more robust data and extensive measures were undertaken to acquire both. Players were initially approached in-game to provide their perspectives and were offered the chance to win a $50 Visa gift card. When no players participated, they were offered $10 just for filling out the survey, yet players still chose not to participate. There were no foreseeable alternatives or venues to get player perspective, thus confederates and coders were used to obtain selected in-game data necessary to analyze the hypotheses.

There also is the question as to whether or not the confederates (also referred to as participant observers) had an effect on gameplay. All confederates had previously played L4D2, were familiar with the game environment, and rated themselves as average skill-level players. Additionally, perceptions and observations of the participant observers (using a group cohesion rating scale and leadership assessment items) were rated by the coders after they viewed the recorded game sessions, which is consistent with reliability checks for group observation (see Wirth et al., 2012). These analyses did not reveal patterns of play by confederates out of character relative to that of non-confederate participates.

Concluding Remarks

Social behavior in online video games and the issue of cyberbullying seem to indicate that prosocial behavior is far more effective at motivating players to perform well within the game, in addition to fostering group cohesion. Nevertheless, cyberbullying behavior was present alongside prosocial behavior, which warrants future investigation to analyze impact and function. One benefit of conducting this study was acquiring in-game group behavioral data in addition to better understanding the appeal of playing online games.

References

Centers for Disease Control and Prevention. (n.d.). Measuring bullying victimization, perpetration and bystander experiences: A compendium of assessment tools. Retrieved from http://www.cdc.gov/violenceprevention/pub/measuring_bullying.html

Cohen, J. (1960). A coefficient of agreement for nominal scales. *Educational and Psychological Measurement, 20,* 37–46.

Dooley, J. J., Pyzalski, J., & Cross, D. (2009). Cyberbullying versus face-to-face bullying: A theoretical and conceptual review. *Journal of Psychology, 217,* 182–188.

Easton, S. S., & Aberman, A. (2008). Bullying as a group communication process: Messages created and interpreted by bystanders. *The Florida Communication Journal, 36,* 46–73.

Entertainment Software Rating Board (ESRB). (2013). Retreived from http://www.esrb.org/ratings/search.jsp

Geidner, N. (2012). Perceived cohesion and individual-level voluntary group participation. Paper presented at the annual meeting of the International Communication Association, Phoenix, AZ. May, 25, 2012.

Greitemeyer, T., & Osswald, S. (2010a). Playing prosocial video games increases empathy and decreases schadenfreude. *Emotion, 10,* 796–802.

Greitemeyer, T., & Osswald, S. (2010b). Effects of prosocial video games on prosocial behavior. *Journal of Personality and Social Psychology, 98,* 211–221.

Hitchens, M. (2011). A survey of first-person shooters and their avatars. *Game Studies, 11(3).*

Hodsen, R., Roscigno, V. J., & Lopez, S. H. (2006). Chaos and the abuse of power: Workplace bullying in organizational and interactional context. *Work and Occupations, 23,* 382–416.

Ivory, J. (2008). The games, they are a-changin': Technological advancements in video games and implications for effects on youth. In D. E. Jamieson & D. Romer (Eds.), *The changing portrayal of adolescents in the media since 1950* (pp. 347–376). New York, NY: Oxford University Press.

Jin, S. A. (2011). "I feel present. Therefore I experience flow:" A structural equation modeling approach to flow and presence in video games. *Journal of Broadcasting & Electronic Media, 55,* 114–136. doi: 10.1080/088.38151.2011.546248

Klimt, C., Hefner, D., Vorderer, P., & Roth, C. (2008). Exploring the complex relationships between player performance, self-esteem processes, and video game enjoyment. Paper presented at the annual meeting of the *International Communication Association,* Montreal, Quebec, Canada.

Kulovitz K.L. (2013). *Cyberbullying in "Left 4 Dead 2": A study in collaborative play.* Dissertation. Retrieved from http://dc.uwm.edu/etd/363 on October 30, 2015.

Kulovitz, K. L., & Mabry, E. A. (2012). Cyberbullying: Perceptions of bullies and victims. In L. A. Wankel & C. Wankel (Eds.), *Misbehavior online in higher education* (pp. 105–126). London: Emerald.

Maillot, P., Perrot, A., & Hartley, A. (2012). Effects of interactive physical-activity video-game training on physical and cognitive function in older adults. *Psychology & Aging, 3,* 589–600. doi: 10.1037/a0026268

Menesini, E., Melan E., & Pignatti, B. (2000). Interactional styles of bullies and victims observed in a competitive and cooperative setting. *The Journal of Genetic Psychology, 161,* 261–281.

Meyers, R. A., & Seibold, D. R. (2012). Coding group interaction. In A. B. Hollingshead & M. S. Poole (Eds.) *Research methods for studying groups and teams: A guide to approaches, tools and technologies* (pp. 329–357). New York, NY: Routledge.

Nocentini, A., Calmaestra, J., Schultze-Krumbholz, A., Scheithauer, H., Ortega, R., & Menesini, E. (2010). Cyberbulling: Labels, behaviors and definition in three European countries. Australian *Journal of Guidance and Counseling, 20,* 129–142.

Olweus, D. (2001). Peer Harassment: A critical analysis and some important issues. In J. Juvonen & S. Graham (Eds.) *Peer harassment in school* (pp. 3–20). New York, NY: Guilford Publications.

Onyett, C. (2009, November). *Left 4 Dead 2* Review: A bloodier, more entertaining, and more complete version of Valve's brutal co-operative shooter. Retrieved from http://www.ign.com/articles/2009/11/17/left-4-dead-2-review

Ozdemir, T., & Eyduran, E. (2005). Comparison of chi-square and likelihood ratio chi-square tests: Power of test. *Journal of Applied Sciences Research 1*, 242–244.

Pearce, N., Cross, D., Monks, H., Waters, S., & Falconer, S. (2011). Current evidence of bets practices in whole-school bullying intervention and its potential to inform cyberbullying interventions. *Australian Journal of Guidance and Counseling, 21*, 1–21.

Schott, G., & Hodgetts, D. (2006). Health and digital gaming: The benefits of a community of practice. *Journal of Health Psychology*, 11, 309–316.

Shafer, D. M. (2012). Causes of state hostility and enjoyment in player versus player and player versus environment video games. *Journal of Communication, 62*, 719–737.

Siitonen, M. (2011). Participant observation in online multiplayer communities. In B. K. Daniel (Ed.), *Handbook of research on methods and techniques for studying virtual communities: Paradigms and phenomena* (pp. 555–567). New York, NY: Information Science Reference.

Spears, B., Slee, P., Owens, L., & Johnson, B. (2009). Behind the scenes and screens: Insights into the human dimension of covert and cyberbullying. *Journal of Psychology, 217*, 189–196.

Wirth, J. H., Feldberg, F., Schouten, A., van den Hoof, B., & Williams, K. D. (2012). Using virtual game environments to study group behavior. In A. B. Hollingshead & M. S. Poole (Eds.), *Research methods for studying group and teams: A guide to approaches, tools, and technologies* (pp. 173–198). New York, NY: Routledge.

NINETEEN

Past Abuse, Cyberstalking, and Help-Seeking Behavior

Elaine L. Davies

Rebecca was a freshman in college when she met Chuck at an on-campus fraternity party. Compared to the boys she had known in high school, he seemed worldly and often bragged of his family's wealth and connections. The two quickly became romantically involved, despite disapproval from her friends. Their one-year intimate relationship was characterized by frequent break-ups, as well as incidents of verbal, emotional, and sexual abuse. She recalled how Chuck would often compare her to his past girlfriends to erode her self-esteem. According to Rebecca, "I felt like it was a reminder that I could be replaced, so I tried so hard to hold on to him." Later, she would add "when I didn't want to do it, he would try to make me feel bad until I gave in or he would just do it … then he would stop calling me, like to punish me … and I'd beg him to take me back." Yet, when summer break came, Rebecca went to her hometown, met a new suitor, and tried to put Chuck behind her. Unfortunately, Chuck was not ready to end his pursuit of Rebecca. Following the final termination of their relationship, Chuck relied on the use of technological devices to stalk and harass Rebecca, as well as to publicly embarrass her. According to Rebecca, he would text and call her with alarming frequency and intensity, use her friend's social media accounts to track her movements, and reveal once-private information about their relationship to a mass audience.

Rebecca's experience is a newer form of interpersonal violence known as cyberstalking. Cyberstalking involves action characterized by the repetitive use of electronic devices to stalk another person that requires a pattern of threatening or malicious behaviors and involves a credible threat (NCLS, 2013; US Attorney General Report, 1999). Thus, to be in violation of cyberstalking laws, the aggressor must make repeated, specific, intentional, and fear-producing threats that lead one to believe an illegal act or an injury will be inflicted on the recipient or his/her family or household.

According to the 2010 Intimate Partner and Sexual Violence Survey, 1 in 6 women and 1 in 19 males will be stalked during their lifetimes (Black et al., 2011). The data within this study reveal the crimes of stalking (both non-cyber, also known as traditional stalking, and cyberstalking) are on the rise. For example, in a 2006 national survey, 5.3 million Americans reported they had been stalked within the last 12 months, but this number rose to 7.1 million in 2010 (Baum, Catalano, & Rand, 2009; Black et al., 2011). Moreover, the number of victims who reported unwanted phone calls, messages, and texts jumped from 66.7% to over 75% during that same time period.

In addition, the extant literature has shown cyberstalking is perpetrated predominately by individuals known to the victim, such as former intimate partners and friends (e.g., Black et al., 2011; Finn, 2004; Mohandie, Meloy, McGowan, & Williams, 2006). Yet, there is a dearth of empirical and theoretical investigation into the precursors of the cyberstalking experience of past intimates. This limited inquiry also has hampered our ability to develop specific, successful coping mechanisms for individuals who have been cyberstalked by a past intimate.

This chapter addresses these concerns by offering an analysis of cyberstalking victims' lived experiences through the lens of Communication Privacy Management (CPM) theory. Specifically, privacy rules central to CPM theory are used to examine how abuse enacted during an intimate relationship helped to guide victims' decisions to close their privacy boundaries with their former partner. Second, CPM will be used to demonstrate how readjustments to victims' privacy boundaries with others in order to receive help from outside others may not always lead to successful outcomes.

Communication Privacy Management Theory

Communication Privacy Management (CPM) theory provides a conceptual framework to illuminate how individuals and groups coordinate the management of private and sensitive information through established boundaries (Petronio, 2002, 2007). The theory predicts communicators will face tensions as they decide the degree of openness or closedness of their privacy boundaries during disclosure. Because all disclosures have both potential benefits and risks, Petronio (2002) argues individuals, pairs, and groups need to develop privacy rules in order to handle any private and co-owned information.

Privacy Rules

Petronio (2002) identified five criteria (culture, gender, context, motivation, and risk-benefit ratio) that guide decisions to reveal/conceal private information, as well as decisions to maintain or to close existing boundaries. Although all criteria are important in the development of privacy rules, this current study focuses on two criteria—*context* and the *risk-benefit ratio.*

The context and risk-benefit criteria provide useful frameworks to demonstrate why individuals who have been involved in abusive relationships may need to readjust their privacy boundaries following relational termination. The *context criterion* is activated when a situation requires individuals to reassess their privacy boundaries. In other words, when individuals face awkward or uncomfortable situations within their environment, they rely on contextual cues in order to evaluate whether or not changes in privacy boundaries are warranted. For example, if a person has had an encounter with a former partner in which threats were made, the victim may choose to close his/her privacy boundaries to the ex-intimate. Similarly, the *risk-benefit criterion* also is used to create or modify privacy rules based on the perceptions of a situation. Rules based on this principle are formulated after people have weighed the dangers and rewards of disclosure or further communication. Because disclosures often have the potential to put people in vulnerable states, individuals must be cognizant of both the rewards and costs of their communication. For example, if individuals believe that they can maintain open privacy boundaries in order to remain friends with their ex-partner, the benefit would be continued social support. On the other hand, the risk may be continued humiliation and denigration.

The two criteria presented here, context and the risk-benefit ratio, provide possible reasons individuals may opt to change or close their privacy boundaries with former partners. Although other reasons are certainly possible, in the next section, I present the results of a previous study to describe how past abuses committed by a relational partner serve as the impetus for these boundary modifications.

Past Abuses Related to Privacy Boundary Modifications

The traditional stalking by past intimates, as well as abusive relational history with a former partner as a precursor to stalking, have been well documented. As stated previously, past relational partners are at the highest risk for traditional stalking (Black et al., 2011). Moreover, research consistently shows stalking often begins during the relationship and escalates following termination (e.g., Logan, Shannon, Cole, & Walker, 2006; Mullen, Pathé, & Purcell, 2000; Spitzberg, 2002). Last, researchers have found strong evidence that abuse enacted during an intimate relationship increases one's propensity for stalking victimization (e.g., Dye & Davis, 2003; Mechanic, Uhlmansiek, Weaver, & Resnick, 2000). Although many of the studies listed above include cyberstalking behaviors in their analyses, no studies have examined directly how victims describe past abuse as a reason to modify how they communicate with their former partner prior to the onset of cyberstalking.

The Cyberstalking Study

The current investigation draws data from a larger study in order to understand the holistic, lived experiences of cyberstalking victims (Davies, 2013). A total of 27 participants engaged in semi-structured interviews to examine the participants' relationships before, during, and after the onset of their cyberstalking experience. The average relationship length was 5.9 years and the average length of time since separation from the aggressor was 1.2 years. Within this sample, 25 reported abuse(s) were enacted during their relationship with the aggressor. Data were analyzed through the use of van Manen's (1990) hermeneutic phenomenology method in order to capture the thick, rich detailed description of the participants' experiences.

Physical abuse. Physical abuse has been conceptualized as an act carried out in order to cause another person to experience physical pain or injury (Straus & Gelles, 1992). This type of abuse has been operationalized as having been hit, slapped, kicked, punched, or beaten.

Past research has shown clear associations between physical abuse and traditional stalking. Studies that rely on self-reports by both aggressors (Burgess et al., 1997; Mohandie, Meloy, McGowan, & Williams, 2006) and victims (Tjaden & Thoennes, 1998, 2000) have found physical violence enacted during an intimate relationship may serve as a warning sign of future stalking.

Within this study, several participants ($N = 8$) reported physical abuse during their relationships. The following exemplars demonstrate how the individuals weighed the risks and benefits of future interaction with their aggressor following incidents of physical assaults.[1]

During Sara's 8-year relationship with Caroline, physical abuse was normative. According to Sara, Caroline's mood swings and violent outbursts had reached an "epic level" after her partner, Caroline, believed she was flirting with a mutual friend. Sara explained:

> *I tried to explain that he's uh, uh, a married man, but she wouldn't hear it, she had it in her head and that's the way it always was, she gets it in her head and I just had to let her rant, but not this time. I was tired of always being her, her, punching bag, both, uh, fig, fig, figuratively and literally.... So she starts throwing things, that's nothing new... so, so, so, I just left as she's throwing my stuff around, but not before she hit me with my laptop. ... I knew I wasn't gonna go back ... but, we had been through so much together and I wanted to keep her in my life. I really wanted us to get married and live that fairytale that everybody lives for.*

Sara's narrative reveals a consistent pattern of physical abuse. Deeper analysis of the incident demonstrates that although the contextual event led Sarah to re-evaluate their relationship, she also struggled with her desire to maintain communication with Caroline.

The persistent physical abuse was also stated by Lucas in his narrative. In his situation, Lucas felt limited contact with his soon-to-be-ex-wife, Mindy, was appropriate following their 5-year relationship, however his fear for Amber, their 4-year-old daughter, required that he carefully weigh his options. In this excerpt, Lucas explained how an incident after he returned from his two-week Army Reserve training led to the final dissolution of their relationship.

> *I came home and she was already stinkin' drunk ... so I asked her where Amber was and she starts in with the whole "you don't love me and I don't love you so just get out but Amber stays with me" thing. I searched the whole house looking for Amber [while] Mindy is at my heels barking at me, slapping my head, just acting a fool. ... So, uh, I said "I'm outta here but trust me no judge will give you Amber" and what does she do? She picks up my softball bat and starts swinging. I had to lock myself in the bathroom to call the police. That's when I knew I had to just stop being with her. ... But yeah, I knew I'd never be rid of her cuz of Amber. ... It was hard cuz I needed her to take care of my baby but she's just outta her mind sometimes and does stupid shit when she's juiced. ... I just wanted some peace and not worry about gettin' beat every time she was bein' all crazy like that.*

Although Lucas and Mindy had been involved in prior physical altercations before, Mindy's use of a weapon and his fear for their daughter's safety were the contextual catalysts behind his resolution to end their relationship.

In both of the situations above, physical abuse served as the driving force for the victims to abandon their relationships and a desire to close any existing privacy boundaries with their aggressor. However, victims in this category struggled with their decision to completely sever all communication. Each participant declared the benefits of continued communication would be to live in an intact family unit, but the risks involved further physical abuse. As a result, when victims decided to initially keep their privacy boundaries porous, these decisions would later provide their aggressors the ability to cyberstalk their past intimates.

Psychological and emotional abuse. Aggressors often use psychological abuse and emotional abuse simultaneously to torment their victims. According to Murphy and Cascardi (1999), the purpose of psychological abuse is "to produce emotional harm ... directed at the target's sense of self" (p. 209). Psychological abusive behaviors include: isolating romantic partners from others; using recurring criticism, threats, verbal aggression, denigration of the victim's self-esteem through control, character assassinations, and humiliation (Roberts, 2005). Similarly, emotional abuse is characterized by expressions of jealousy, anger, and verbal abuse aimed at the victim (Simonelli & Ingram, 1998). The extant research has shown control and jealousy are strong predictors of traditional stalking (e.g., Davis, Ace, & Andra, 2000; Spitzberg, 2002).

The results of the current study reveal many of the behaviors listed above were present in the victims' recollections. Victims frequently recalled incidents of incessant criticism and derogation, yet described jealousy as the motivating factor for relational termination and contributed to their decision to close their privacy boundaries. In the first exemplar, Hailey explains Marco's sharp criticisms and threats were often brought on by his resentment of her friendly demeanor.

> *He always said I paid too much attention to other people … on our one month anniversary he thought I was flirting with our girl waitress. Oh, come on, … he harped on that for days, he just went on and on about how I was a filthy lesbo deep down inside and if I kept on like that everyone would know it too. … So, I learned real quick not to talk to anyone. I had no friends because Marco said all my old friends were fake … horrible, slutty people. Eventually, I ended up taking in a stray cat so I would have her to talk to, but he was always so mean to her, kicking her, spraying her with water, calling her names. … But he took Pippsie away because I forgot to feed her before I left for work. He said it was my fault and that she was dead and I didn't deserve her. I just fell apart and called my mom. She came over and told me if he can do that to an innocent animal, there's no telling what he'd do to me. Mom convinced me to come home with her for one night … so he could cool off. … I mean I missed him a lot, but that night was horrible. He must've texted 100 times that night, each worse than the last. He even said that if I didn't come back right now I was gonna end up like Pippsie and no one would ever find me.*

Hailey's situation demonstrates how Marco isolated her from her friends, criticized her behavior through denigration, and used intimidation as a means to control her, but the mysterious disappearance of her cat served as the context to end the relationship. Despite his final threat to inflict physical harm her, she felt the benefit to communication would be to ease his mind. Yet, she also sensed he would use the incident at a future date to inflict more psychological harm.

Shana also explained that during her 5-year relationship with Andrew, his constant denigration of her was done both at home and in public. Shana recollected an incident at her company Christmas party led to their breakup. However, Andrew's threats after she had moved out were the impetus for her to close her privacy boundaries.

> *He was so jealous of my boss. I think he wanted his money, or car, or power, I don't know, but that night, he (Andrew) went right up and asked him (her boss) flat out if I had, um, I had slept with him. … (The boss) was so confused … then Andrew said "if you want this lazy, fat cow, you can have her." I almost prayed he (Andrew) would kill me so I wouldn't die of embarrassment. After that night, I couldn't go back to work … it just felt so weird, so I got fired. I really liked that job, but it was just too weird. Then after weeks of applying for jobs and Andrew constantly nagging me about my make-up and clothes, like he'd say "you're wearing that?"… He would post pictures of me on social media and ask his friends to comment about how, uh, I guess, inappropriate I looked … finally, I left.*

Following the breakup, Andrew's communication with her fluctuated between solicitous and hostile. Shana explained that at first, he seemed pleased that she had moved in with a friend and found a new job; however, when he made a veiled threat of sexual assault, Shana decided to permanently seal her privacy boundaries. She states, "*I was so confused until he went after my roommate saying she was a whore just like me and that he wouldn't be surprised if someone broke into our house and raped both of us.*"

The exemplars in this section illustrate how aggressors engage in psychological and emotional abuse through the use of expressed jealousy, incessant criticism, and denigration during the relationship. These acts were communicated face-to-face, as well as through the use of technology. Although these negative behaviors were described as extremely hurtful by the participants, the threats towards others served as the catalysts for victims to finally close their privacy boundaries with their ex-partners.

The initial struggles to disengage from the relationship reported by Hailey and Shana were not uncommon. Similar to the victims who described physical abuse, the participants in this section discussed their initial choice to maintain communication with their past intimates following relational termination. Yet, it is important to note here that all victims who reported any of three types of abuse also stated their attempts to modify or close their privacy boundaries led to the perception of increased hostility and incidents of cyberstalking by their aggressor.

The results presented here were the study participants' own descriptions of actual events.

In nearly all interviews, participants were able to articulate at least one episode of emotional, physical, or emotional abuse, with several reporting multiple types of abuse. Taken as a whole, the results of this study demonstrate abuses enacted during an intimate relationship are often pre-cursors to traditional stalking and cyberstalking and often intensify following termination. Therefore, victims must also use the context and risk-benefit ratio privacy rules in decisions to reveal or conceal information as they seek help to end their ordeals.

Help-Seeking Behaviors Following Incidents of Cyberstalking

Successful coping with any traumatic event requires victims to utilize a variety of personal skills and the involvement of trained professionals. Yet, past stalking research has provided mixed results concerning the effectiveness of individuals' efforts and the involvement of legal and medical assistance to end the ordeal (Hall, 1998; Logan, Walker, Stewart, & Allen, 2006).

One possible theoretical explanation for the victims' feelings of dissatisfaction can be found in the tenets of Communication Privacy Management Theory. Specifically, individuals must reveal private information and open their privacy boundaries to outside others only to have that information mismanaged. In this section, the results will demonstrate how victims opened their privacy boundaries to professionals in an effort to end their cyberstalking experience, but in many cases, the information disclosed led to further denigration of the victim.

Customer service representatives. Placement of blocking mechanisms ($N = 24$) was the most frequently used tactic specifically employed by the victims in their attempts to stop further communication with their former partners. The act of blocking may be defined as the ability to restrict incoming electronic communications from cell phones, email, and social media. Although several participants did not need assistance to successfully put blocking mechanisms in place, others had to rely on professionals. For example, Shana contacted her land line telephone company provider to block Andrew.

> *Getting the calls and texts to stop on my cell was really easy, I just followed the instructions that came with the phone, but the home phone was harder. I had to call [the telephone company] and waited on hold forever, was transferred again and again and again; first they sent me to disconnection services, then to sales only to be asked if I wanted to to, uh, to bundle my services ... then to technical assistance, back to sales ... when I finally got through to one person who took the time to actually listen, the problem was taken care of and actually suggested that I change my number and make it unlisted.*

Jason also opened his privacy boundaries with customer service representatives to share his story in an effort to receive assistance. Contrary to Shana's experience, the clerks showed little sympathy for his plight.

> *I originally went to [cell phone provider] to change my number which was kinda, kinda embarrassing 'cause the dude was like, "girlfriend problems, huh, wish I had that problem, ha ha," I didn't think it was funny, he even made sure that I knew how to re-add her, just in case, when I told him there was like no way that I was going back to her; he just smiled and said "sure buddy." I got so pissed off that when I got home I called to cancel my plan only to be told I'd get hit with an astronomical cancellation fee. The first thing the lady said was, "It says in my notes that you're the guy with the girlfriend problems." Then I was told my story didn't qualify under the provisions of my contract. I still remember that exact phrase.*

Jason and Shana's experiences highlight how the revelation of private information can be mismanaged when one must deal with insensitive outside others. Both explained how they were shown little compassion in their initial reactions with employees.

Assistance from law enforcement. In order to enlist the aid from law enforcement and judicial services to end their cyberstalking experience, many participants ($N = 13$) explained the need to open their privacy boundaries with outside others to start the process of obtaining an order of protection. Within this theme, the majority of victims stated they were unable to receive assistance from these professionals and nearly all reported dissatisfaction with how they were treated when they filed their complaint. However, victims who were persistent in navigating the legal system were able to secure some form of protection.

Claudia's experience required her to open her privacy boundaries to a number of individuals, but she has not received help. She claimed her attempts to cease all communication with her ex-husband, Carl, have been difficult because they share a daughter. During the divorce proceedings, Claudia shared with her attorney the intimidating messages Carl had sent to her sister. In this excerpt, Claudia explained her perception of the lack of aid and support she received from her lawyer and a police officer.

> *Mr. Johnson [her attorney] said Carl's threats were not made directly to me, so they were of little relevance, so I showed him the ones he sent me and he [the attorney] said, "well that's interesting, but these messages can be ... interpreted in lots of ways, so it would be best if you just share custody with him." I remember thinking what's it gonna take for him [the attorney] to do something with all the money I was payin' him. I mean, he [Carl] still sends me texts all the time saying awful things, blaming me 'cause Jacqueline wants to come home every time she's with him. It's not my fault he's a bully and she sees that, right? I asked a cop friend of my brother's if I could do anything and he said*

"not really unless he makes a … clear threat and even then these things are hard to prove." I guess I've just given up.

Like Claudia, Michael's situation, with his former partner, Taylor, is on-going. Although his parents have convinced him to go to the authorities, to date, the authorities have done little to end the stalking and cyberstalking. In this exemplar, Michael recounted his experience with two jurisdictions of police:

> *My mom … made me go to the cops at my school, but they say they can't do anything 'cause no threat has been made, but they called the [his hometown] cops 'cuz I guess that's where it all started. The [school] cops said the [hometown] cops promised to stop by her house and have a talk with her. A couple of days later, I get a call from [name of the hometown Chief of Police] who says that he received a report from some of his cops and that I need to come in to talk to them. So my mom and me had to go to the cops and they like begin saying things like she's [Taylor]very upset and to give her some time to get over it and try to talk me out of pursuing the matter any further. So, they were no help, but my mom wouldn't let it go. She went to the State's Attorney's office and convinced them to issue a restraining order. So, I go in to show it to the cops and they pretty much said that that they couldn't do much because they have better things to do and it's a free country, crap like that, so if we're somewhere public and she show's up, I can call them, but it's not a priority for them.*

The inability to receive satisfactory assistance in these cases is consistent with past research that demonstrates individuals often perceive cyberstalking as less severe because there is a lack of physical contact between the aggressor and victim (Alexy et al., 2005; Lee, 1998; Spitzberg & Hoobler, 2002). This misperception was also seen in the narratives of individuals who turned to medical and psychological professionals in their attempts to cope with their cyberstalking experiences.

Assistance from medical/psychological professionals. In cases where victims reported feelings of grave fear and mental/emotional distress as a result of technological stalking, psychological and medical professional services were utilized. Like the individuals who sought help from the legal system, these victims had to open their privacy boundaries when discussing their situation in their efforts to obtain treatment. For example, Rebecca recalled after her ex-intimate released a video of her involvement in a wet T-shirt contest, her mother took her to see their family physician.

> *So I tell the doctor everything that happened and [he] says that what I'm like going through is, like normal, I mean, seriously, like having everyone in the world see you half naked on the fucking internet is, like fucking normal. He tells my mom it will … pass or something, and like all that I need to, is … not get so worked up over this. He didn't even look at me … when he like said it. That just made it worse and I like started crying, like real bad, so he like, tells my mom, that I should see a shrink … but mom says that we don't have the money … and since I'm not in school, we don't like have insurance, or something retarded like that. It was, like she, and he (the doctor) was uh, blaming me for everything that happened. He just couldn't understand how he could help.*

On the other hand, some victims report they were able to receive treatment after they opened up to their physicians about their situation. In this exemplar, Alicia recounted the assistance she received for a "breakdown" after she discovered her husband had been using tracking software to monitor her movement. She explained:

> *I really remember being in the hospital. ... I broke down again and told him [the emergency room physician] that my husband was ... stalking me and it all came out ... He introduced me to Dr. Walters [her psychologist] who has become my saving grace. We talk about all kinds of things like how Dennis has taken advantage of me and how I need to keep a journal ... about of all the times he contacts me. Doc says it's important to not only my health, but also so I can give the police enough ammunition. I have a detective that I'm working with, but it's hard ya know, somedays I can handle him [Dennis] and other days, I feel like the crazy person ... all over again.*

The two exemplars provide contradictory evidence of how opening one's privacy boundaries to outside others served victims who sought treatment following incidents of technological stalking. Although these narratives demonstrate the variety of psychological harms that result from technological stalking, Rebecca's recollections also show how victims are often re-traumatized when they reveal private information. According to Rebecca, her mother's reaction to the discovery that her daughter had engaged in "shameless activities" was to blame the victim. Thus, when Rebecca shared private information with her parent, the information was used as a weapon to belittle her daughter. Moreover, the doctor's inability to address Rebecca directly and his attempts to minimize her suffering only further exacerbated her distress. On the other hand, Alicia's information was well received. Her ability to open her boundaries with the first physician to treat her allowed her to feel comfortable enough to share her story with the recommended psychiatrist and law enforcement. Past research shows that when individuals disclose full and accurate information with their doctors, they receive better health outcomes.

Conclusion

This chapter has examined the results from a larger cyberstalking study to demonstrate how past abuses committed during an intimate relationship may put one at risk for cyberstalking. The use of Communication Privacy Management theory may help to explain how, following a particular event, victims will re-evaluate their communication patterns with former partners. The results show victims look at a variety of factors, such as past abuse, familial obligations, and self-serving motivations, in order to assess if privacy boundary modifications with their ex-intimate are necessary. Specifically, the results indicate the context of a situation plays a large role in one's decision to modify or close privacy boundaries after carefully weighing the risks and benefits of continued interaction.

Victims also rely on contextual cues and the risk-benefit ratio when seeking assistance from outside sources. The results indicate that due to the severity of their situations, participants had to open their privacy boundaries in order to obtain protection from their cyberstalkers' attacks. In instances when the private information revealed to the professionals was well managed, victims reported better resolution and higher satisfaction with the help they received. On the other hand, those whose private information was mismanaged expressed deep dissatisfaction and heightened psychological distress. Therefore, the takeaway messages from this chapter is for individuals to be aware of warning signs from past abuses within intimate relationships that may lead to cyberstalking and to seek assistance from those who respect the gravity of the situation.

Notes

1 All names are pseudonyms.

References

Alexy, E. M., Burgess, A. W., Baker, T., & Smoyak, S. A. (2005). Perceptions of cyberstalking among college students. *Brief treatment and crisis intervention, 5*(3), 279–289. doi:10.1093/brief treatment/mhi020

Baum, K., Catalano, S., & Rand, M. (2009). *National crime victimization study: Stalking victimization in the United States.* Report submitted to the Bureau of Justice Statistics (NCJ 224527). Washington DC: US Department of Justice.

Black, M. C., Basile, K. C., Breiding, M. J., Smith, S. G., Walters, M. L., Merrick, M. T., Chen, J., & Stevens, M. R. (2011). *The National Intimate Partner and Sexual Violence Survey (NISVS): 2010 summary report.* Atlanta, GA: National Center for Injury Prevention and Control, Centers for Disease Control and Prevention.

Burgess, A. W., Baker, T., Greening, D., Hartman, C. R., Burgess, A. G., Douglas, J. E., & Halloran, R. (1997). Stalking behaviors within domestic violence. *Journal of Family Violence, 12,* 389–403.

Davies, E. L. (2013). *Technological stalking of relational partners: A hermeneutic phenomenological approach to understanding victims' lived experiences through a communication privacy management theory lens.* (Unpublished doctoral dissertation). University of Missouri, Columbia, Missouri.

Davis, K. E., Ace , A., & Andra , A. (2000). Stalking perpetrators and psychological maltreatment of partners: Anger-jealousy, attachment insecurity, need for control, and break-up context. *Violence and Victims, 15,* 407–425.

Dye, M. L., & Davis, K. E. (2003). Stalking and psychological abuse: Common factors and relationship-specific characteristics. *Violence and Victims, 18,* 163–180. doi: 10.1891/vivi.2003.18.2.163

Finn, J. (2004). A survey of online harassment at a university campus. *Journal of Interpersonal Violence, 19*(4), 468–483. doi:10.1177/0886260503262083.

Hall, D. M. (1998).The victims of stalking. In J. R. Meloy (Ed.), *The psychology of, stalking: Clinical and forensic perspectives* (pp. 113–137). San Diego, CA: Academic Press.

Lee, R. (1998). Romantic and electronic stalking in a college context. *William and Mary Journal of Women and the Law, 4,* 373–466.

Logan, T. K., Shannon, L., Cole, J., & Walker, R. (2006). The impact of differential patterns of physical violence and stalking on mental health and help-seeking among women with protective orders. *Violence Against Women, 12,* 866–886. doi:10.1177/1077801206292679

Logan, T. K., Walker, R., Stewart, C., & Allen, J. (2006). Victim service and justice system representative responses about partner stalking: What do professionals recommend? *Violence and Victims, 21,* 49–66.

Mechanic, M. B., Uhlmansiek, M. H., Weaver, T. L., & Resick, P. A. (2000). Impact of severe stalking experienced by acutely battered women: An examination of violence, psychological symptoms and strategic responding. *Violence and Victims, 15,* 443–458.

Mohandie, K., Meloy, J. R., McGowan, M. G., & Williams, J. (2006). The RECON typology of stalking: Reliability and validity based upon a large sample of North American stalkers. *Journal of Forensic Sciences, 51,* 147–155. doi:10.1111/j.1556-4029.2005.00030.x

Mullen, P. E., Pathé, M., & Purcell, R. (2000). *Stalkers and their victims.* Cambridge, UK: Cambridge University Press.

Murphy, C. M., & Cascardi, M. (1999). Psychological abuse in marriage and dating relationships. In R. L. Hampton (Ed.), *Family violence prevention and treatment* (2nd ed.). Beverly Hills, CA: Sage.

National Conference of State Legislatures. (2013). State cyberstalking and cyberharassment laws. Retrieved from: http://www.ncsl.org/research/telecommunications-andinformation-technology/cyberstalking-and-cyber harassment-laws.aspx

Petronio, S. (2002). *Boundaries of privacy: Dialectics of disclosure.* Albany, NY: State University of New York Press.

Petronio, S. (2007). Translational research endeavors and the practices of communication privacy management. *Journal of Applied Communication Research, 35,* 218–222.

Roberts, K. A. (2005). Women's experience of violence during stalking by former romantic partners: Factors predictive of stalking violence. *Violence Against Women, 11*(1), 89–114. doi:10.1177/1077801204271096

Simonelli, C. J., & Ingram, K. M. (1998). Psychological distress among men experiencing physical and emotional abuse in heterosexual dating relationships. *Journal of Interpersonal Violence, 13,* 667–681. doi: 10.1177/088626098013006001

Spitzberg, B. H. (2002). The tactical topography of stalking victimization and management. *Trauma, Violence, & Abuse, 3,* 261–288. doi:10.1177/1524838002237330

Spitzberg, B. H., & Hoobler, G. (2002). Technological stalking and the technologies ointerpersonal terrorism. *New Media and Society, 4(1),* 71–92. doi:10.1177/14614440222226271

Straus, M. A., & Gelles, R. J. (1992). *Physical violence in American families: Risk factors and adaptations to violence in 8,145 families.* New Brunswick, NJ: Transaction Publishers.

Tjaden, P. G., & Thoennes, N. (1998). *Stalking in America: Findings from the National Violence Against Women Survey.* (NCJ 169592). Washington, DC: National Institute of Justice and Centers for Disease Control and Prevention.

Tjaden, P. G., & Thoennes, N. (2000). *Full report of the prevalence, incidence, and consequences of violence against women: Findings from the National Violence Against Women Survey.* (NCJ 183781). Washington DC: US Department of Justice, National Institute of Justice.

U.S. Attorney General Report (1999). *Cyberstalking. A new challenge for law enforcement and industry.* (Electronic Version) Retrieved from http://www.usdoj.gov.criminal/cybercrime/cyberstalking.htm

van Manen, M. (1990). *Researching lived experience: Human science for an action sensitive pedagogy.* London: Althouse Press.

TWENTY

How Do Computer-Mediated Channels Negatively Impact Existing Interpersonal Relationships?

Amy J. Johnson, Eryn Bostwick, and Chris Anderson

The ubiquitous nature of computer-mediated communication and mobile phone technology has provided many ways to maintain interpersonal relationships that originated face-to-face. In fact, a line of research illustrates how these channels can be used to aid interaction between both long-distance and geographically close friends (Johnson & Becker, 2011), romantic relationships (Maguire & Connaughton, 2011), and family members (Johnson, Haigh, Becker, Craig, & Wigley 2008). However, there is also research related to the down side of these mediated channels for interpersonal relationships, including the benefits and costs of engaging in conflict over CMC channels (Baym, 2010), the impact of mobile phone use on interpersonal relationships (Miller-Ott, Kelly, & Duran, 2012), the dangers of idealizing long-distance relationships with whom one communicates mainly through mediated means (Stafford, 2005), and potential disappointments when expectations for how one communicates through CMC are not met (Adams, 1998). This study explores the current research on potential drawbacks of using computer-mediated channels to maintain existing interpersonal relationships.

A major research focus of computer-mediated communication has been the impact of CMC technology on individuals, where CMC use has been associated with an array of negative outcomes (Brandtzæg, 2012). A meta-analysis of 18 research studies that used a variety of measures of loneliness identified a significant relationship between Facebook use and loneliness (Song et al., 2014). Frequent users of social network sites (SNS) often have higher rates of narcissism (Ljepava, Orr, Locke, & Ross, 2013), and Neo and Skoric (2009) found that students who had a compulsion to instant message felt inconvenienced by face-to-face communication. Given the pervasive nature of technology, identifying how CMC affects individuals' relationships with family, friends, and romantic partners instead of just themselves is equally important. While most forms of CMC are viewed as a great way to keep in touch, there is a dark side to the integration of these technologies into a person's interpersonal relationships.

The next sections focus on three potentially negative impacts of CMC use on interpersonal relationships: (1) Unmet expectations associated with CMC use; (2) CMC and negative relational processes; and (3) negative impacts of cell phone usage. Ten focus groups were conducted to further explore negative impacts of CMC on existing relationships, and findings and conclusions from a thematic analysis of these focus groups are presented.

Unmet Expectations of CMC Use in Interpersonal Relationships

There can be idealized views of how computer-mediated communication may benefit interpersonal relationships. How these ideals coincide with reality will determine whether these communication channels have potentially positive or negative effects on relationships.

There are several lines of research that have examined violated expectations that occur due to interaction over computer-mediated communication channels. For example, in the past, friendships tended to end by "fading away" (Rose & Serafica, 1986). However, now one can unilaterally end a Facebook "friendship" by "unfriending" the individual, which no longer allows the individual to have access to one's profile information only available to "friends." A less face-threatening method of controlling the information one receives on a social media network is to "hide" an individual's information from one's newsfeed. Pena and Brody (2014) found people reported that they were more likely to intend to hide someone's information from their newsfeed than unfriend a person. Bevan, Ang, and Fearns (2014) found of those individuals who knew they were unfriended, about 14% said they had contacted the person who unfriended them. Closeness to the individual, whether individuals contacted the person who unfriended them, and being female were associated with perceived violation importance.

Additionally, due to how individuals portray themselves and the feedback they receive from each other, relationship development may occur faster over CMC than through face-to-face communication, a phenomenon known as hyperpersonal interaction (Walther, 1996). Unfortunately, this relationship development may be based on biased perceptions, leading to unmet expectations. For example, Brody (2013) extends hyperpersonal interaction to existing long-distance friendships that did not originate online. He found that how often one communicated through computer-mediated channels had more of an impact on friendship satisfaction and commitment for those friends who had not seen each other face-to-face recently.

Another area where expectations regarding CMC use may affect interpersonal relationships is the provision of social support. For example, Stefanone, Kwon, and Lackaff (2012) had individuals send requests for instrumental social support to their Facebook friendship networks. The great majority (80%) of these requests went unmet; however, closeness of the individuals and how often they communicated on Facebook predicted greater likelihood to provide instrumental support. Johnson et al. (2013) found the closeness of a Facebook friend did not predict whether social support was offered to an individual or whether this support was perceived as effective if it occurred. Only biological sex (women were more likely to provide effective social support) and Facebook reciprocity (those who communicated with each other more often through Facebook were more likely to provide effective social support) predicted support provision and effectiveness (see also Rozzell et al., 2014). These findings may be related to Ellison, Steinfield, and Lampe's (2007) findings that social network sites are particularly relevant to bridging social capital, which focuses on weak ties, rather than bonding social capital, which focuses more on strong ties.

CMC and Negative Relational Processes

When communicating via computer-mediated channels, the opportunity for a variety of conflicts and issues arises. The Internet provides a unique opportunity for interactional partners to carefully craft an image of themselves (Toma & Hancock, 2010). Because of this ability, individuals are easily able to deceive others by presenting themselves differently than they actually appear. For example, Toma, Hancock, and Ellison (2008) found that this type of deception can occur in up to 81% of online dating profiles. Hall and Pennington (2013) also found people are dishonest on Facebook.

Social media sites seem to cause a variety of conflicts in interpersonal relationships. Not only can people create idealized versions of themselves on their Facebook page or lie about their behavior, but Miller-Ott and Kelly (2013) also found individuals use Facebook to spread rumors and gossip about others. Research also suggests people post rumors, post embarrassing photos of others, and engage in cyberbullying through Facebook (Miller-Ott & Kelly, 2013). According to Stern and Taylor (2007), these types of behaviors, as well as misinterpreting information posted by others, can lead to interpersonal conflict.

Utz and Beukeboom (2011) also found Facebook could contribute to jealousy and conflict. Seeing a romantic partner post on another person's wall or interacting with other potential partners is associated with increased feelings of jealousy. This was more prominent for individuals with low self-esteem. Also, 30% of participants admitted to monitoring their relational partner's behavior via Facebook, compared to 10–17% of participants who admitted to doing so offline. Simply sending or accepting a friend request, or remaining or becoming friends with a romantic interest, is associated with increases in jealousy (Drouin, Miller, & Dibble, 2014).

Sometimes computer-mediated channels are used to actually engage in conflict communication. There are mixed opinions regarding this process (Baym, 2010). Many individuals report believing that handling conflict through CMC is not appropriate, but others explained that in some circumstances it might be preferable. For example, if a relational partner tends to get angry when confronted, communication via CMC might be best. Also, people who are uncomfortable engaging in conflict face-to-face might be more willing to use CMC channels, which would be more effective than avoiding the conflict all together (Baym, 2010). Perry and Werner-Wilson (2011) claim couples indicated liking to use CMC to solve problems because it allows them to take time to reflect and think about what they want to say, which ultimately led to decreased escalation of conflict. Taken together, this research suggests CMC channels present plenty of opportunity for conflict to arise. However, it also suggests there are times when conflict communication via CMC is not only acceptable, but also preferable.

Disadvantages of Cell Phones

Kang and Jung (2014) found that in America cell phones are used to fulfill needs for safety, self-actualization, self-esteem, and belongingness. Text messaging has become a major channel for communication in intimate relationships. In fact, according to Luo (2014), some individuals communicate with a romantic partner via text up to 90% of the time. Madianou (2014) posits that the "taken for grantedness" of smartphones gives us a moral obligation to communicate and constantly be available, which can cause issues for some interpersonal relationships.

This constant access to information is related to increased anxiety in cell phone users (Lepp, Barkley, & Karpinski, 2014). Other individuals indicated although cell phones help them feel connected to others, they also made people seem more shallow and self-centered, or people end up paying more attention to their phones than other people (Mihailidis, 2014).

Researchers have sought to understand how people negotiate what type of cell phone use is appropriate in interpersonal relationships. Kirby Forgays, Hyman, and Schreiber (2014) refer to this as *cell phone etiquette* and found the average person thinks of text messages as more appropriate than phone calls in most situations; however, younger people were more likely to communicate via text message, while older participants were more likely to call or use an email or Facebook message. They also found younger participants were more likely to be in constant contact with one another and more likely to get irritated if they did not get a text message response from a conversational partner within a day (Kirby Forgays et al., 2014).

Miller-Ott, Kelly, and Duran (2012) focused on how cell phone rules are related to relational satisfaction. Rules associated with relational issues (e.g., rules prohibiting partners from starting relational arguments over the phone) were positively related to relationship satisfaction, while those associated with repetitive contact and monitoring one's partner were negatively associated with relational satisfaction.

Another aspect of cell phone use that can lead to conflict is deception. Wise and Rodriguez (2013) found most participants admitted to sending deceptive text messages, although they deceived strangers more than intimate partners. The majority of participants also believed they were lied to via text. Soleil Archambault (2011) studied cell phone use in Mozambique and found that 47% of females and 32% of males fought with their partners because of real or perceived deception on their phones.

The next section will present the methods and results of ten focus groups that were conducted to explore how individuals report CMC use negatively impacts their interpersonal relationships. General research questions driving the analysis include the following: What unmet expectations do individuals report for CMC use in interpersonal relationships and how did these unmet expectations impact the relationships? What negative processes (e.g., deception, conflict) did individuals report regarding CMC use in interpersonal relationships? What conflicts did individuals report regarding the use of CMC channels, including smartphones, with their interpersonal relational partners?

Method

Participants

Participants were recruited from undergraduate communication courses from a large southern university. In exchange for participation, students were offered course credit. Ten focus groups were conducted with 4 to 11 participants in any one group with a total of 73 participants across all ten groups. Participants had little variation over age (range: 18–31, $M = 20$) and a roughly even gender division (33 male, 40 female). Sixty-six students reported their ethnicity (Caucasian, n = 50; other identifications n = 16). Sixty-three percent (n = 46) of the sample identified as not having a current romantic relationship and 23.3% (n = 17) identified as being in a serious relationship (casual dating 9.6%, n = 7; married 4.1%, n = 3).

Procedures

Each focus group met for approximately 45–60 minutes. Participants first completed informed consent forms and were reminded any answers they gave would remain confidential, and they were asked to keep anything they or others discussed in the focus group private. A facilitator used a semi-structured group interview while probing deeper for inconsistent comments, clarification, and understanding (Kvale & Brinkmann, 2009; see Appendix for questions). All focus groups were audio recorded and transcribed. Following the question and answer period, participants completed a short demographic questionnaire.

Analysis

A thematic analysis was used to examine the data. Themes were defined as patterned responses that capture important responses to the research question. We took what Braun and Clarke (2006) call a semantic approach, and only identified explicit content, while avoiding interpreting beyond what the participants said.

In order to identify the themes, the first and second authors separately read through five focus group transcripts and made notes about topics that came up in conversation related to the research questions. Then, the first and second author each read through the transcripts again with the goal of identifying codes, or labeling the data that related back to the research question. Once these codes were produced, the co-authors compared the codes found in each transcript and identified similarities between them to create more general themes. We required that these themes be present in at least 3 of the 5 focus groups. Lastly, the authors met to discuss and compare the themes they found individually. In any case where one author had a theme that the other did not, they examined the data together to discuss whether the theme was appropriate. Once they agreed that they had sufficiently identified all themes related to the research questions, they found examples representative of each theme. The second author then read through the remaining five transcripts to ascertain that no new themes emerged and to select further examples representative of each theme.

Results

The analysis revealed six major themes for how CMC use can damage interpersonal relationships: (1) using CMC can lead to misunderstandings due to lean features of the channel, (2) failing to reciprocate computer-mediated communication can cause conflict, (3) posts on CMC channels can lead to conflict, (4) people being constantly connected to media can lead to conflict, (5) certain norms regarding how to use CMC should be followed or negative outcomes may result, and (6) unmet expectations regarding CMC use can cause negative outcomes for interpersonal relationships. Each theme and examples of responses reflecting the theme are presented below.

The first theme dealt with misunderstandings that can occur when communicating via CMC. Responses focused on not being able to rely on certain cues, such as vocal tone, when communicating via CMC and therefore misinterpreting a message. For example, one participant said, "I've had more miscommunications happen because I can't tell the tone of a certain word, or, even like what does the length of a message mean? Rather than—you know, whereas, in face-to-face, I can tell instantly, the feedback is very instant." Similarly, a participant from another focus group said, "Like when they're right there, you can kind of pick up on like little actions they do a lot easier—like, if they roll their eyes, or like if how—what they're doing with their hands…that kind of gives you more insight into the conversation…" Lastly, another participant stated, "I think when you're not face-to-face with a person, it's easy to lose context of like what they're saying through just words and not seeing, like their nonverbal communication." Participants emphasized this type of problem can be eliminated in face-to-face communication, or at least channels that allow access to a variety of different cues, such as Skype or FaceTime.

The second theme that emerged was related to failing to reciprocate communication, such as not responding to a text message or Facebook post. For example, one participant said,

> I think that's like, interesting, because now on Facebook, you can, like see if someone read your message…Um, and also, text message is an option now and they can see if we read it. So, it causes conflict just because, like, 'why didn't you reply? You saw it and then didn't.'

Another participant expressed similar sentiments:

> Yeah, one of my best friends has had another son recently; told him congratulations; he didn't respond to me. I was like, 'well never mind then, I don't want to talk to you either.'

Some participants suggested the impact of not responding to a message depends on the channel. According to one participant,

> I don't know, whenever like—if someone doesn't reply to a post on Facebook or something, I find it a lot easier to just write off—it's like, "o, they didn't see it—you know–they didn't." It's kind of like, "oh, whatever." It's easy to excuse, whereas like if I send them a text or call them, or something, and they still don't, like, answer, then I'm like, "hey, man"—I'll try and find them in person and be like, "did you not just get this?"

Lastly, one participant said sometimes you assume people are ignoring you, but really they might not be paying attention to the medium you used to contact them. For example, one participant said, "… when I'm like away from my, like, dad, he thinks because I'm on it so much when I'm with him, he does not understand …why I don't like respond that second. And so, I get in trouble for that, sometimes …Because, I just kind of—I forget, sometimes."

The third theme reflected posting inappropriate content online. This theme had two subthemes: first, posting certain pictures or comments can lead to conflict; second, posting inappropriate remarks about other people that you might not otherwise say face-to-face can be problematic. Many participants talked about friends or family members posting pictures or tagging them in something without their approval, causing conflict. For example, one participant said, "…my ex-boyfriend in high school, he got really upset with me because one of my other exes, who ended up being my best friend through high school, told me happy birthday on Facebook."

Many other participants talked about how posts can cause conflict between friends or family. For example, one participant said, "Um, when my family members found out I was with my current boyfriend, they would comment on the picture on Facebook and, like, 'oh, you're with him; he sucks; he's awful,' and stuff like making that a public thing."

The second subtheme dealt with posting content or comments that one would not say face-to-face, leading to conflict. Some participants said this occurs because people feel more comfortable posting online, while others discussed using posts to be passive-aggressive. For example, one participant said,

> I've observed a lot of Twitter beef, um, like, you know—now, people can tweet something and like, not anonymously, but they won't name the person in particular that they're talking about, but everyone knows, "oh my God, she's talking about Sarah."

The fourth theme dealt with constant connectivity and how people seem to pay more attention to the CMC channels than one another. Within this theme some participants talked about how people have become so connected to their phones they no longer interact with certain interpersonal partners. For example, one participant said, "I find it interesting how people use their phone, sometimes, as a way to, like, um, sort of avoid social situations sometimes...Uh, and, this ability to sort of withdraw becomes more elaborate as you get better phones and more apps, and stuff." Other participants focused on how constant connectivity has caused people to focus on their phones instead of conversations when they are in groups with friends or family members. For example, one participant said, "Um, I think also with phones, you know, being—er, smart phones having everything that a computer has, it distracts people from just ordinary conversation and so if you're on, like, a date, or if you're hanging out with friends, a lot of people tend to just—if an awkward situation comes up, just look at their phones and kind of just lose interest." Another participant echoed this statement and said,

> Um, well, what I thought of—if you're, say you have a big family, like, reunion, or dinner and there are people are on their phones and they're not talking. Like, that's one thing that family has been, like, really hard about—like, no phones at the table, because, like it's family time.

Lastly, some participants said this constant connectivity has led to a decrease in interpersonal communication skill. For example, one participant said, "I guess with my little sister, even talking to her verbally is kind of annoying, because now she's—kind of speaks the way she writes on social media, and I'm just like, 'what are you saying? I don't understand. Speak like a human.'"

The fifth theme that emerged from the data related to how people use media to communicate and whether their practices conform to communication norms. Within this theme, people talked about certain media being appropriate in certain situations and times, and inappropriate in others. One participant said, "Um, I think it depends on the seriousness of the situation. Basically, like, say there's...a death in the family, you'll probably want to, you know, do a mediated channel [rather] than a CMC...because you know want to personally tell that person." Some other participants said if certain types of contact are made at certain times, it signals importance. For example, one person said, "I don't mind people who are not close with texting me, but there are definitely certain times, even within any social media, where if I get a phone call—if I get a phone call at two in the morning, I'm going to assume something bad just happened."

Lastly, some participants said CMC should be used for certain purposes, but many people violate those norms. For example, one participant said, "Um, I feel like ideally it would be... more of, like, a way to connect, instead of a way to like, kind of brag about what you're doing." Another participant made a similar comment: "[People use] social media, I think...to be kind of hateful, or to gossip, or to put meaningless things on there, like what they have for dinner, instead of actually using it to connect with, like their friends or family."

The sixth theme focused on conflict that occurs when expectations are not met. For example, participants indicated getting upset when people do not conform to expected norms of a medium. For example, one participant said, "And, well, especially like emojis and stuff, like... when I text my friends, I don't put emojis in. If I'm texting a girl, if you don't—if you don't use it, it's almost like—almost kind of rude if you don't." Lastly, some participants talked about getting upset because someone used CMC to talk to the "wrong" people. One participant said, "I think it's definitely a problem sometimes when...you have one friend who asks you, 'well, why did you comment on that person's profile, or that person's picture. Why did you like that person's information; I thought you weren't friends with them.'" Another participant echoed this statement and said, "...Like, SnapChat, I had a girl come up to me and say, 'why is she your best friend on SnapChat,' or something like that. And it's just like, 'well, I just snapchatted a few more than you.'"

Discussion

The purpose of this study was to explore the current literature on the dark side of CMC use for interpersonal relationships. Three general themes were found in the existing literature and were echoed in the focus group findings.

First, expectancy violations regarding how individuals utilize computer-mediated communication in interpersonal relationships have been associated with negative impacts on these relationships. From defriending on Facebook to idealization in long-distance friendships to requesting support from others through social media sites, negative expectancy violations have been shown to occur. Our focus group findings emphasize this process as an important part of the potentially dark side of CMC communication for interpersonal relationships. Individuals reported misunderstandings due to unmet expectations related to *how* they communicated with CMC, *when* they communicated by CMC, and *with whom* they communicated by CMC. Individuals reported unmet expectations concerning how they believed individuals should communicate via CMC and the reality of how individuals communicated through these channels, causing frustration and conflict with their interpersonal partners.

The second general theme in the literature related to negative relational processes occurring *through* computer-mediated channels, such as deception and conflict. From posting overly positive portrayals of oneself on social media sites to inciting conflict on one's Facebook or Twitter feed, research has shown that not all communication that occurs on computer-mediated channels is straightforward. Other research has focused on jealousy from CMC use, and disadvantages and advantages of engaging in conflict through CMC. Findings from our focus groups reflected these negative relational processes through CMC. Individuals reported that their interpersonal partners posted content they did not want posted, leading to conflict (or did not post content they wanted posted). Romantic partners reported jealousy related to Facebook posts from ex-romantic partners, and participants reported passive aggressive behavior where individuals would deal with conflict with another by posting a supposedly innocuous statement on a social media site. Focus group respondents emphasized that often negative relational processes involved behavior individuals would not engage in face-to-face.

A third theme in the literature focused on the disadvantages of smartphones in particular for romantic relationships; however, in the focus groups a more general theme of the dark side of constant connectivity in general, which smartphones epitomized, was emphasized. Specifically, participants discussed the continuous problem of having an interpersonal partner's divided attention. Rules for using smart phones in interpersonal situations such as family dinners (similar to Kelly and Miller-Ott's, 2014, findings regarding rules for cell phones and romantic relationships) were enacted. Utilizing computer-mediated communication to avoid face-to-face communication and thus avoid opportunities to interact interpersonally or to avoid awkward moments within communication encounters with interpersonal partners were recalled.

Although not discussed in the literature review, a major theme from the focus groups was the downside of CMC *channels* for certain types of communication with interpersonal partners. Respondents repeatedly emphasized that the lean nature of the media and the lack of interpersonal cues led to misunderstandings and conflict that might have been avoided if they had been discussed face-to-face. These findings echo Walther's (1992) social information processing (SIP) theory, which says individuals can overcome deficiencies in lean computer-mediated channels over time through learning to compensate for the lack of nonverbal cues. Although individuals in this study did report utilizing methods, such as emojis, to counteract a lack of nonverbal cues, even the use of this method was not without potential strife when individuals followed different norms. Most of the research regarding SIP has focused on groups formed via computer-mediated communication (e.g., Walther, 1993). How these limitations of computer-mediated channels continue to affect existing interpersonal relationships is an interesting area for future research.

One issue in the current study is that the generalizability to interpersonal relationships beyond college students is limited. However, the college students in our study talked about a wide range of interpersonal relationships from friends, geographically close and long-distance romantic relationships, parents, siblings, and extended family members, helping to expand the scope of these findings.

Conclusion

This study has focused on the dark side of computer-mediated communication for existing interpersonal relationships, such as friendships, family relationships, and romantic relationships. Common themes of unmet expectations, negative relational processes over CMC, and difficulties due to the medium and constant connectivity are discussed and illustrated through the findings of focus groups. These results suggest many fruitful directions for future research to help clarify how computer-mediated communication channels lead to both positive and negative outcomes for existing interpersonal relationships.

References

Adams, R. G. (1998). The demise of territorial determinism: Online friendships. In R. G. Adams & G. Allan (Eds.), *Placing friendship in context* (pp. 153–182). Cambridge, UK: Cambridge University Press.

Baym, N. K. (2010). *Personal connections in the digital age*. Cambridge, UK: Polity.

Bevan, J. L., Ang, P., & Fearns, J. B. (2014). Being unfriended on Facebook: An application of Expectancy Violation Theory. *Computers in Human Behavior, 33,* 171–178. doi: 10.1016/j.chb.2014.01.029

Brandtzæg, P. B. (2012). Social networking sites: Their users and social implications—A longitudinal study. *Journal of Computer-Mediated Communication, 17*, 467–488. doi: 10.1111/j.1083-6101.2012.01580.x

Braun, V., & Clarke, V. (2006). Using thematic analysis in psychology. *Qualitative Research in Psychology, 3*, 77–101. doi: 10.1191/1478088706qp063oa

Brody, N. (2013). Absence—and mediated communication—makes the heart grow fonder: Clarifying the predictors of satisfaction and commitment in long-distance friendships. *Communication Research Reports, 30*, 323–332. doi: 10.1080/08824096.2012.837388

Drouin, M., Miller, D. A., & Dibble, J. L. (2014). Ignore your partners' current Facebook friends: Beware the ones they add! *Computers in Human Behavior, 35*, 483–488. doi: 10.1016/j.chb.2014.02.032

Ellison, N. B., Steinfield, C., & Lampe, C. (2007). The benefits of Facebook "friends": Social capital and college students' use of online social network sites. *Journal of Computer-Mediated Communication, 12*, 1143–1168. doi: 10.1111/j.1083-6101.2007.00367.x

Hall, J. A., & Pennington, N. (2013). Self-monitoring, honesty, and cue use on Facebook: The relationship with use extraversion and conscientiousness. *Computers in Human Behavior, 29*, 1556–1564. doi: 10.1016/j.chb.2013.01.001

Johnson, A. J., & Becker, J. A. H. (2011). CMC and the conceptualization of "friendship": How friendships have changed with the advent of new methods of interpersonal communication. In K. B. Wright and L. M. Webb (Eds.), *Computer-mediated communication in personal relationships* (pp. 225–243). New York, NY: Peter Lang.

Johnson, A. J., Haigh, M. M., Becker, J. A. H., Craig, E. A., & Wigley, S. (2008). College students' use of relational management strategies in email in long-distance and geographically close relationships. *Journal of Computer-Mediated Communication, 13*, 381–404.

Johnson, A. J., Lane, B., Tornes, M., King, S., Wright, K. B., Carr, C. T., Piercy, C., & Rozzell, B. (2013). *The social support process and Facebook: Soliciting support from strong and weak ties.* Paper presented to the Interpersonal Division of the International Communication Association for their annual conference in London, England, June 2013.

Kang, S., & Jung, J. (2014). Mobile communication for human needs: A comparison of smartphone use between the US and Korea. *Computers in Human Behavior, 35*, 376–387. doi: 10.1016/j.chb.2014.03.024

Kelly, L., & Miller-Ott, A. (2014). *Divided attention: Romantic interactions in an age of continuous availability by cell phones.* Paper presented at the 100th Annual National Communication Association Conference, Chicago, IL.

Kirby Forgays, D., Hyman, I., &Schreiber, J. (2014). Texting everywhere for everything: Gender and age differences in cell phone etiquette and use. *Computers in Human Behavior, 31*, 314–321. doi: 10.1016/j.chb.2013.10.053

Kvale, S., & Brinkmann, S. (2009). *InterViews: Learning the craft of qualitative research interviewing* (2nd ed.). Los Angeles, CA: Sage Publications.

Lepp, A., Barkley, J. E., & Karpinski, A. C. (2014). The relationship between cell phone use, academic performance, anxiety, and satisfaction with life in college students. *Computers in Human Behavior, 31*, 343–350. doi: 10.1016/j.chb.2013.10.049

Ljepava, N., Orr, R. R., Locke, S., & Ross, C. (2013). Personality and social characteristics of Facebook non-users and frequent users. *Computers in Human Behavior, 29*, 1602–1607. doi: 10.1016/j.chb.2013.01.026

Luo, S. (2014). Effects of texting on satisfaction in romantic relationships: The role of attachment. *Computers in Human Behavior, 33*, 145–152. doi: 10.1016/j.chb.2013.01.014

Madianou, M. (2014). Smartphones as polymedia. *Journal of Computer-Mediated Communication, 19*, 667–680. doi: 10.1111/jcc4.12069

Maguire, K. C., & Connaughton, S. L. (2011). A cross-contextual examination of technologically mediated communication and social presence in long-distance relationships. In K. B. Wright and L. M. Webb (Eds.), *Computer-mediated communication in personal relationships* (pp. 244–265). New York, NY: Peter Lang.

Mihailidis, P. (2014). A tethered generation: Exploring the role of mobile phones in the daily life of young people. *Mobile Media & Communication, 2*, 58–72. doi: 10.1177/2050157913505558

Miller-Ott, A. E., & Kelly, L. (2013). Communication of female relational aggression in the college environment. *Qualitative Research Reports in Communication, 14*, 19–27. doi: 10.1080/17459435.2013.835338

Miller-Ott, A. E., Kelly, L., & Duran, R. L. (2012). The effects of cell phone usage rules on satisfaction in romantic relationships. *Communication Quarterly, 60*, 17–34. doi: 10.1080/01463373.2012.642263

Neo, R. L., & Skoric, M. M. (2009). Problematic instant messaging use. *Journal of Computer-Mediated Communication, 14*, 627–657. doi:10.1111/j.1083-6101.2009.01456.x

Pena, J., & Brody, N. (2014). Intentions to hide and unfriend Facebook connections based on perceptions of sender attractiveness and status updates. *Computers in Human Behavior, 31,* 143–150. doi: 10.1016/j.chb.2013.10.004

Perry, M. S., & Werner-Wilson, R. J. (2011). Couples and computer-mediated communication: A closer look at the affordances and use of the channel. *Family & Consumer Sciences Research Journal, 40*, 120–134. doi: 10.1111/j.1552-3934.2011.02099.x

Rose, S., & Serafica, F. C. (1986). Keeping and ending casual, close, and best friendships. *Journal of Social and Personal Relationships, 3*, 275–288. doi: 10.1177/0265407586033002

Rozzell, B., Piercy, C., Carr, C. T., King, S., Lane, B., Tornes, M., Johnson, A. J., & Wright, K. B. (2014). Notification pending: Online social support from close and nonclose relational ties. *Computers in Human Behavior, 38*, 272–280. doi: 10.1016/j.chb.2014.06.006

Soleil Archambault, J. (2011). Breaking up 'because of the phone' and transformative potential of information in Southern Mozambique. *New Media and Society, 13*, 444–456. doi: 10.1177/1461444810393906

Song, H., Zmyslinski-Seelig, A., Kim, J., Drent, A., Victor, A., Omori, K., & Allen, M. (2014). Does facebook make you lonely? A meta-analysis. *Computers in Human Behavior, 36*, 446–452. doi: 10.1016/j.chb.2014.04.011

Stafford, L. (2005). *Maintaining long-distance and cross-residential relationships.* Mahwah, NJ: Lawrence Erlbaum.

Stefanone, M. A., Kwon, K. H., & Lackaff, D. (2012). Exploring the relationship between perceptions of social capital and enacted support online. *Journal of Computer-Mediated Communication, 17*, 451–466. doi: 10.1111/j.1083-6101.2012.01585.x

Stern, L. A., & Taylor, K. (2007). Social networking in Facebook. *Journal of the Communication, Speech & Theatre Association of North Dakota, 20*, 9–20.

Toma, C. L., & Hancock, J. T. (2010). Looks and lies: The role of physical attractiveness in online dating self-presentation and deception. *Communication Research, 37*, 335–351.doi: 10.1177/0093650209356437

Toma, C. L., Hancock, J. T., & Ellison, N. B. (2008). Separating fact from fiction: An examination of deceptive self-presentation in online dating profiles. *Personality and Social Psychology Bulletin, 34*, 1023–1036. doi: 10.1177/0146167208318067

Utz, S., & Beukeboom, C. J. (2011). The role of social network sites in romantic relationships: Effects on jealousy and relationship happiness. *Journal of Computer-Mediated Communication, 16*, 511–527. doi: 10.1111/j.1083-6101.2011.01552.x

Walther, J. B. (1992). Interpersonal effects in computer-mediated interaction: A relational perspective. *Communication Research, 19*, 52–90. doi: 10.1177/009365092019001003

Walther, J. B. (1993). Impression development in computer-mediated interaction. *Western Journal of Communication, 57*, 381–398.

Walther, J. B. (1996). Computer-mediated communication: Impersonal, interpersonal, and hyperpersonal interaction. *Communication Research, 23*, 3–43. doi: 10.1177/009365096023001001

Wise, M., & Rodriguez, D. (2013). Detecting deceptive communication through computer-mediated technology: Applying interpersonal deception theory to texting behavior. *Communication Research Reports, 30*, 342–346. doi: 10.1080/08824096.2013.823861

Appendix

Focus group questions[1]

1) Ideally, how would you like to use computer-mediated channels to communicate with your friends, family, or romantic partners? You can give specific examples.

2) How does reality match up with this ideal?

3) Have you ever been disappointed when a friend, romantic partner, or family member did not use CMC channels to communicate with you; for example, respond to a Facebook posting?

4) Have you ever had conflict with your family, friends, or romantic partners related to using CMC channels to communicate?

5) Thinking of smart phones more generally, have you ever had conflict with your family, friends, or romantic partners related to using a smart phone (own or other use)?

6) How do you decide whether to communicate face-to-face or through a mediated or computer-mediated channel with a certain family, friend, or romantic partner?

7) Have you ever engaged in conflict with a family member, friend, or romantic partner using computer-mediated channels? Describe this experience. Would it have been more effective to have engaged in this conflict face-to-face?

8) Have you ever engaged in conflict with a family member, friend, or romantic partner using text messages? Describe this experience. Would it have been more effective to have engaged in this conflict face-to-face?

9) Have you ever engaged in a conflict verbally over your cell phone? Describe this experience. Would it have been more effective to have engaged in this conflict face-to-face?

10) Does anyone have any additional relevant experiences that you would like to share?

[1]During the interview, the facilitator listened for inconsistent comments and probed for understanding. Each focus group was asked the same set of core questions; however, the facilitator developed follow-up questions.

Context Five

The Dark Side of Blended-Communication Contexts

TWENTY-ONE

"The More Things Change…": Technologically Mediated Abuse of Intimate Partner Violence Victims

Jessica J. Eckstein

According to Black et al. (2011), nearly half of all adults in the U.S. have been psychologically abused by a partner and physical abuse is experienced by 35.6% of women and 28.5% of men. With increasing frequency, intimate partner violence (IPV) victims are able to utilize electronic resources and social support. However, technology is also harnessed by perpetrators. Previously limited to in-person, third-party, and telephone/mail contact, today's media abet perpetrators with new ways to harass, stalk, violate, and assault their victims (Southworth, Dawson, Fraser, & Tucker, 2005). Research on IPV should now not only include the ways technology is used, but must examine the currently unknown consequences of mediated violence. In the current study, I take an initial step in researching technologically mediated abuse (TMA) as experienced by IPV victims. After first examining the limited technological-abuse literature, I integrate results and discussion of a mixed-methods study of victims' TMA.

Laying Foundations

Intimate partner violence (IPV) involves intentional physical (i.e., implementing objects or body to corporally injure), emotional (i.e., targeting identity to hurt or control), and/or verbal (i.e., using profanity or offending to attack) tactics used to harm a romantic relational partner (Straus, Hamby, & Warren, 2003). Due to a current lack of research on the topic in romantic relational contexts, it is unknown to what extent diverse types of IPV differ in terms of TMA. On the one hand, technological immediacy allows people to voice instantaneous reactions privately and publicly following disagreements, suggesting that particular media may be used by perpetrators of conflict-based IPV relationships in goal-directed situations. On the other hand, media amplify the means through which abusers can control, intrude, monitor, and otherwise attack victims. Ultimately, the extent to which intimates in IPV contexts are stalked (or abused) remains unknown.

Coercive control, theorized to distinguish IPV relationship types, may be a form of abuse particularly practiced with new technologies. Operationalized by Lehmann, Simmons, and Pillai (2012) as "multidimensional" and "repetitive" behaviors functioning because of victims' beliefs in abusers' reward-punishment potential, *coercive control* is distinctly comprised of two simultaneous processes (p. 916). Coercion, whereby implications or embodiments of menace drive compliance, is a factor that rarely requires actual force to be performed, because the mere threat of it is the primary facilitator (Dutton & Goodman, 2005). Control, or resulting obedience, is guaranteed via resource management, behavioral micro-regulation, and limiting support options. IPV victims experience coercive control largely through surveillance methods—today, almost entirely technologically facilitated.

Another key consideration for IPV victims, because of technological immediacy, is *intrusion*, or "external control or interference that demands attention, diverts energy…and limits choices" (Wuest, Ford-Gilboe, Merritt-Gray, & Berman, 2003, p. 600). Intrusion involves direct abuse and control, indirect health issues from abuse, treatment costs, and/or ongoing lifestyle adjustments due an abuser's invasion. Clearly, IPV victims experience intrusion in ways never before documented. Being able to identify the many possibilities of TMA—beyond potential tools—is an important step in addressing abusers' tactics.

RQ1: What personal and relational IPV victim characteristics are associated with TMA?

RQ2: In what ways do IPV victims report experiencing TMA?

Method

Participants recruited for this IRB-approved study via social network emailing and diverse (i.e., both violence and non-violence related) web forums included anyone having experienced physically and/or psychologically abusive behavior from a former romantic partner. Completed surveys comprised a final sample of 495 (157 men and 338 women) self-identified victims aged 18 to 74 years (M = 36.68, SD = 13.61). Differently-sexed (94.1%, n = 466) and similarly-sexed (5.9%, n = 29) couples included 160 female- (32.3%) and 335 male- (67.7%) perpetrators. Previous research indicates hetero-/homosexual victims differ primarily on social support resources and societal stigma received—not on particulars of IPV experiences (Hardesty, Oswald, Khaw, & Fonseca, 2011). Therefore, as no significant sexuality differences were found on any variables in this study, all victims were collapsed in the final sample.

After accessing the survey via direct web-link to an SSL-encrypted site erasing IP addresses, people indicated informed consent before completing measures assessing IPV for a larger research project. Participants who responded to an open-ended question indicating ways their partner "used technology to threaten, accuse, or hurt" them during their relationship are denoted as the TMA subsample (n = 187; 67 men, 120 women).

Quantitative measures included demographic questions and the following victimization measures: (a) 19 items (α = .94) on physical tactic frequency (0 = *Never*–6 = *Always*, Total sample M = 2.13, SD = 1.10; TMA subsample M = 2.23, SD = 1.15) from the CTS2 physical subscale (Straus et al., 2003) and the Partner Abuse Scale-Physical (Hudson, 1997); (b) 25 frequency (1 = *Never*–7 = *Always*) items (α = .94) from the Index of Psychological Abuse (Sullivan, Parisian, & Davidson, 1991), Total sample M = 4.15, SD = 1.24; TMA subsample M = 4.29, SD = 1.23; (c) coercive control, from the non-overlapping control-relevant IPA items (Total sample M = 4.21, SD = 1.34; TMA subsample M = 4.32, SD = 1.30); and (d) 10 items (α = .92) of technological abuse frequency (1 = *Never*–7 = *Always*), from a scale created to measure how "forms of technology were used" to abuse (Total sample M = 2.53, SD = 1.58; TMA subsample M = 2.57, SD = 1.59). No significant differences were found between the Total and TMA groups on any victimization variables.

Findings

To address the research questions, I integrated both quantitative and qualitative data analyses with an implications discussion.

TMA-Associated Variables

RQ1 queried the role of personal and relationship factors in victims' TMA. Currently intrusive practices from the former abuser were first assessed. Descriptive data showed 53.5% of participants never interacted with their abuser. For victims still un/voluntarily in contact with their abuser, communication was primarily via phone (52.3%, e.g., call, voice/text message) and/or computer (30.8%, e.g., instant message, email, social networking site). Subsequent findings suggest that "contact" actually served as a form of surveillance for many victims.

Next, bivariate correlations among relationship variables and abuse experiences for the total sample and the TMA subsample show all types of abuse positively correlated with one another in both groups. For all victims, physical abuse positively corresponded to time spent out of the relationship; more physical victimization was reported by those longer out of IPV relationships. Men reported significantly less physical abuse than did women. Finally, across the total sample (but not the TMA subsample), the more victims were established in new relationships, the more physical victimization they reported experiencing from former partners, with whom they had less contact.

Psychological abuse positively correlated with longer IPV relationships, amount of time victims stayed after abuse first began, and more developed relationships with current romantic partners. More coercive control was found in relationships where abuse onset began sooner and among victims currently in more developed new relationships. Men and women did not significantly differ in psychological victimization or coercive control.

Higher levels of technological abuse were significantly associated with: shorter IPV relationship duration, quicker initial abuse onset, less time stayed after abuse began, less time since the relationship ended, less development in new relationships, and younger victims. It may be that younger victims experience more TMA; age was associated only with TMA—not with any other abuse types. If young adults are less likely to view technological communication as harassment than in-person methods of the same behavior (Short & McMurray, 2009), TMA may be even more common than perceived by victims in this study. In the total sample (but not the TMA subsample), men reported significantly less technological victimization than did women.

Technology's potential to reduce relational control and to "force" connection (i.e., reduce autonomy) with a partner have been discussed in non-abusive contexts; current findings suggest abusers particularly harness the technology for intrusion and control purposes. The actual content of perceived-TMA is thus important for victim-response measures.

TMA Content

To explore how TMA is used (RQ2), qualitative free-responses were analyzed for typological variety of technology used by abusers. Using open-coding thematic analysis, 12 TMA categories emerged from 187 volunteered responses. Unedited exemplars are presented with participants' identification number. Phi-coefficients and chi-square analyses show category co-use likelihood and sex-differences, respectively (see Table 21.2). Final analyses determined how particular TMA categories related to differences among abuse types reported.

Emergent categories were grouped into three larger themes based on perceived abuser goals and methods: Emotional-Psychological, Structural-Systemic, and Tangible-Physical. Most of the TMA subsample reported more than one TMA category, with sex differences found in only four categories; women reported more emotional attacks and stalking, whereas men reported more identity theft and slander.

Emotional-Psychological TMA. Emotional or psychological abuse was the primary goal in many TMA categories. Obsessive relational intrusion, emotional attacks, stalking, and intrusion to others were all significantly positively intercorrelated. Further, emotional control also positively related to slander/libel, suggesting a high degree of co-use of emotional-psychological tactics.

Stalking. Typical examples of this TMA form included "surveillance" behaviors such as "Record[ing] my phone calls on our home phone without me knowing it" (#242) or "Monitored my computer use, web access, and read my email w/o my knowledge" (#277). Stalking also included more obtrusive e-monitoring, as with one victim "required to get online with AIM when [her partner was] home" and not allowed to "get on [the] computer without him knowing" (#53); their joint-AIM account allowed her partner to monitor her from any location.

Another form of explicit tracking emphasizing control was mentioned by victims when not immediately reachable by their partners: "If I didn't have my phone on me then it would always be assumed I was cheating on him or just out with another guy" (#229). Consistent with general population stalking statistics, women in this subsample were significantly more likely than men to report technological stalking behaviors, a finding attributable to women's higher fear and threat-perceptions of stalking behavior (compared to men's). In non-technological contexts, even when victimization fits legal definitions of stalking, it is typical for men to not define themselves as "victims" of stalking (Spitzberg & Cupach, 2014).

The fact that victims' knowledge of being unobtrusively monitored was usually due to the abuser telling them lends support to perceptions that stalking-TMA was used more to control and/or psychologically intrude on their lives than it was to merely monitor (see Spitzberg & Cupach, 2014). Stalking as a form of emotional harassment, rather than merely as a tool to stay up-to-date on others' whereabouts, works only if victims become aware of the method's invasiveness (Salter & Bryden, 2009).

Slander/libel. Another psychological TMA form was that of *slander* (i.e., spoken defamation) *or libel* (i.e., written defamation), used to spoil victims' reputations, either directly where he/she would observe it (e.g., My wife posted all over the internet that I was hiding money, cheating on her, and abusing her when I wasn't," #98) or indirectly via third-party notification. For example, one woman's abuser "blind copied certain emails to the kids that would make me look bad and also court documents to…stress them out so that they would turn against me" (#4). Another victim noted his wife had, after their relationship ended, "hacked into my yahoo account and was emailing my church sister telling them I am a liar" (#408).

Reputational damage also occurred through indirect implication of victims' wrongdoing, such as when an abuser "inflicted injuries on herself of which she took digital photos, and accused me of causing them" (#250) or "sending herself an email [from victim's hacked account] to violate a restraining order" (#285). Interestingly, those who reported slander were significantly less likely to co-report experiences with ORI and stalking. This finding could be attributed to sex differences, as males reported significantly more slander than did females, who reported more stalking (see Table 21.2). In some ways, this aligns with previous findings where abused men felt more emotionally victimized by the "system" than by wives' actual attacks (Eckstein, 2010).

ORI. TMA via *obsessive relational intrusion* occurred after actual or perceived relational breaks. For example, one woman's ex-boyfriend would "instant message me numerous times in a day, even if I had an away message up, and write mean things saying that I'm a bitch for not answering him…One time he threw rocks at my dorm window screaming my name and when I didn't respond, he messaged me 15 times in a row; my roommate and I sat there counting" (#153). Another woman's ex-partner "created new screen names and email addresses to try to contact me (I've blocked everything I know how to block), as well as tried to find out info about where I lived" (#216). After moving away and changing her phone number, another woman's ex-partner tracked her down and resorted to relational interaction alternatives, using "Facebook and MySpace as ways to send me emails about our fights" (#170). Males and females did not differ in victimization via this TMA category.

Intrusion to others. TMA also occurred when abusers attempted to irk and/or manipulate victims by invading others' privacy or pursuing relationships with people in the victim's life. Targets of intrusion could be *victim affiliates* (see Spitzberg & Cupach, 2014) directly valued in the victim's life or mere social network members. Close targets were typically family members, as "when I left him he called my sister's house phone and threatened to cause harm to my sister and her children, so I went back" (#57) or "He would call my home late at night, disrupting my entire family while they slept because I wouldn't answer my cell phone after telling him that I no longer wanted to argue with him that night" (#151). Others demonstrated the role social networks could play, as one man noted his girlfriend "Used my cell phone to send demeaning and inappropriate messages to random people in my address book" (#169).

These tactics *to* others affected victims by harming valued people in their lives, as in the example of an abuser who "texted my best friend who he thought was responsible for the downfall of our relationship and said the cruelest thing: 'Hey asshole, I hear your dad likes to drink antifreeze, why don't you go kill yourself too, you fat sack of shit.' (Her father killed himself by drinking antifreeze when she was 5)" (#173). In some cases, intrusion to others was intended to directly hurt those involved with the victim, as in one extreme of alienation attempts from an abuser who (never intending her victim to find out) used technology to "tell people I WAS DEAD so her friends and classmates wouldn't try to find out about me; as well as to make up an excuse for visiting me (she told them she was 'going to my funeral' when really she was just spending vacation with me)" (#231).

Emotional attack. When TMA involved intentionally inflicted affective pain based on intimate, personal knowledge, it constituted an emotional attack. Emotional attack TMA was reported significantly more by women than by men (see Table 21.2). For example, as one woman reported, "He knew I despised porn, so he would leave it up as a screen saver or wallpaper" (#365). Stored films or images were used to disturb victims, as in the case of one woman whose partner "would videotape violent sex acts and send them" (#17) to her; a separate instance involved a woman whose abuser put an extremely graphically violent cartoon in a file under her name on her family computer (#335).

Harnessing both the verbal-attack and public potential of social media, another victim reported "comments via facebook making fun of me or telling me i should go kill myself" (#91). Another victim reported ongoing insults from her partner supplemented by "giv[ing] me a website to go to that would tell me about a new weight loss product or exercise, constantly was telling me I was fat" (#198). In some cases, abusers substituted public-embarrassment with mere frequency of tactics, such as in the case of an onslaught of verbal assault: "At one point, I hung up on him and turned off my phone so I could sleep. He called back every minute and left belittling messages every minute stating the time until my voice mail was full. He then proceeded to text message me insults until my texts were at their limit" (#214). In non-technological IPV contexts, it is relatively easy to distinguish between verbal assault tactics and intrusiveness (and between verbal versus emotional abuse), as the former is limited to in-person exchanges. In the presence of technology, verbal assaults can be compounded by and with intrusiveness and thus, coercive control.

Emotional control. This TMA was categorized by abusers' goals to manipulate and exert power using affective personalized knowledge (as opposed to *emotional attack*, where emotional hurt is the primary goal). The following examples were typical: "[He] lied about his relationship status on Facebook (i.e., saying he was single if we were still dating but fighting or saying he was in a relationship if we were on a break), turned off his phone when he knew i needed his help" (#65). Using his boyfriend's personal contacts against him, another perpetrator "would flirt with other guys and have cyber sex with them to get back at me when he was upset" (#161). Other abusers used threats of technology to control (e.g., "threaten to make calls to my family, if i ever left him" #311), to physically threaten (e.g., "After his pedo pals would take torture pictures of their victims he would threaten me with them that would happen to my kids 'if I made trouble'" #421), or to blackmail (e.g., "Threatened to put up nude pictures of me in embarrasing [*sic*] situations on the internet" #325).

Confirming the coding of this category as coercively controlling were victims' quantitative scores on the coercive control scale—higher if they reported this TMA category (M = 4.54, SD = 1.35, compared to those who did not: M = 4.10, SD = 1.21), $t(185) = 2.39, p < .05$ or $F(1, 185) = 5.69, p < .05$. Emotional control was more likely co-reported by victims experiencing intrusion *by* others (a structural-systemic theme category).

Structural-Systemic TMA. A second theme included TMA using cultural, social, and/or government entities to implement legal or psychological intrusions. This theme included use of *proxies* who may be "co-opted pursuers" and/or knowingly or unknowingly "professionalized" human or instrumental tools of the abuser (Spitzberg & Cupach, 2014, p. 122). Post-coding analyses of this theme showed intrusion by others, identity theft, and economic control all positively related in co-occurrence.

Identity theft. Abusers used TMA by stealing or using previously-shared personal information to interact *as* the victim (e.g., "used my computer while i was at class, got on my chat messenger and tried to get a friend to say innapropriate [*sic*] things to him so he could 'bust' me" #16) or to enact transactions/decisions in the *role of* the victim: (e.g., "used my identity... closed bank accounts..., canceled credit cards, changed address for bills, canceled health insurance, auto insurance...would cancel or change my accounts" #82). Men were significantly more likely to report this TMA than were women.

Those reporting identity theft were more likely to be technologically abused (M = 3.19, SD = 1.63) than those who did not report this category (M = 2.48, SD = 1.57), $t(183) = 2.06$, $p < .05$) or $F(1, 183) = 4.24, p < .05$. Identity theft victims were less likely to report co-experiences with emotional attacks, but more likely to experience psychological abuse (M = 4.89, SD = 1.09, compared to those who did not report identity theft: M = 4.21, SD = 1.22), $t(185) = 2.66$, $p < .01$ or $F(1, 185) = 7.06, p < .01$. Coercive control was also higher among those experiencing identity theft (M = 4.95, SD = 1.04, compared to those who did not: M = 4.23, SD = 1.31), $t(185) = 2.62, p <.01$ or $F(1, 185) = 6.85, p < .01$. Perhaps due to the financial implications as well as life-controlling aspects of identity theft, those who reported this category were also more likely to report their abuser's co-use of economic control (see Table 21.2).

Economic control. This category occurred whenever technology was used to impair financial resources (e.g., "He had my disability & child support checks direct deposited into his acct. w/o my permission" #17) or to limit the earning power of the victim (e.g., "[He] would manipulate [phone] conversations...start arguments on the phone late night when I needed to sleep for long work hours the following day" #214). Unsurprisingly, those reporting economic control experienced higher coercive control (M = 4.54, SD = 1.35, compared to those not reporting: M = 4.10, SD = 1.20), $t(185) = 2.39, p < .05$.

Others reported abusers would "mess with" or otherwise prevent technology use in order to "keep me from performing my job" (#297). Expectedly, economic control was significantly related to co-reports of technological destruction. Both identity theft and economic control would not be possible without the use of technological, largely institutional/organizational, proxies.

Intrusion by others. Abusers were often aided by third-parties – in some cases, members of shared social networks: "his friends are harassing me...prank calls!" (#6); "having other girls that he would cheat on me with call/text me and harass me" (#88). Proxies' harassment or abuse may be conscious or unknowing and "systems" are frequently used, as in using a business to put "personal info on the Internet, specifically mortgage refi sites, to get them to call me or my family a lot" (#438). Proxy intrusion was reported more by those experiencing higher levels technological abuse (M = 3.04, SD = 1.86, compared to those not intruded by others: M = 2.46, SD = 1.50), $t(183) = 1.99, p < .05$ or $F(1, 183) = 3.96, p < .05$.

Tangible-Physical TMA. A final theme included categories where tangible technology objects directly or indirectly harmed victims' physiological and psychological well-being, accomplished via technology as a physical instrument to effect negative health outcomes. Technological destruction was significantly related to with isolation and physical attacks.

Technology destruction. Intentional demolition of machinery was largely used to limit access to it. For example, abusers "cut off computer wires" (#31) or "would always break the house phone so that I couldn't use it to call for help…went through about 18 in one year" (#128). In addition to limiting outside access, this category also included technology destroyed to financially or emotionally harm, as with one man's report: "[She] completely smashed my entire computer system…since my computer was and still is central to how I earn a living" (#418). Because simply removing access also accomplishes physical/social isolation, abusers' choice of technology destruction is notable for its clear co-motive of physical violence threat. Unsurprisingly, technological destruction and TMA physical attacks significantly co-occurred and those reporting technological destruction experience higher technological abuse levels ($M = 3.68$, $SD = 2.07$, compared to those who did not: $M = 2.46$, $SD = 1.50$), $t(17.74) = 2.36$, $p < .05$ or $F(1, 183) = 9.44$, $p < .01$.

Physical attack. Technological objects used to physically assault or to affect victims' health were coded in this category. For example, one abuser used nightly instant messaging to "threaten to break up with me if I signed off to go to bed, knowing very well that I have a medical condition that makes sleep a necessity" (#39). Although some victims questioned its TMA classification, the most common physical attack report involved using a cell phone as a weapon: "He would throw the cell phone at me all the time" (#47), "She threw a phone at my head" (#263). Victims reporting TMA physical attacks also experienced more physical victimization ($M = 3.22$, $SD = 1.28$) than those who did not report this category ($M = 2.16$, $SD = 1.11$), $t(185) = 3.29$, $p < .01$ or $F(1, 185) = 10.84$, $p < .01$. It is unsurprising that technologies' proximal ubiquity increases likelihood of employing them for physical violence.

Isolation. Finally, TMA occurred by preventing access to technology that could provide necessary outside-communication: "If I got a call on the house phone, he would change the number" (#17); "[I] wasn't allowed to have a Facebook or MySpace even though she had both. [I] wasn't allowed to communicate to female friends over the phone via txt or online" (#152); "[He] didn't pay the phone bill so I could not have contact with others. We lived in the country, no neighbors for about a mile" (#428). Another woman reported, "No technology [was] available in the home. [I had] isolation away from people except for a regular phone line at home. He was very sophisticated and would frequently send messages via pager like 'I love you.' These were very twisted in meanings; on the surface they meant one thing to ordinary people, but to him it meant a forced assault" (#355).

Higher levels of psychological abuse were experienced by victims who reported isolating TMA ($M = 4.86$, $SD = 1.13$, compared to those who did not: $M = 4.21$, $SD = 1.22$), $t(185) = 2.47$, $p < .05$ or $F(1, 185) = 6.11$, $p < .05$. It is possible that this was merely co-occurring, as isolated victims were those more likely to be psychologically abused. More likely, however, is that technological isolation served as a facilitator of psychological victimization involving coercive control—or at least the elimination of counter-balancing support networks; those reporting isolating TMA experienced more coercive control ($M = 4.82$, $SD = 0.96$, compared to those not isolated: $M = 4.25$, $SD = 1.32$), $t(36.76) = 2.54$, $p < .05$ or $F(1, 185) = 4.13$, $p < .05$.

Older parallels. A subset of individuals (n = 29, 15.5% of TMA subsample) voluntarily indicated that although absent technology at that time, they were certain their abuser would have used anything available. The same victims then described tactics identical to those victimized by TMA: "My relationship was 13 years ago, email, texting and even cell phones weren't as prevalent…However, he did rack up a $1000 regular phone bill one month checking in on me when he was out of town" (#77). Even in cases where technology was clearly used, some victims compared their experiences to today's advances and did not view their experiences as TMA: "My abuse was prior to most of today's technology. I've no doubt that [TMA] would apply if technology had been available then. He did use hidden cameras, remote listening devices on the home phone (bugged the phone) & monitored the car's odometer" (#330). For IPV in particular, where perpetrators overlap abuse types, the societal boon of technology is clearly a bane for IPV victims.

Conclusions

Quantitative results demonstrated both personal and relational characteristics associated with TMA (RQ1). Technological abuse was a primary facilitator of psychological abuse and coercive control experienced both during and after IPV relationships. Qualitative results revealed additional nuance in the ways (compared to means) technology is used to victimize. Thematic analyses of free-response reports found here inform a future Technologically Mediated Abuse Scale and reveal IPV victims' *ways of experiencing* TMA (RQ2). Emerging categories constructed three themes based on perceived abuser-goals and technologies used: Emotional-Psychological, Structural-Systemic, and Tangible-Physical. Minimal sex differences and frequent category co-use confirmed thematic distinctions based on theories of coercive control.

To gain a more complete picture of TMA, future studies may benefit from using checklists instead of asking participants to self-identify specific past experiences; however, mixed-methods approaches are ideal, because evolving technologies are impossible to comprehensively anticipate in solely checklist-based measures. These methods would allow both researchers and practitioners to move beyond studies of prevalence and description to discover the ways TMA affects victims. For example, what forms of TMA are most "successful" at controlling, harming, stalking, discrediting and/or isolating victims? These findings could aid victims seeking treatment and inform policies surrounding e-privacy, invasiveness, and the ways it is monitored (or not) by larger structures.

To date, TMA research has focused primarily on media, leaving largely unexamined the *extent* to and *ways* in which technologies are used by abusers. Findings from this study begin to fill this void. It is clear that abusers use technology in ways that not only supplement, but *increase* the invasive ubiquity—particularly in intrusiveness and coercive control—of already existing abusive means.

References

Black, M. C., Basile, K. C., Breiding, M. J., Smith, S. G., Walters, M. L., Merrick, M. T., … Stevens, M. R. (2011). *The National Intimate Partner and Sexual Violence Survey (NISVS): 2010 summary report.* Atlanta, GA: National Center for Injury Prevention and Control, Centers for Disease Control and Prevention. Retrieved from www.cdc.gov/violenceprevention/pdf/nisvs_report2010-a.pdf

Dutton, M. A., & Goodman, L. A. (2005). Coercion in intimate partner violence: Toward a new conceptualization. *Sex Roles, 52*(11), 743–756. doi: 10.1007/s11199-005-4196-6

Eckstein, J. (2010). Masculinity of men communicating abuse victimization. *Culture, Society & Masculinities, 2*(1), 62–74. doi: 10.3149/csm.0201.62

Hardesty, J. L., Oswald, R. F., Khaw, L., & Fonseca, C. (2011). Lesbian/bisexual mothers and intimate partner violence: Help seeking in the context of social and legal vulnerability. *Violence Against Women, 17*(1), 28–46. doi: 10.1177/1077801209347636

Hudson, W. W. (1997). *The WALMYR assessment scales scoring manual.* Tallahassee, FL: WALMYR Publishing.

Lehmann, P., Simmons, C. A., & Pillai, V. K. (2012). The validation of the Checklist of Controlling Behaviors (CCB): Assessing coercive control in abusive relationships. *Violence Against Women, 18*(8), 913–933. doi: 10.1177/1077801212456522

Salter, M., & Bryden, C. (2009). I can see you: Harassment and stalking on the Internet. *Information & Communications Technology Law, 18*(2), 99–122. doi: 10.1080/13600830902812830

Short, E., & McMurray, I. (2009). Mobile phone harassment: An exploration of students' perceptions of intrusive texting behavior. *Human Technology, 5*(2), 163–180.

Southworth, C., Dawson, S., Fraser, C., & Tucker, S. (2005). A high tech twist on abuse: Technology, intimate partner stalking, and advocacy. *National Network to End Domestic Violence.* Retrieved from nnedv.org/downloads/SafetyNet/NNEDV_HighTechTwist_PaperAndApxA_English08.pdf

Spitzberg, B. H., & Cupach, W. R. (2014). *The dark side of relationship pursuit: From attraction to obsession and stalking.* New York, NY: Routledge.

Straus, M. A., Hamby, S. L., & Warren, W. L. (2003). *The Conflict Tactics Scales handbook: Revised Conflict Tactics Scales (CTS2), CTS: Parent-Child Version (CTSPC).* Los Angeles, CA: Western Psychological Services.

Sullivan, C. M., Parisian, J. A., & Davidson, W. S. (1991). *Index of psychological abuse: Development of a measure.* Paper presented at the annual conference of the American Psychological Association, San Francisco, CA.

Wuest, J., Ford-Gilboe, M., Merritt-Gray, M., & Berman, H. (2003). Intrusion: The central problem for family health promotion among children and single mothers after leaving an abusive partner. *Qualitative Health Research, 13*(5), 597–622. doi: 10.1177/1049732303013005002

Table 21.1. Total sample and TMA subsample findings regarding victimization, personal, and relational characteristics.

Bivariate Correlations				Men		Women							
Variable	1	2	3	4	5	6	7	8	9	10	*M (SD)*	*M (SD)*	*t (df)*
1. Physical Abuse													
Total Sample	–										1.86 (0.90)	2.25 (1.16)	4.09 (384.49) ***
TMA Subsample	–										1.85 (0.83)	2.45 (1.24)	3.95 (178.76) ***
2. Psycho-logical Abuse													
Total Sample	.55 ***	–									4.02 (1.23)	4.22 (1.25)	1.69 (492)
TMA Subsample	.56 ***	–									4.06 (1.20)	4.42 (1.22)	1.93 (185)
3. Coercive Control													
Total Sample	.44 ***	.89 ***	–								4.08 (1.33)	4.27 (1.34)	1.43 (492)
TMA Subsample	.47 ***	.88 ***	–								4.20 (1.36)	4.40 (1.26)	1.02 (185)
4. Tech-nological Abuse													
Total Sample	.36 ***	.55 ***	.58 ***	–							2.18 (1.29)	2.70 (1.68)	3.76 (381.80) ***
TMA Subsample	.26 ***	.45 ***	.51 ***	–							2.32 (1.35)	2.71 (1.70)	1.57 (183)

Bivariate Correlations				Men		Women							
Variable	1	2	3	4	5	6	7	8	9	10	*M (SD)*	*M (SD)*	*t (df)*
5. IPV Relat. Duration													
Total Sample	.02	.16 ***	.06	-.12 **	–						8.35 (7.43)	5.71 (6.42)	3.83 (268.11) ***
TMA Subsample	.04	.16 *	.03	-.16 *	–						7.72 (6.66)	6.42 (6.84)	1.26 (185)
6. Time Before Onset													
Total Sample	-.07	-.08	-.10 *	-.11 *	.44 ***	–					1.88 (3.08)	0.97 (1.73)	3.46 (201.46) **
TMA Subsample	-.04	-.02	-.04	-.10	.46 ***	–					2.13 (3.48)	0.92 (1.48)	2.71 (79.63) **
7. Time Stayed After Onset													
Total Sample	.04	.19 ***	.08	-.11 *	.93 ***	.12 **	–				6.14 (6.13)	4.52 (5.67)	2.81 (284.57) **
TMA Subsample	.08	.19 *	.05	-.13	.92 ***	.10	–				5.30 (4.88)	5.43 (6.46)	0.15 (184)
8. Time Since IPV Relat.													
Total Sample	.11 *	.01	-.02	-.33 ***	.12 **	.10 *	.11 *	–			6.39 (6.21)	5.97 (7.00)	0.64 (491)
TMA Sub-sample	.16 *	-.02	-.06	-.43 ***	.14 *	.08	.12	–			6.45 (6.43)	7.52 (7.46)	0.99 (185)

Bivariate Correlations				Men		Women							
Variable	1	2	3	4	5	6	7	8	9	10	*M (SD)*	*M (SD)*	*t (df)*
9. New Relat. Commitment													
Total Sample	.10 *	.12 *	.11 *	-.09 *	.00	-.00	.01	.45 ***	–		2.05 (1.19)	2.16 (1.16)	0.98 (486)
TMA Subsample	.08	.04	.06	-.15 *	-.05	.00	-.06	.48 ***	–		1.98 (1.17)	2.30 (1.26)	1.69 (140.58)
10. Current Abuser-Interaction													
Total Sample	-.09 *	-.00	-.02	.02	.16 ***	.11 *	.13 **	-.25 ***	-.19 ***	–	2.38 (1.61)	2.12 (1.59)	1.69 (491)
TMA Subsample	-.09	.00	.01	.04	.20 **	.09	.18 *	-.24 **	-.19 *	–	2.30 (1.54)	2.17 (1.62)	0.54 (184)
11. Current Age													
Total Sample	-.02	.08	-.04	-.33 ***	.60 ***	.27 ***	.57 ***	.54 ***	.19 ***	-.01	43.45 (13.39)	33.50 (12.53)	8.03 (490) ***
TMA Subsample	.01	.04	-.10	-.41 ***	.55 ***	.27 ***	.50 ***	.56 ***	.16 *	-.02	43.43 (13.44)	37.16 (12.50)	3.20 (184) **

Note. Total Sample N = 495 (157 men, 338 women). TMA Subsample N = 187 (67 men, 120 women). t-tests are independent samples. * $p < .05$. ** $p < .01$. *** $p < .001$.

Table 21.2. Technologically-Mediated Abuse Open-Ended Categories

TMA Category	Sex differences[a]			Phi (φ) score likelihood associations of dual category co-use										
	Men	Women	χ^2	ORI	Emot. Attack	Stalking	Intrus By Others	Slander	Econ. Control	ID Theft	Intrus To Others	Isolation	Phys. Attack	Tech. Destruct
Emotional Control	35	61	.25	.10	–.05	.03	.17 *	.17 *	–.06	.08	.14	–.01	–.17 *	–.14
(N = 96, 58.9%)	*56.5%*	*60.4%*												
Obsess. Rel. Intrus.	19	41	1.64	–	.23 **	.20 *	–.01	–.17 *	–.03	–.01	.19 *	–.10	–.18 *	–.19 *
(N = 60, 36.8%)	*30.6%*	*40.6%*												
Emotional Attack	15	41	4.58 *		–	–.12	–.04	.06	–.01	–.17 *	.17 *	–.05	–.02	–.13
(N = 56, 34.4%)	*24.2%*	*40.6%*												
Stalking	11	33	4.35 *			–	–.09	–.22 **	–.05	.09	–.10	–.02	–.08	–.17 *
(N = 44, 27.0%)	*17.7%*	*32.7%*												
Intrusion By Others	18	18	2.81				–	.21 **	.20 *	.23 **	.20 *	–.14	–.05	–.05
(N = 36, 22.1%)	*29.0%*	*17.8%*												
Slander	20	10	12.79 ***					–	.00	.19 *	.29 ***	–.06	–.08	–.12
(N = 30, 18.4%)	*32.3%*	*9.9%*												
Economic Control	14	13	2.62						–	.18 *	.05	.00	–.01	.21 **
(N = 27, 16.6%)	*22.6%*	*12.9%*												
Identity Theft	16	9	8.45 **							–	.16 *	–.03	–.06	–.04
(N = 25, 15.3%)	*25.8%*	*8.9%*												
Intrusion To Others	11	13	.73								–	–.03	–.06	–.04
(N = 24, 14.7%)	*17.7%*	*12.9%*												
Isolation	5	19	3.53									–	–.06	.19 *
(N = 24, 14.7%)	*8.1%*	*18.8%*												
Physical Attack	3	10	1.34										–	.26 **
(N = 23, 1.2%)	*4.8%*	*9.9%*												
Technol. Destruct.	6	12	.19											–
(N = 18, 11.0%)	*9.7%*	*11.9%*												

Notes. N = 187 participants (120 women, 67 men). [a]Delineations show sexes reporting each TMA category and "within sex" percentages.
* $p < .05$. ** $p < .01$. *** $p < .001$.

TWENTY-TWO

Turbulence, Turmoil, and Termination: The Dark Side of Social Networking Sites for Romantic Relationships

Jesse Fox and Courtney Anderegg

Social networking websites (SNSs) have become an integral medium for communicating within and about interpersonal relationships (boyd & Ellison, 2008; Stafford & Hillyer, 2012). SNSs have been lauded for their ability to unite distal friends, maintain relational ties, facilitate relationship development, and promote social capital (e.g., Ellison, Vitak, Gray, & Lampe, 2014; Fox, Warber, & Makstaller, 2013; McEwan, 2013). Although considerable research has elected to focus on the benefits of using SNSs, it is also important to examine the dark side of computer-mediated communication (DeAndrea, Tong, & Walther, 2011). For example, SNS use has been tied to decreases in psychological well-being (Chen & Lee, 2013), and scholars have noted negative psychological outcomes when users experience rejection on SNSs (e.g., Bevan, Ang, & Fearns, 2014; Tokunaga, 2011a, 2014).

One area where the dark side of SNSs may be most prevalent is that of romantic relationships. Research has begun to acknowledge the role that SNSs play in the initiation, escalation, maintenance, and dissolution of romantic relationships (e.g., Carpenter & Spottswood, 2013; Fox, Jones, & Lookadoo, 2013; Fox & Warber, 2013; Marshall, 2012; Papp, Danielewicz, & Cayemberg, 2012). SNSs provide evidence of online and offline activities, which allows a romantic partner to covertly engage in information seeking and uncertainty reduction (Fox & Anderegg, 2014; Tokunaga, 2011b). It also introduces new sources of potential conflict, may create undesirable uncertainty, and gives other social network members greater access to information about the couple (Fox, Osborn, & Warber, 2014).

In this chapter, we examine the affordances of SNSs in terms of how they initiate, promote, or intensify destructive romantic relationship communication. We elaborate various dark side behaviors and experiences on SNSs related to romantic relationships, including social comparison, negative relational maintenance, romantic jealousy, and partner monitoring. Additionally, we discuss relational issues exacerbated by SNSs, including technological incompatibility, destructive secret tests, and cyberstalking.

Affordances of SNSs

SNSs have specific social affordances that enable the actions one can take within the site (Fox & Moreland, 2015; Treem & Leonardi, 2012). *Affordances* are the perceived properties of a technology that enable specific actions (Norman, 1988). These affordances determine how social information is conveyed and transmitted throughout the network, which influences how users receive, interpret, and are affected by this information. Thus, affordances have important implications for how dark side behaviors manifest differently on SNSs compared to other communication channels.

One draw of SNSs is their ability to link individuals in one common virtual space. The affordance of *connectivity* or *association* enables network members, no matter how disparate or geographically distant, to recognize each other's presence and view each other's content through a common node or "friend." *Visibility* means that information that was not easily accessible or publicized previously is now shared among the network (Treem & Leonardi, 2012). Connectivity and visibility enable individuals to view information about their romantic partners that they may not have regular access to, such as seeing pictures and posts from previous relationships, which may foster relational uncertainty, jealousy, or suspicions. Further, given that social network members often have a significant influence on an individual's romantic relationships (Hogerbrugge, Komter, & Scheepers, 2013; Sprecher, 2011), these two affordances may maximize the network's influence on—or meddling in—a romantic relationship, as there is more fodder for gossip and speculation about the nature or health of the relationship.

Persistence, editability, and *replicability* are tied to the digital nature of text, pictures, videos, and other content. Information shared online may be accessible long after the initial post and difficult to remove permanently (Treem & Leonardi, 2012). Persistence and replicability also make it difficult to hide transgressions, relational indiscretions, or otherwise suspicious behaviors if they are posted online. Even if content is removed, others may have stored it or shared it among other networks. Furthermore, several editing tools enable digital information to be manipulated, from simple cropping to intensive reconfiguration. In this way, artificial or deceptive material could be created to cause turmoil in a relationship.

Individual sites also have specific affordances that may foster negative experiences. One particular Facebook feature, the ability to go "Facebook official" or "FBO" (i.e., link to one's partner in the relationship status), affords partner-specific connectivity (Fox & Warber, 2013; Papp et al., 2012). Although this opportunity may seem like a way to promote togetherness, partners often have differing perceptions of the meaning and timing of this relationship status (Fox & Warber, 2013), which can lead to tension, uncertainty, and conflict (Fox et al., 2014). Other SNSs like Whisper, Secret, and Yik Yak are designed to afford *anonymity* (Wang, Wang, Wang, Nika, Zheng, & Zhao, 2014). In these environments, posters feel confident they will not be identified, which may facilitate cyberaggression (Wright, 2014). Thus, it is important to consider that the same affordances that allow us to share experiences and memories also have the potential to challenge, complicate, or damage romantic relationships.

Technological Incompatibility

Although similar attitudes and behaviors regarding technology use can facilitate relationships (Ledbetter, 2014), relationship difficulties can also emerge as a result of *technological incompatibility*, or any problematic discrepancy in technology use between partners. This incompatibility may be based on the amount or timing of use, type of connections maintained, or content shared on a site (Fox et al., 2014; Fox & Moreland, 2015). For example, Bailey may feel uncomfortable with Thomas's insistence on posting all of their intimate honeymoon pictures publicly on Instagram, because Bailey prefers to keep his social media presence professional.

Any such discrepancies in SNS use may create discord or conflict in romantic relationships. Indeed, negative perceptions of how a romantic partner uses social media can diminish feelings of relational intimacy (Hand, Thomas, Buboltz, Deemer, & Buyanjargal, 2013). Some couples have divergent expectations for romantic relationship maintenance via SNSs, and different practices by partners can create conflict (Fox & Moreland, 2015). Some romantic partners struggle to establish boundaries for privacy on SNSs and argue about what is acceptable to publicize to the network; in extreme cases, this can lead to relationship termination (Fox et al., 2014).

One possible explanation for these discrepancies is differences in romantic partners' attachment style. Attachment Theory suggests that our tendencies to be anxious or avoidant toward others has significant implications for how individuals experience, enact, and communicate within romantic relationships (Hazan & Shaver, 1987; Simpson, 1990). Several studies have found that attachment styles predict various negative relational behaviors on SNSs (e.g., Fox, Peterson, & Warber, 2013; Fox & Warber, 2014; Marshall, 2012; Marshall, Benjanyan, Di Castro, & Lee, 2013). In general, those who are high in anxious attachment rely more on SNSs, put significant stock in their content, and experience more negative emotions as a result. Avoidant individuals typically prefer not to communicate with their partners via SNSs unless they can be used to create distance from the partner. Differences in attachment style may lend themselves to technological incompatibility and lead to conflict. Thus, it is important that couples assess their behaviors and relational expectations and negotiate acceptable SNS practices within the relationship. Although technological incompatibility could incite conflict on SNSs, users also need to be mindful of negative maintenance behaviors enacted online.

Negative Relationship Maintenance

Relationship maintenance refers to the behaviors that an individual engages in to keep a romantic relationship in its current state (Canary & Stafford, 1994; Dindia, 2003). Relationship maintenance behaviors, such as displaying positivity to one's partner, disclosing personal information, and attempting to integrate friends and family into the relationship, are largely seen as positive behaviors (Stafford & Canary, 1991). Several behaviors, however, qualify as negative relational maintenance, such as jealousy induction (Dainton & Gross, 2008) and interpersonal electronic surveillance (Tokunga, 2011b). Importantly, the use of negative maintenance behaviors has been found to decrease levels of relationship satisfaction (Dainton & Gross, 2008).

Relationship maintenance behaviors—both positive and negative—are often enacted online via SNSs (McEwan, 2013). Indeed, relationship maintenance is one of the most important reasons for why individuals use Facebook; however, it is not the *amount* of Facebook use, but the *type* of Facebook use that has the greatest impact on romantic relationships online. Dainton and Berksoski (2013) found that tests of infidelity (a negative maintenance behavior), assurances (a positive maintenance behavior), and levels of jealousy on Facebook predicted almost 50% of the variance explained in relational satisfaction. Thus, negative relational maintenance behaviors on SNSs may take a significant toll on romantic relationships.

Several researchers claim that whether or not individuals engage in negative maintenance behaviors may depend on the initial state of the relationship (e.g., Dainton & Gross, 2008; Goodboy & Bolkan, 2011). For instance, if Rhonda is fearful that her partner Cedric is interested in other women, she may monitor Cedric's interactions on Facebook. However, if Rhonda uncovers suspicious posts on Cedric's profile or if Cedric finds out about Rhonda's monitoring, the relationship may become even more dysfunctional than before. As such, negative maintenance behaviors are often not successful in maintaining relationships, but instead propel them toward dissolution. Similar to negative maintenance behaviors, negative social comparisons can also take place on SNSs and be detrimental to a relationship.

Social Comparison

Upward social comparison occurs when an individual identifies someone of higher status or other desirable traits and then reflects on one's own shortcomings in contrast. Several studies have found that SNSs are a common context for detrimental social comparisons, and they lead to diminished self-perceptions, negative emotions, and depressive symptoms (Feinstein, Hershenberg, Bhatia, Latack, Meuwly, & Davila, 2013; Fox & Moreland, 2015; Haferkamp & Krämer, 2011; Lee, 2014).

SNSs are also a context in which relational comparisons may be made. According to Interdependence Theory (Kelley & Thibaut, 1978), there are two types of comparisons that an individual makes in the context of relationships: comparing the existing relationship to others' relationships, or comparing the choice to remain in the relationship with other opportunities. An individual may consider the comparison level (CL) for the relationship (Kelley & Thibaut, 1978) by comparing it to other couples' relationships as portrayed on SNSs. SNSs also enable individuals to explore their comparison level for alternatives (CLalt), or other options besides remaining in the current relationship (Kelley & Thibaut, 1978).

Another form of social comparison is judging the self against a partner's perceived alternatives. Because of the affordances of connectivity and visibility, considerably more information about a partner's romantic alternatives is made available on SNSs. Rebecca might use Facebook to see how attractive or successful her boyfriend's ex-girlfriends are, or she might scope out his single female friends and compare herself to them. Given the number of bases for comparison typically available on SNSs, it is likely that at least one of these will evoke negative reactions. Another example of using SNSs in a negative manner in relationships is employing destructive secret tests.

Destructive Secret Tests

Throughout the various stages of a relationship (i.e., initiation, maintenance, and termination),[1] individuals may need to reduce or reconcile uncertainty that they may have about their partner or the future of the relationship (Berger & Calabrese, 1975). Different strategies may be used to reduce uncertainty, but often the goal is the same: to acquire information about a romantic relational partner. These strategies, or *secret tests*, allow individuals to gain insight and reduce uncertainty about the relationship (Baxter & Wilmot, 1984).

SNSs provide a unique platform in which individuals are able to conduct secret tests. Tests can range from positive and hopeful (e.g., trying to determine the seriousness of a relationship) to detrimental and damaging (e.g., trying to catch a partner engaging in inappropriate behavior). Fox, Peterson, and Warber (2013) found that secret tests are executed both positively and negatively via SNSs. Partners often took advantage of the affordances of SNSs (e.g., the ability to make comments to one's partner visible to the network or the visibility of one's connections and communication with other network members) to test the definition and boundaries of their relationship and the intentions, commitment, and fidelity of their partners.

One commonly used test is the separation test, in which the individual attempts to disconnect from or avoid the partner (Fox, Peterson, & Warber, 2013). On SNSs, this test is executed by deliberately ignoring messages, tags, and posts from the partner to see how he or she will react. If the partner also avoids contact, the individual might take this as a sign that the partner is uninterested in the relationship. Perhaps the most frequently used negative test is one that attempts to invoke partner jealousy by openly flirting with another person through posts or "liking" an ex-partner's content. Because SNSs make these interactions visible to the network, these actions are used to bait the partner and evoke a reaction. A third, relatively infrequent type is the triangle test, where a third party would be asked to help test a relational partner's fidelity (Fox, Peterson, & Warber, 2013). For example, if Louise doesn't trust her girlfriend Amy, she may ask another friend to post something flirtatious on Amy's page to see if Amy flirts back. Although secret tests can be used to benefit a relationship, relational partners often use SNSs to tempt their partners with opportunities for infidelity or create relational turmoil. As we can see, secret tests often stem from or invoke romantic jealousy, which has been a common focus of research on SNSs in romantic relationships.

Romantic Jealousy

Due to the amount of information available on SNSs, it is possible that they may stir up jealousy in relationships (Bevan, 2013), particularly if those individuals are anxiously attached (Marshall et al., 2013). Previous studies have found that higher levels of Facebook use or involvement with Facebook predict greater relational jealousy (Elphinston & Noller, 2011; Muise, Christofedes, & Desmarais, 2009) and dissatisfaction (Elphinston & Noller, 2011). Other studies have shown that certain content on a partner's SNS profile has the potential to trigger jealous or angry reactions (Muise, Christofedes, & Desmarais, 2014; Muscanell, Guadagno, Rice, & Murphy, 2013).

Experiences of jealousy and uncertainty in relationships may be a vicious cycle when both partners use SNSs (Fox & Warber, 2014; Muise et al., 2009). Individuals may seek out their partner's profile to alleviate relational concerns, but the content they find may trigger greater uncertainty or jealousy. As a result, the individual may then engage in ongoing surveillance, which may exacerbate feelings of uncertainty or jealousy. Thus, particularly for individuals high in trait jealousy (Utz & Beukeboom, 2011), SNSs may be a consistent trigger if the partner's romantic history or current interactions are visible, which could encourage partner monitoring on these sites.

Partner Monitoring

Social networking sites provide a novel way for partners to gather information about each other (Fox & Anderegg, 2014). Indeed, monitoring another person is one of the most common reasons people use SNSs (Joinson, 2008). Tokunaga (2011b) identified four characteristics of SNSs that promote *interpersonal electronic surveillance* (IES) of one's romantic partner. First, information is readily accessible through these sites. It is easy to join an SNS and access the profiles of your connections or your connections' connections. Second, information on SNSs is often comprised of various media such as textual messages, photographs, links, and audio or video clips. Given that pictures are considered more credible than words on SNS profiles (Van Der Heide, D'Angelo, & Schumaker, 2012), this capability may be particularly relevant to partners with suspicions. Third, SNSs allow the archiving of profile information (i.e., they afford persistence; Treem & Leonardi, 2012). Partners may conduct IES of the target's past posts, photos, or interactions with others to gather more data. Fourth, given that neither geographical proximity nor social interaction is necessary to obtain this information, data may be gathered more surreptitiously. Many SNSs, including Facebook and Twitter, do not provide feedback regarding which network members have accessed one's profile. Thus, the target may never know that he or she is under surveillance by the partner.

In addition to Tokunaga's (2011b) characteristics, a fifth characteristic also makes SNSs optimal for partner surveillance: the multiplicity of sources available. It is not only the target who is contributing to his or her profile page, but also other network members. According to Warranting Theory, information that comes from sources other than the self is seen as more credible (Walther & Parks, 2002). Also, information that comes from multiple sources (e.g., several network members, or both comments and pictures) would also be perceived as more credible (Flanagin & Metzger, 2007). Facebook in particular makes this "friendsourcing" easy: not only can friends mention the target or upload media about the target, but they can also tag the target in posts, check-ins, or photos and have that information appear on the target's page as well.

Given these affordances, it is unsurprising that several studies have shown that Facebook is commonly used to monitor one's romantic partner or ex-partner (e.g., Elphinston & Noller, 2011; Fox & Warber, 2014; Lyndon, Bonds-Raacke, & Cratty, 2011; Marshall, 2012; Marshall et al., 2013; Tokunaga, 2011b; Tong, 2013). Because Facebook allows both self-generated and other-generated information to be tied to one's profile, there are multiple sources of information conveniently amalgamated in one easily accessible location. Perhaps the greatest source of information is photographs, which may reveal considerable detail about where a partner is, who the partner is with, and what the partner is doing. Thus, Facebook often serves as an indirect source for knowledge about romantic partners and may inform feelings or decisions about the relationship at every stage, even after dissolution.

Research indicates that potential relationship threats often arise on SNSs: attractive new friends may emerge, questionable photographs from a weekend event may be shared, or flirty comments from an enviable other may appear on the partner's page (Fox, Warber, & Makstaller, 2013; Marshall et al., 2013). Without SNSs, many of these behaviors would still occur, but they would remain hidden from the partner. It is the expression enabled through SNSs, as well as the act of distributing this information online versus offline, that creates distress that may have otherwise been avoided. Interestingly, despite knowledge of the potential relational consequences, many individuals acknowledge that they "creep" (i.e., inspect a person's page without his or her knowledge in order to gain information) on their partner's and others' profiles to obtain information the partner might otherwise try to conceal (Fox, Warber, & Makstaller, 2013; Muise et al., 2014). This behavior, however, can escalate from minimally invasive to far more threatening.

Cyberstalking and Obsessive Relational Intrusion

Although the terms "creeping" and "Facebook stalking" already indicate that there is something discomforting about having someone surreptitiously monitoring one's SNS profile, the casual social monitoring promoted by SNSs can escalate to a problematic or even dangerous level. Continuous surveillance and unwanted pursuit of a romantic interest is known as *obsessive relational intrusion* (ORI; Spitzberg & Cupach, 2003). SNSs are optimally designed to facilitate ORI because (1) targets often share a vast amount of personal information on these sites; (2) perpetrators can monitor this information easily, privately, and as frequently as they like; and (3) SNSs provide many different channels through which the perpetrator can reach the target (Chaulk & Jones, 2011). Some ORI behaviors on SNSs include posting unwanted material to the target's profile; sending unwanted private messages; or tagging the target in posts or pictures.

Recent research suggests that cyberstalking via SNSs is not uncommon (Dreßing, Bailer, Anders, Wagner, & Gallas, 2014). Among users of a German SNS, Dreßing and colleagues found more than 40% had been cyberstalked and 6.3% experienced problematic cyberstalking. Most often, the victim and perpetrator were ex-romantic partners. Despite the fact that these interactions were taking place virtually, there were still significant negative outcomes for victims, including anger, depression, and sleep disturbances. Thus, one of the darkest aspects of SNSs is that they may enable persistent and potentially harmful unwanted attention, interference, or stalking from former romantic partners. These behaviors often occur in the wake of relationship dissolution, perhaps because SNSs are often one of the last lingering connections between ex-partners.

Relationship Dissolution and SNSs

Given both the public nature of the relationship and the integration of the couple's digital presence on SNSs, relationship dissolution in the age of social media is a particularly messy process (Gershon, 2011). If couples have been together for a long period of time, it is likely that they have developed a conjoined presence on the sites they both use (e.g., old posts and pictures may populate the profile). Thus, it is unsurprising that individuals typically report cleaning up their SNS profile by removing the digital detritus of the relationship (Fox, Jones, & Lookadoo, 2013). Although this may be a painful process, this purging may also serve as a coping ritual.

Because SNSs offer easy access to one's own network as well as the significant other's, they often serve many functions in the wake of a breakup and may allow some dark side behaviors to emerge. Lyndon and colleagues (2011) identified three manners in which individuals use Facebook negatively in the wake of a breakup: *venting* (e.g., directly making negative comments about an ex-partner or relationship), *covert provocation* (e.g., passive aggressive posting on the wall to make the ex jealous or angry), and *public harassment* (e.g., spreading rumors about or posting embarrassing photos of the ex-partner). Another recent study explored the different ways in which individuals react to a breakup on Facebook (Fox, Jones, & Lookadoo, 2013). Most commonly, people felt pressured by their SNS presence to pretend that they were unaffected by the breakup. Often, users exaggerated positive activities after the breakup, trying to prove to their network (and often the ex as well) that they were doing better than ever. Although people may be able to grieve the relationship normally offline, the pressure to maintain face and hide one's true emotional state on SNSs may cause greater distress. Facebook users also were found to publicly bash the ex-partner—or to allow friends to bash the ex-partner—on one's page after a breakup (Fox, Jones, & Lookadoo, 2013). In these cases, Facebook was weaponized in a battle to "win" the breakup publicly, either by hurting the ex's reputation or getting shared network members to take sides. Often, this created more animosity between ex-partners.

After a breakup, uncertainty about the relationship's future may remain. In the wake of termination, it is not uncommon for ex-partners to remain "friends" on Facebook (Fox & Warber, 2014; Marshall, 2012; Marshall et al., 2013; Tokunaga, 2011b). This lingering connection and access to post-breakup experiences may foster feelings of uncertainty after dissolution (Fox, Jones, & Lookadoo, 2013; Tong, 2013). Thus, it is unsurprising that individuals often monitor their exes on SNSs long after the relationship is over (Fox & Warber, 2014; Marshall, 2012; Marshall et al., 2013; Tong, 2013).

Post-breakup SNS monitoring is not without consequence. Marshall (2012) found that individuals who monitor their ex-partner's Facebook page after a breakup reported greater levels of distress and negative feelings, greater longing for the ex-partner, and less emotional recovery from the breakup. Thus, even when the individual is not using an SNS for negative expression or self-disclosure about the breakup, SNSs may still have negative consequences for individuals post-dissolution.

Another recent line of research has examined SNSs as a potential instigator or trigger for relationship termination. According to a survey by the American Academy of Matrimonial Lawyers (2010), 81% of divorce lawyers reported an increase in the use of SNSs as evidence in divorce proceedings. To address this possible relationship, Valenzuela, Halpern, and Katz (2014) examined SNS use in U.S. married couples over time. After controlling for several social and economic factors, they observed an association between the adoption of Facebook and increasing divorce rates. Further, SNS use was negatively correlated with perceptions of marital quality and happiness, and positively correlated with relationship trouble and contemplating divorce. Although these data are survey based and thus no causal conclusions can be drawn, they indicate that SNSs may introduce or exacerbate the dark side of romantic relationships.

Conclusion

As Stafford and Hillyer (2012) note, our understanding of the role of technologies in personal relationships is nascent. Unfortunately, people tend to adopt technologies and integrate them into their lives without stopping to question whether their impact is mostly beneficial or detrimental in particular contexts. Considerable research indicates that SNSs may have negative effects on relationships in terms of stirring up jealousy and conflict. Romantic partners should critically evaluate how SNSs function in their relationship, as they may need to set boundaries in terms of SNS use to capitalize on its benefits while avoiding or mitigating its downsides. Although SNSs have often been shown to have positive effects in relationships, there is great potential for the dark side to emerge in romantic relationships, and it is up to users to manage that balance.

Notes

1 According to Knapp (1978), relationship initiation is defined as the first interaction between two individuals. Maintenance is defined as the behaviors enacted to keep the relationship in a specified state (Dindia, 2003) and termination is defined as the relationship's end, whether incremental or due to a critical event (Baxter, 1984).

References

American Academy of Matrimonial Lawyers. (2010). Big surge in social networking evidence says survey of nation's top divorce lawyers. Retrieved from http://www.aaml.org/about-the-academy/press/press-releases/e-discovery/big-surge-social-networking-evidence-says-survey-

Baxter, L. A. (1984). Trajectories of relationship disengagement. *Journal of Social & Personal Relationships, 1*, 29–48. doi: 10.1177/0265407584011003

Baxter, L. A., & Wilmot, W. W. (1984). "Secret tests": Social strategies for acquiring information about the state of the relationship. *Human Communication Research, 11*,171–201. doi: 10.1111/j.1468-2958.1984.tb00044.x

Berger, C. R., & Calabrese, R. J. (1975). Some explorations in initial interaction and beyond: Toward a developmental theory of interpersonal communication. *Human Communication Research, 1*, 99–112. doi: 10.1111/j.1468-2958.1975.tb00258.x

Bevan, J. L. (2013). *The communication of jealousy.* New York, NY: Peter Lang.

Bevan, J. L., Ang, P. C., & Fearns, J. B. (2014). Being unfriended on Facebook: An application of expectancy violation theory. *Computers in Human Behavior, 33*, 171–178. doi: 10.1016/j.chb.2014.01.029

boyd, d. m., & Ellison, N. B. (2008). Social network sites: Definition, history, and scholarship. *Journal of Computer-Mediated Communication, 13*, 210–230. doi: 10.1111/j/1083-6101.2007.00393.x

Canary, D. J., & Stafford, L. (1994). Maintaining relationships through strategic and routine interaction. In D. J. Canary & L. Stafford (Eds.), *Communication and relational maintenance* (pp. 3–22). San Diego, CA: Academic Press.

Carpenter, C. J., & Spottswood, E. L. (2013). Exploring romantic relationships on social networking sites using the self-expansion model. *Computers in Human Behavior, 29*, 1531–1537. doi: 10.1016/j.chb.2013.01.021

Chaulk, K., & Jones, T. (2011). Online obsessive relational intrusion: Further concerns about Facebook. *Journal of Family Violence, 26*, 245–254. doi: 10.1007/s10896-011-9360-x

Chen, W., & Lee, K. H. (2013). Sharing, liking, commenting, and distressed? The pathway between Facebook interaction and psychological distress. *Cyberpsychology, Behavior, & Social Networking, 16*, 728–734. doi:10.1089/cyber.2012.0272

Dainton, M., & Berksoski, L. (2013). Positive and negative maintenance behaviors, jealousy, and Facebook: Impacts on college students' romantic relationships. *Pennsylvania Communication Annual, 69*, 35–50.

Dainton, M., & Gross, J. (2008). The use of negative behaviors to maintain relationships. *Communication Research Reports, 25*, 179–191. doi: 10.1080/08824090802237600

Dindia, K. (2003). Definitions and perspectives on relational maintenance communication. In D. J. Canary & M. Dainton (Eds.), *Maintaining relationships through communication: Relational, contextual, and cultural variations* (pp. 1–30). Mahwah, NJ: Erlbaum.

DeAndrea, D. C., Tong, S. T., & Walther, J. B. (2011). Dark sides of computer-mediated communication. In W. R. Cupach & B. H. Spitzberg (Eds.), *The dark side of close relationships II* (pp. 95–118). New York, NY: Routledge.

Dreßing, H., Bailer, J., Anders, A., Wagner, H., & Gallas, C. (2014). Cyberstalking in a large sample of social network users: Prevalence, characteristics, and impact upon victims. *Cyberpsychology, Behavior, & Social Networking, 17*, 61–67. doi: 10.1089/cyber.2012.0231

Ellison, N. B., Vitak, J., Gray, R., & Lampe, C. (2014). Cultivating social resources on social network sites: Facebook relationship maintenance behaviors and their role in social capital processes. *Journal of Computer-Mediated Communication, 19*, 855–870. doi: 10.1111/jcc4.12078

Elphinston, R. A., & Noller, P. (2011). Time to face it! Facebook intrusion and the implications for romantic jealousy and relationship satisfaction. *Cyberpsychology, Behavior, & Social Networking, 14*, 631–635. doi: 10.1089/cyber.2010.0318

Feinstein, B. A., Hershenberg, R., Bhatia, V., Latack, J. A., Meuwly, N., & Davila, J. (2013). Negative social comparison on Facebook and depressive symptoms: Rumination as a mechanism. *Psychology of Popular Media Culture, 2*, 161–170. doi: 10.1037/a0033111

Flanagin, A. J., & Metzger, M. J. (2007). The role of site features, user attributes, and information verification behaviors on the perceived credibility of Webbased information. *New Media & Society*, 9(2), 319–342.

Fox, J., & Anderegg, C. (2014). Romantic relationship stages and behavior on social networking sites: Uncertainty reduction strategies and perceived norms on Facebook. *CyberPsychology, Behavior, & Social Networking, 17*, 685–691. doi: 10.1089/cyber.2014.0232

Fox, J., Jones, E. B., & Lookadoo, K. (2013, June). Romantic relationship dissolution on social networking sites: Social support, coping, and rituals on Facebook. Paper presented at the 63rd Annual Conference of the International Communication Association, London, UK.

Fox, J., & Moreland, J. J. (2015). The dark side of social networking sites: An exploration of the relational and psychological stressors associated with Facebook use and affordances. *Computers in Human Behavior, 45*, 168–176. doi: 10.1016/j.chb.2014.11.083

Fox, J., Osborn, J. L., & Warber, K. M. (2014). Relational dialectics and social networking sites: The role of Facebook in romantic relationship escalation, maintenance, conflict, and dissolution. *Computers in Human Behavior, 35*, 527–534. doi: 10.1016/j.chb.2014.02.031

Fox, J., Peterson, A., & Warber, K. M. (2013). Attachment style, sex, and the use of secret tests via social networking sites in romantic relationships. Paper presented at the Multi-Level Motivations in Close Relationship Dynamics Conference of the International Association for Relationship Research, Louisville, KY.

Fox, J., & Warber, K. M. (2013). Romantic relationship development in the age of Facebook: An exploratory study of emerging adults' perceptions, motives, and behaviors. *CyberPsychology, Behavior, & Social Networking, 16*, 3–7. doi:10.1089/cyber.2012.0288

Fox, J., & Warber, K. M. (2014). Social networking sites in romantic relationships: Attachment, uncertainty, and partner surveillance on Facebook. *CyberPsychology, Behavior, & Social Networking, 17*, 3–7. doi: 10.1089/cyber.2012.0667

Fox, J., Warber, K. M., & Makstaller, D. C. (2013). The role of Facebook in romantic relationship development: An exploration of Knapp's relational stage model. *Journal of Social & Personal Relationships, 30*, 772–795. doi:10.1177/026540751246837

Gershon, I. (2011). *The breakup 2.0: Disconnecting over new media.* Ithaca, NY: Cornell University Press.

Goodboy, A. K., & Bolkan, S. (2011). Attachment and the use of negative relational maintenance behaviors in romantic relationships. *Communication Research Reports, 28*, 327–336. doi: 10.1080/08824096.2011.616244

Haferkamp, N., & Krämer, N. C. (2011). Social comparison 2.0: Examining the effects of online profiles on social-networking sites. *Cyberpsychology, Behavior, & Social Networking, 14*, 309–314. doi: 10.1089/cyber.2010.0120

Hand, M. M., Thomas, D., Buboltz, W. C., Deemer, E. D., & Buyanjargal, M. (2013). Facebook and romantic relationships: Intimacy and couple satisfaction associated with online social network use. *Cyberpsychology, Behavior, & Social Networking, 16*, 8–13. doi: 10.1089/cyber.2012.0038

Hazan, C., & Shaver, P. (1987). Romantic love conceptualized as an attachment process. *Journal of Personality & Social Psychology, 52*, 511–524. doi: 10.1037/0022-3514.52.3.511

Hogerbrugge, M. J., Komter, A. E., & Scheepers, P. (2013). Dissolving long-term romantic relationships: Assessing the role of the social context. *Journal of Social & Personal Relationships, 30*, 320–342. doi: 10.1177/0265407512462167

Joinson, A. N. (2008). Looking at, looking up, or keeping up with people? Motives and use of Facebook. In *Proceedings of the SIGCHI Conference on Human Factors in Computing Systems* (pp. 1027–1036). Florence, Italy: ACM.

Kelley, H. H., & Thibaut, J. (1978). *Interpersonal relations: A theory of interdependence.* New York, NY: Wiley.

Knapp, M. L. (1978). *Social intercourse: From greeting to goodbye.* Needham Heights, MA: Allyn & Bacon.
Ledbetter, A. M. (2014). Online communication attitude similarity in romantic dyads: Predicting couples' frequency of e-mail, instant messaging, and social networking site communication. *Communication Quarterly, 62*, 233–252. doi: 10.1080/01463373.2014.890120
Lee, S. Y. (2014). How do people compare themselves with others on social network sites? The case of Facebook. *Computers in Human Behavior, 32*, 253–260. doi: 10.1016/j.chb.2013.12.009
Lyndon, A., Bonds-Raacke, J., & Cratty, A. D. (2011). College students' Facebook stalking of ex-partners. *Cyberpsychology, Behavior, & Social Networking, 14*, 711–716. doi: 10.1089/cyber.2010.0588
Marshall, T. C. (2012). Facebook surveillance of former romantic partners: Associations with postbreakup recovery and personal growth. *Cyberpsychology, Behavior, & Social Networking, 15*, 521–526. doi: 10.1089/cyber.2012.0125
Marshall, T. C., Bejanyan, K., Di Castro, G., & Lee, R. A. (2013). Attachment styles as predictors of Facebook-related jealousy and surveillance in romantic relationships. *Personal Relationships, 20*, 1–22. doi: 10.1111/j.1475-6811.2011.01393.x
McEwan, B. (2013). Sharing, caring, and surveilling: An actor–partner interdependence model examination of Facebook relational maintenance strategies. *Cyberpsychology, Behavior, & Social Networking, 16*, 863–869. doi: 10.1089/cyber.2012.0717
Muise, A., Christofides, E., & Desmarais, S. (2009). More information than you ever wanted: Does Facebook bring out the green-eyed monster of jealousy? *CyberPsychology & Behavior, 12*, 441–444. doi: 10.1089/cpb.2008.0263
Muise, A., Christofedes, E., & Desmaris, S. (2014). "Creeping" or just information seeking? Gender differences in partner monitoring in response to jealousy on Facebook. *Personal Relationships, 21*, 35–50. doi: 10.1111/pere.12014
Muscanell, N. L., Guadagno, R. E., Rice, L., & Murphy, S. (2013). Don't it make my brown eyes green? An analysis of Facebook use and romantic jealousy. *Cyberpsychology, Behavior, & Social Networking, 16*, 237–242. doi: 10.1089/cyber.2012.0411
Norman, D. A. (1988). *The psychology of everyday things.* New York, NY: Doubleday.
Papp, L. M., Danielewicz, J., & Cayemberg, C. (2012). "Are we Facebook official?" Implications of dating partners' Facebook use and profiles for intimate relationship satisfaction. *Cyberpsychology, Behavior, & Social Networking, 15*, 85–90. doi: 10.1089/cyber.2011.0291
Simpson, J. A. (1990). Influence of attachment styles on romantic relationships. *Journal of Personality & Social Psychology, 59*, 971–980. doi: 10.1037/0022-3514.59.5.971
Spitzberg, B. H., & Cupach, W. R. (2003). What mad pursuit? Obsessive relational intrusion and stalking related phenomena. *Aggression & Violent Behavior, 8*, 345–375. doi: 10.1016/S1359-1789(02)00068-X
Sprecher, S. (2011). The influence of social networks on romantic relationships: Through the lens of the social network. *Personal Relationships, 17*, 1–15. doi: 10.1111/j.1475-6811.2010.01330.x
Stafford, L., & Canary, D. J. (1991). Maintenance strategies and romantic relationship type, gender, and relational characteristics. *Journal of Social & Personal Relationships, 8*, 217–242. doi: 10.1177/0265407591082004
Stafford, L., & Hillyer, J. D. (2012). Information and communication technologies in personal relationships. *Review of Communication, 12*, 290–312. doi: 10.1080/15358593.2012.685951
Tokunaga, R. S. (2011a). Friend me or you'll strain us: understanding negative events that occur over social networking sites. *Cyberpsychology, Behavior, & Social Networking, 14*, 425-432. doi: 10.1089/cyber.2010.0140
Tokunaga, R. S. (2011b). Social networking site or social surveillance site? Understanding the use of interpersonal electronic surveillance in romantic relationships. *Computers in Human Behavior, 27*, 705–13. doi: 10.1016/j/chb.2010.08.014
Tokunaga, R. S. (2014). Relational transgressions on social networking sites: Individual, interpersonal, and contextual explanations for dyadic strain and communication rules change. *Computers in Human Behavior, 39*, 287–295. doi: 10.1016/j.chb.2014.07.024
Tong, S. T. (2013). Facebook use during relationship termination: Uncertainty reduction and surveillance. *Cyberpsychology, Behavior, & Social Networking, 16*, 788–793. doi: 10.1089/cyber.2012.0549
Treem, J., & Leonardi, P. (2012). Social media use in organizations: Exploring the affordances of visibility, editability, persistence, and association. *Communication Yearbook, 36*, 143-189.
Utz, S., & Beukeboom, C. J. (2011). The role of social network sites in romantic relationships: Effects on jealousy and relationship happiness. *Journal of Computer-Mediated Communication, 16*, 511–527. doi: 10.1111/j.1083-6101.2011.01552.x
Valenzuela, S., Halpern, D., & Katz, J. E. (2014). Social network sites, marriage well being and divorce: Survey and state-level evidence from the United States. *Computers in Human Behavior, 36*, 94–101. doi: 10.1016/j.chb.2014.03.034

Van Der Heide, B., D'Angelo, J. D., & Schumaker, E. M. (2012). The effects of verbal versus photographic self-presentation on impression formation in Facebook. *Journal of Communication, 62*, 98–116. doi: 10.1111/j.14602466.2011.01617.x

Walther, J. B., & Parks, M. R. (2002). Cues filtered out, cues filtered in: Computer-mediated communication and relationships. In M. L. Knapp & J. A. Daly (Eds.), *Handbook of interpersonal communication* (3rd ed., pp. 529–563). Thousand Oaks, CA: Sage.

Wang, G., Wang, B., Wang, T., Nika, A., Zheng, H., & Zhao, B. Y. (2014). Whispers in the dark: Analysis of an anonymous social network. In *Proceedings of the ACM Internet Measurement Conference* (pp. 1–13). Vancouver, BC: ACM.

Wright, M. F. (2014). Predictors of anonymous cyber aggression: The role of adolescents' beliefs about anonymity, aggression, and the permanency of digital content. *Cyberpsychology, Behavior, & Social Networking, 17*, 431–438. doi: 10.1089/cyber.2013.0457

Twenty-Three

Shedding Light on Dark Structures Constraining Work/Family Balance: A Structurational Approach

Jenny Dixon and Corey J. Liberman

Although the great majority of organizational communication scholarship focuses on constructive dialogic exchanges, it is equally prudent to assess the processes associated with destructive communication within the organizational environment. Lutgen-Sandvik and Sypher (2009) define destructive communication as that which is "...evidenced through incivility, harassment, and abuse of power...[where] injustice and incivility often underscore communicative interactions... and illegal, unethical, and reprehensible interactions sadly become the norm" (p. 1). Phenomena wrought with destructive communication include workplace bullying, backstabbing, sexual and gender harassment, and racial harassment. From a critical perspective, these communicative behaviors have drastic implications for organizational constituents across organizational units and can, and oftentimes do, come to impinge on work-related practices and interpersonal relationships. An additional, and perhaps unexpected, site for considering the dark side of organizational communication is in the attempted negotiation of work/life balance, where variables such as difference, imbalance, perception, and comparison permeate the interaction process.

For more than 30 years, organizations have implemented programs, such as flextime and on-site daycare, to help employees balance work and family obligations (Hochschild, 1997). The goal of these initiatives was to prevent organizational problems caused by a failure to adequately balance work and family (i.e., absenteeism, turnover) (Hochschild, 1997). Over time, these initiatives gained complexity to account for the changing work environment, such as the increasing number of dual-earner households and single parents (Clark, 2000). One particularly accommodating mechanism for assisting employees with balancing work and family is family and medical leave. Unpaid leave is guaranteed, under certain conditions, by the Family and Medical Leave Act (Society for Human Resource Management, 2015), and the necessity of paid leave is beginning to be recognized at the state level (A Better Balance, n.d.). However, organizational communication literature unearths a disparity between the presence of policy and its availability as constituted through workplace communication (Dixon & Dougherty, 2014; Kirby & Krone, 2002).

Originally, work/life balance accommodations were considered only familial or domestic aspects of non-work life, neglecting to consider a myriad of affected constituents. The rise of work-life initiatives parallels the increase in women joining the workforce as women's domestic roles were (and largely still are) maintained despite entry into the workforce. However, by the 1990s programs began to include considerations for non-work life beyond family obligations (e.g., on-site workout facilities) and research began to account for work/life balance (Kirby, Golden, Medved, Jorgenson, & Buzzanell, 2003), as opposed to the perhaps limited consideration of work and family. This conceptual expansion of work/life balance—while aptly accounting for the diversity of non-work goals and obligations—may cloud considerations of work/life balance that are specific to family (e.g., the divergence between how family is understood at the managerial level and the varying family structures represented by workplace employees) (Cohen, 2014). Despite a liberal conceptualization of family within the family communication (Baxter & Braithwaite, 2006) and organizational communication (Buzzanell & Lucas, 2006) disciplines, workplaces largely maintain work/life balance accommodations that privilege employees from traditional family structures (Dixon & Dougherty, 2014), ultimately creating a large population of disenfranchised individuals. The ways in which work/life balance are negotiated interpersonally in the workplace—despite divergent notions of what it means to have family obligations—make for a rich and important opportunity to examine the dark side of organizational communication. The following literature review considers (a) how traditional notions of family are communicated in the workplace, (b) how these traditional notions present unique barriers for some working adults, and (c) the managerial barriers that also account for struggles in balancing work and family life.

Literature Review

An examination of the work/life balance literature suggests that many of the struggles in negotiating work and family stem from divergent notions of what *family* means. Furthermore, problematic structures occur when ambiguity and inconsistency exist regarding whether—and usually the extent to which—work is expected to take priority. Therefore, it is important to consider the communication processes with which family identity is advocated in the workplace and the barriers to advocacy that often emerge.

Negotiating Family

Family communication scholars have historically adopted liberal definitions of family, including choice (Floyd, Mikkelson, & Judd, 2006) and social custom (Edwards & Rothbard, 2000) as possible determinants. However, academic conceptualizations of family have not sufficiently intervened in the workplace. Those who may not benefit from existing work/life balance initiatives include child-free singles and gay and lesbian couples (Buzzanell & Lucas, 2006). Working adults caring for voluntary kin (i.e., family of choice, which describes family members not bound by biological or legal ties; Braithwaite et al., 2010) likely also experience barriers. From these few examples, narrow understandings of what it means to have a family can be a foundation for workplace inequalities. As Buzzanell & Lucas (2006) explain, "when only a certain form of family is privileged via daily discourses, all other forms are marginalized" (p. 341). The negotiated meaning of family as communicated in the workplace is vital for self-advocacy and attempting work/family balance.

In many workplace cultures, communication about family while at work is informally expected of employees. The ability to talk about family at work is important for workplace satisfaction and company loyalty (Allen, 2001). Additionally, receiving social support from coworkers reduces perceptions of stress related to work/life balance conflict (Carlson & Perrewé, 1999). What might be less salient is the importance of communication about family at work for employees with families outside the "spouse and child" construct. Indeed, "the very messages of what constitutes and does not constitute family both enable and constrain material choices and affect individuals' negotiation or adoption of benefits, workplace practices, organizational policies, and government legislation" (Buzzanell & Lucas, 2006, p. 337–338). As Buzzanell and Lucas (2006) suggest, employees with families that have not been discursively constructed into legitimacy face extra challenges regarding work/life balance and are socially partitioned from their traditional "spouse and child" counterparts.

Working adults belonging to families that have not been discursively constructed as family in the workplace often find that their work/life needs are secondary, if acknowledged at all. For example, child-free employees are less likely to ask for accommodation (Young, 1999). Additionally, inability to communicate about family at work due to concerns of social stigma may result in the invisibility of one's family balance needs altogether. As employees feel they cannot advocate for their own work/life balance needs, they are often burdened with picking up the slack for employees using family-friendly policies (Dixon & Dougherty, 2014). LGBT and single working adults often feel the expectation to yield to the needs of those with traditional families (Dixon & Dougherty, 2014), again socially constructing the "us versus them" divide within the workplace: a divide that can have both organizational and relational ramifications.

The idea that some families are communicated into legitimacy through interpersonal communication at work and some are not, makes for an interesting site for examining the dark side of organizational communication. It might be assumed that employees with families that have not been socially constructed within the workplace have either (a) attempted unsuccessfully to advocate for their work/life balance needs or (b) felt unable to communicate about family out of concern that there would be negative consequences. Regardless of the reason for this, the lack of acknowledgment of certain family structures provides an obvious disconnect between policy and practice. Of course, the ability to take advantage of work/life balance accommodations is not limited to one's ability to advocate for his/her family needs to peers. Managers and other gatekeepers of work/life balance policies also play an indispensable role.

Managerial Barriers

Managers and other workplace authority figures play a significant role in shaping the creation and usability of work/life balance policy. Employees often find that management uses mixed messages when communicating about work/life balance policy. Specifically, work/life balance programs are commonly adopted, and yet workplace culture and managerial demands warrant such programs to be neglected in favor of workplace productivity (Kirby, 2000).

The role of management in administering work/family balance policy gains complexity when considering managers' personal attitudes about family. Tracy and Rivera (2010) explored women's organizational challenges by talking to male managers and executives (gatekeepers of work/family balance policy) about gender roles and work/life balance. When asked about work/life balance policy, respondents continually referenced their personal beliefs and their own family structures. Importantly, most executives in this study held traditional notions of gender roles in relation to work/family balance (Tracy & Rivera, 2010). The results of this study were somewhat straightforward: traditional notions of family structure (i.e., spouse and child) are reinforced within the organizational confines.

In addition to mixed messages regarding how work and family should be prioritized, working adults often find that they need to be strategic in their approach to taking time off from work (Medved, 2004). It is possible that managers create a strategic ambiguity regarding the balance of work and family and that employees, likewise, use strategic ambiguity to attain the accommodations they need (Eisenberg, 1984). The problem, however, is that strategic ambiguity can quickly become equivocal if/when there is a disconnect between the managers creating and enforcing the said rules and the employees attempting to abide by them.

Contradictory and ambiguous messages about work/life balance lend to confusion regarding the amount of accommodation to which one is entitled and the extent one should advocate for her or his work/life balance needs. Considerations of the communicated negotiation of family, as well as the ambiguous messages from work/life policy gatekeepers make for an important site for examining the dark side of organizational communication. A helpful way to frame this exploration is through the propositions of Structuration Theory.

Structuration Theory

Anthony Giddens's (1984) theory of structuration has been used as a guiding framework for many empirical studies in the fields of sociology and communication over the past three decades. The major argument undergirding this theory is that social action is not random, nor is it unexpected. Rather, all communication episodes are, at least to an extent, predictable. According to Giddens, this predictability emerges because of the routine, communicative behaviors in which social actors engage. In brief, Giddens argues that social beings, in any dialogic situation, have a myriad of different communicative choices. This is what Giddens (1984) calls agency. As an example, assume that an employee, in an attempt to manage his work/life balance, asks for time off to see his daughter's gymnastics competition. The employee in question has many potential ways of posing this question, including: "I have truly been a model employee for the past month, so would it be alright if I took this Friday afternoon off?"; "I know that you allowed Bradley to take off a few weeks ago for his daughter's recital so I am hoping that you will allow me to take off to see my daughter's competition"; "Do you mind if I take a few hours off this Friday afternoon to see my daughter's gymnastics competition?"; or "Would it be alright with you if I took this Friday afternoon off? I will be sure to make up the hours." His direct supervisor also has several different responses at her disposal, including: "Absolutely yes"; "Absolutely no"; "Yes, but you will have to get your work completed prior to her competition"; "Yes, but you will have to work on the weekend"; or "No, but I will allow you to take time off next week instead."

What is crucial to Giddens's (1984) structuration theory is that even though social actors have agency to communicate as they wish, they also must be cognizant that they are, in fact, socially operating amidst others (what Giddens calls a *system*) within the confines of already existing rules (what Giddens calls *structures*). These rules are not written, but rather are expectations or guidelines for what might be deemed as appropriate behaviors or actions. Thus, in an organization whose system's members both anticipate and appreciate time off for family events, and whose recurring structures indicate that supervisors should grant family leave requests, the appropriate/expected/routinized response would be more representative of "absolutely yes" as compared to "no, but I will allow you to take time off next week instead." It is, therefore, the relationship among agency, system, and structure that produces predictability, which, in the end, is extremely salient for structuration theory.

The final, and perhaps most important, element of Giddens's (1984) theory of structuration is what he termed the duality of structure. In essence, this maintains that while structures (the rules for interaction) come to shape our communicative and behavioral choices, these structures are the results of dialogic practices as well. In other words, this duality of structure argues that rules are both a prerequisite for, and the effect of, communication. One of the arguments at the start of this chapter was that many organizations still provide work/life balance accommodations that largely privilege employees from traditional families. Using the results from Kirby and Krone's (2002) study, one important theme that emerged regarding work/family balance and existing policies associated with it was "preferential treatment" for traditional families. That is, not only were those from non-traditional families less able to reap certain organizational benefits (e.g., time off), but they were also less likely to ask for certain benefits in the first place. This duality of structure would argue that giving preferential treatment to traditional families creates a social environment where certain employees are more reluctant to seek benefits and, over time, such silence becomes the norm—what, in this chapter, is framed as the dark side of organizational communication. To better examine the dark side of organizational communication through the context of negotiating work and life balance, the following research question was explored:

> RQ: What destructive practices occur (if any) in workplaces as perceived by employees attempting to balance work and non-work obligations?

Method

To gain additional insight into the barriers of balancing work and life obligations, a small, exploratory study was conducted to see what destructive practices might be reported. Specifically, a small thematic analysis further illuminated the disparity among policy, practice, and communication, thereby anecdotally showcasing the "darkness" behind current organizational structures and informing means of alleviating these daily constraints. The following sections explain the sampling, interview, and analysis procedures used to understand work/life balance in the workplace.

Data Collection

Utilizing the philosophical underpinnings of the *active interview,* data were collected with the assumption that knowledge was constructed through the act of attaining it (Holstein & Gubrium, 1995; Kvale, 2009). In other words, interviews functioned to co-create meaning. In an active interview, the interviewer "activates narrative productions, [and] suggests narrative positions, resources, and orientations" (Kvale, 2009, p. 158). Interviewees also had an active role: becoming a researcher by composing meaning through a co-constructive process with the interviewer (Holstein & Gubrium, 1995; Kvale, 2009).

Procedure

Participation was limited to adults over the age of 18, who work for pay outside the home. Five participants were recruited for this study—each representing affiliation with a different family structure, including a gay man living with a roommate; a straight, divorced, mother of two whose children are grown and out of the house; a straight, child-free, polyamorous woman; a straight, married, and child-free woman; and a straight, divorced father of a young daughter. Though the number of participants for this study is quite low, each provided a rich and invaluable perspective on the dark structures of work/life balance. The ensuing paragraph offers a brief description of each participant.

Oscar is a 31-year-old case manager for a state agency that provides medical care to low-income families. In addition to his work obligations, Oscar attends nursing school and lives with his roommate and their combined roster of pets (three dogs and a cat). Brenda is a 61-year-old mortgage-banking manager. She has been at her current position for 12 years and has unique insight into the changing landscape of the workplace with regard to work/life balance. As a single mother of two adult children, Brenda can also draw comparisons between balancing childcare and the work/life balance obstacles that remain after children have grown and left the home. Kelly is a 31-year-old director of a college admissions office. Her unique perspective on work/life balance includes advocating for her non-work obligations, despite her and her spouse's choice to be child free. Jamie is a 36-year-old veterinary technician. She explains that she is very fortunate to be in a workplace where schedules are negotiated by coworkers who are invested in making sure each employee gets the time off that he/she needs. Her work/life balance includes considering making time for polyamorous relationships. Finally, Michael is a 43-year-old tenured college professor. Though Michael enjoys the flextime available for a tenured professor, balancing his work obligations with his need to care for his young daughter—of whom he has secondary custody—presents a unique perspective.

To ensure a range of viewpoints, participants were recruited through purposive sampling in which no two individuals belonging to the same family structure (e.g., single parent, married and child free, etc.) were represented. Participants were provided a consent form and offered $10.00 for their time and contributions. Two of the five participants chose not to accept compensation. It was at this point that participants were asked to select a pseudonym. All except one opted to have one selected. Interviews took place via phone interview or Skype. Skype interviews allowed for a rich opportunity to understand participants' descriptions of their families. For example, when asked who he considers to be a part of his immediate family, Oscar picked up a cuddly grey cat—Carl—and set him in his lap within the frame of the video feed. Interviews lasted between 23 and 55 minutes. All interviews were recorded.

Interviews began with the solicitation of demographic information, including age, biological sex, ethnicity, current job title, and time spent at current job. Then, participants were asked questions that sought to get at the lived experience of negotiating work/life balance. Interview prompts included: "Name a time (if any) when you had to decide between tending to a work obligation or a family obligation. How did you arrive at your decision?" and "Please describe what successful work/family balance looks like." To get at the discursively constructed notion of work/family balance, participants were asked questions such as, "Please tell me about a time (if any) when you heard coworkers or management criticizing someone else for taking time off of work for family-related reasons. In what ways (if any) do you feel this situation caused you to rethink whether you would take time off of work for similar reasons?" Finally, each participant was asked to describe his/her family. Special care was taken to not shape participants' answers to this question with predetermined criteria (e.g., people living in your home).

Analysis of the transcribed interviews took place with the hope of locating cogent and informative themes, which serve to describe the dark side of work/life balance. First, open coding consisted of delineating the data into conceptual categories (Corbin & Strauss, 2008); then, axial coding served to build structure around the axis of each category by locating examples in the data (Charmaz, 2006). Analysis of notes taken during the data collection process (Charmaz, 2006) also contributed to the development of three themes: (un)acceptable absences, deception and suspicion, and passive observation.

Results

The research question sought to explore destructive organizational practices as perceived by employees seeking to navigate work/life balance. Across the five interviews, participants noted destructive structures that favored the work/life balance needs of employees with traditional families. Furthermore, participants described aspects of the work/life balance negotiation process to include suspicion and deception. Finally—though not in itself a "dark structure"—participants discussed using passive and active information-seeking strategies to develop understanding of the work/life balance norms of the workplace.

(Un)Acceptable Absences

The idea that some reasons for leaving work were considered more important and worthy of accommodation than others was a dominant theme in the interview data. It could be speculated that common, non-work topics of discussion created and reinforced acceptable family structures and, therefore, acceptable reasons to take time off from work. Reflecting organizational communication literature, results showed that communication about family at work centered primarily on talk of children. As explained by Jamie:

> A lot of people talk about kids… A coworker that I work most closely with has two infant grandchildren that she is very involved with. That's probably the primary, you know, babysitting and birthday parties and school starting, and who learned to use the potty this week and I don't participate in that a whole lot.

It is likely that the centrality of children in workplace communication affects the negotiated meaning of family. This is not to suggest that talking about children in the workplace is an antecedent of the dark side of communication. However, reflection is needed for how the prominent topic of children may foster biases in determining legitimate and illegitimate reasons for taking time off from work.

The most prominent structure to occur in the interview data was the perceived inequality in what was considered to be a legitimate reason to leave work early or to take a day off. Specifically, participants spoke to a double standard in which they were discouraged from taking time off from work for reasons other than attending to the needs of children. Oscar, who balances work with taking college classes, explained:

> Oscar: I would kind of like to go over to the math tutoring lab in the middle of the day and take half an hour to get a question answered, you know?

Interviewer: Would it be feasible?

Oscar: Yeah, and I mean technically that's the policy, but if someone wants to take her kid to the doctor it's okay, but if I want to go visit with the tutor for half hour or an hour to get my questions answered it's like, "well, it's not a good reason," and I'll need to schedule paid time off.

The first author's exchange with Oscar points to both the perceived allowances given to employees balancing work with childcare and lapses in the integrity of workplace policy. Kelly mentioned a similar inequality when trying to take time off to receive medical care:

> I have had some recent health issues, just in the past four months, and I know if it was someone's child, it would be, "take time, go, absolutely." For me, it was like "you can't do this at another time of the day? You can't see the doctor at 7[p.m.]?" I take vacation time but it's seen as I should be doing doctor's visits and MRIs and CAT scans on my own time. But when people who have children have serious illnesses, it's "absolutely, no questions asked, totally fine."

Importantly, Kelly's experience does not draw a specific instance in which a coworker was allowed to take time off to attend to a child's serious illness. However, the negotiation of work/life balance proves to be a dark organizational structure when employees feel that some non-work obligations are more important and worthy of interruptions in productivity than others and makes less traditional family structures seem much less relevant and important.

Suspicion and Deception

The data also unearthed a theme of suspicion and deception. Explanations of suspicion and deception took many forms. Brenda discussed anticipating deception, saying, "There are always people who want to take advantage of the system." Brenda went on to explain that deception was pervasive and yet not tolerated. All but one participant discussed knowing of a particular coworker who was the source of workplace gossip on account of perceived attempts at taking advantage of work/life policies.

In addition to organizational structures fraught with suspicion that one or more coworkers are taking advantage of "the system," participants discussed making the decision of whether to lie about their availability to work. Oscar explained:

Oscar: I get asked kind of regularly to work extra...and I get paid for it of course, but there have been times that I [makes air quotation marks] wasn't in Tulsa [lowers hands] because I would get calls to see if I would come fill in for somebody and I didn't want to.

Interviewer: You had to choose whether to come back to Tulsa to work those hours?

Oscar: Yeah, well, I mean I *was* in Tulsa, but I lied about it. That was how I got out of work because, they would call me to see if I would fill in and I really didn't want to do it.

Interviewer: I see. Do you think that they ask you more than the average employee to do things like that?

Oscar: They used to, but I started turning them down a lot lately.

Here, Oscar described how he had become a go-to person for coming in to work on short notice. As a result, he decided to lie and say he was unavailable.

Just as some participants elected to lie in their negotiations of work/life balance, others relayed making the conscious decision to be truthful. Jamie told a story of when she was called into work on her birthday. She explained to her boss that she had gone to work for the last 10 days and would be unavailable to come into work. Though truthfulness is hardly a dark structure of organizational communication, the potential risk involved in advocating for one's self presents a structure of insecurity.

Passive Observation

As previously mentioned, four of the five participants had a token coworker in mind who was either known for her or his inclination to "abuse the system" or otherwise "take too much time off." Though the authors feel that passive observation of work/life norms within a workplace is not a dark structure, it is symptomatic of the gap between policies in place and the unequivocal freedom to use them.

In addition to having coworkers in mind who function outside the acceptable work/life balance parameters, two participants discussed passively observing coworkers negotiating work/life balance amidst one or more non-work obligations. As stated by Michael,

> I had a colleague who was up for tenure and was going through a fairly similar situation to what I'm going through this year in terms of a close family member being sick and I remember hearing she—meaning my colleague—was having a hard time with administration...she was not getting much sympathy.

It makes sense that, when preparing to apply for tenure, Michael would consider the experiences of those who applied in the years prior. Again, Michael's considerations of the work/life negotiations of a colleague in a metaphorical balancing act similar to his own is not an element of the dark structures of communicating work/life balance, but rather a strategy employed to manage ambiguities regarding how sympathetic workplaces may or may not be. Problems such as lack of managerial support, fear of poor performance evaluations, fear of ostracism from coworkers, and the failure of policy to address key issues prevent the effective management of work/family conflict. Passive observation appears to be an enlightened coping strategy for a potential dark structure.

The three themes that emerged from the data help to paint a picture of the dark structures inhibiting fair and functional work/life balance. Disparities in what it means to have a family and attend to family obligations resulted in *(un)acceptable absences*, or perceived inequalities among employees with varying work/life balance needs. Additionally, *suspicion and deception* was a consistent theme, illuminating dishonesty and mistrust as a pervasive aspect of securing time off from work. Finally, *passive observation* was depicted as a routine strategy for charting strategies for work/life balance negotiation. Implications of these findings are provided below.

Discussion

The results suggest that care of the self, with regard to staying healthy, as well as care of the self with regard to seeking a college degree, were not viable reasons for leaving work, whereas leaving work to care for a sick child is allowable and narratively expected. This perception that some non-work obligations are deemed more important than others reflects the work of Dixon and Dougherty (2014) who found that working adults belonging to non-traditional family structures are expected to yield to the needs of their counterparts who belong to "opposite-sex spouse and child" family structures.

Importantly, participants did not share explicit instances in which they were unequivocally denied the opportunity to take time off. For example, when requesting time off to go to the doctor, Kelly was asked if it would be possible to go in the evening. Similarly, Oscar shared that he would be allowed to take time to go to a tutoring session, but that he would have to request the time off through formal processes that his coworkers would not have to follow. These examples suggest subtle inequalities rather than overt double standards. Based on Giddens's (1984) structuration theory, the recurring and subtle inequalities that occur within organizations become operating norms. By adhering to an organization's structures, according to Giddens, employees begin to reinforce a sense of routinization, where certain discursive forms become dominant, while others become recessed (or ignored). The result reflects the partitioning of voices heard and topics discussed on one hand and voices silenced on the other.

As indicated in the present study, as well as in existing literature, workplace members have a grammar for discerning acceptable and unacceptable absences. Participants frequently referred to the granting of acceptable leave, and the criteria by which the purpose for a leave is considered acceptable, as "the system." Each participant was able to speak of a coworker—past or present—who was rumored to regularly (attempt to) "cheat the system." Some participants discussed how they, themselves, lied to avoid going to work. This capacity to "cheat the system" gave way to the second theme: *suspicion and deception.*

Participants were *suspicious* of coworkers who they felt took time off of work under false pretenses. Also, some participants discussed using *deception* when negotiating time away from work. According to Oscar, lying might be a viable strategy and one that he has effectively used before in order to refuse a superior's request to work overtime. He also felt the need to deceive his boss because he felt he was becoming the "go-to" person when the office needed an extra worker. This supports existing research, indicating that people feel the need to be strategic when negotiating time off (Medved, 2004). It could be argued that *(un)acceptable* absences give way to *suspicion and deception.* These subtle double standards and deceptive means of coping with them, could be alleviated, at least in part, with work/life balance policies that are customized to the individual employee (Buzzanell & Lucas, 2006).

Finally, as Michael explained during his interview, much about one's attempt to balance work and family is informed by *passive observation.* Similar to *suspicion and deception*, *passive observation* may be a symptom of the disparity between organizational policy and workplace norms (Kirby & Krone, 2002). This passive information-seeking strategy can be explained using Structuration Theory: if structures exist that bar individuals from asking for time off or that impinge on one's decision to talk about family-related issues that extend beyond children, and employees model what they passively witness, then the very structures that exist are reinforced, thus producing the duality of structure about which Giddens (1984) warns.

Of course, the small sample size of five participants leaves our results far from generalizable. However, this chapter has brought us one step closer to understanding how a dominant discourse within an organizational environment, can ultimately evolve and become the structure on which one's social decisions are predicated. To think that an individual might engage or disengage in a social behavior because he/she identifies with a less traditional family structure might, to the layperson, seem ludicrous. However, to the informed scholar and practitioner it is emblematic of the dark side of organizational communication fostered by a predominant ideological framework that favors and privileges certain employees while restricting the social and behavioral practices of others. It is important to now figure out how, especially considering the influx of individuals self-identifying as part of non-traditional family structures, a new ideology can emerge: one that will provide the platform for a universal voice.

References

A Better Balance. (n.d.). The need for paid family leave. Retrieved from http://www.abetterbalance.org/web/our issues/familyleave

Allen, T. (2001). Family-supportive work environments: The role of organizational perceptions. *Journal of Organizational Behavior, 58*(3), 413–435. doi: 10.1006/jvbe.2000.1774

Baxter, L. A., & Braithwaite, D. O. (2006). Introduction: Metatheory and theory in family communication research. In D. O. Braithwaite and L. A. Baxter (Eds.), *Engaging theories in family communication: Multiple perspectives* (pp. 1–16). Thousand Oaks, CA: Sage.

Braithwaite, D. O., Bach, B. W., Baxter, L. A., DiVerniero, R., Hammonds, J. R., Hosek, A. M., Willer, E. K., & Wolf, B. M. (2010). Constructing family: A typology of voluntary kin. *Journal of Social and Personal Relationships, 27*(3), 388–407. doi:10.1177/0265407510361615

Buzzanell, P. M., & Lucas, K. (2006). Employees "without" families: Discourses of family as an external constraint to work-life balance. In L. H. Turner & R. West (Eds.), *The family communication sourcebook* (pp. 335–352). Thousand Oaks, CA: Sage.

Carlson, D. S., & Perrewé, P. L. (1999). The role of social support in the stressor-strain relationship: An examination of work/family conflict. *Journal of Management, 25*(4), 513–540. doi: 10.1177/014920639902500403

Charmaz, K. (2006). *Constructing grounded theory: A practical guide through qualitative analysis.* Thousand Oaks, CA: Sage.

Clark, S. C. (2000). Work/family border theory: A new theory of work/family balance. *Human Relations, 53*(6), 747–770. doi: 10.1177/0018726700536001

Cohen, P. (2014). *The family: Inequality and social change.* New York, NY: W. W. Norton & Company.

Corbin, J., & Strauss, A. (2008). *Basics of qualitative research* (3rd ed.). Thousand Oaks, CA: Sage.

Dixon, J., & Dougherty D. S. (2014). A language convergence/meaning divergence analysis exploring how LGBTQ and single employees manage traditional family expectations in the workplace. *Journal of Applied Communication Research, 42*(1), 1–19. doi: 10.1080/00909882.2013.847275

Edwards, J. R., & Rothbard, N. P. (2000). Mechanisms linking work and family: Clarifying the relationship between work and family constructs. *Academy of Management Review, 25*(1), 178–199. doi: 10.5465/AMR.2000.2791609

Eisenberg, E. M. (1984). Ambiguity as strategy in organizational communication. *Communication Monographs, 51,* 227–242.

Floyd, K., Mikkelson, A. C., & Judd, J. (2006). Defining the family through relationships. In Lynn H. Turner and R. West (Eds.), *The family communication sourcebook* (pp. 21–39). Thousand Oaks, CA: Sage.

Giddens, A. (1984). *The constitution of society: Outline of the theory of structuration.* Malden, MA: Polity Press.

Hochschild, A. R. (1997). *The time blind: When work becomes home and home becomes work.* New York, NY: Metropolitan.

Holstein, J. A., & Gubrium, J. F. (1995). *The active interview* (Vol. 37). Thousand Oaks, CA: Sage.

Kirby, E. L. (2000). Should I do as you say or do as your do? Mixed messages about work and family. *Electronic Journal of Communication, 10.* Retrieved from: http://www.cios.org/EJCPUBLIC/010/3/010313.html

Kirby, E. L., Golden, A. G., Medved, C. E., Jorgenson, J., & Buzzanell, P. M. (2003). An organizational communication challenge to the discourse of work and family research: From problematics to empowerment. *Communication Yearbook, 27,* 1–43.

Kirby, E. L., & Krone K. J. (2002). The policy exists but you can't really use it: Communication and the structuration of work/life policies. *Journal of Applied Communication Research, 30*(1), 50–77. doi: 10.1080/00909880216577

Kvale, S. (2009). *InterView: Learning the craft of qualitative research interviewing* (2nd ed.). Thousand Oaks, CA: Sage.

Lutgen-Sandvik, P., & Sypher, B. D. (2009). *Destructive organizational communication: Processes, consequences, and constructive ways of organizing.* New York, NY: Routledge Press.

Medved, C. E. (2004). The everyday accomplishment of work and family: Exploring practical actions in daily routines. *Communication Studies, 55*(1), 128–145. doi: 10.1080/10510970409388609

Society for Human Resource Management. (2015). FMLA: Eligibility: Who is a covered family member under the FMLA? Retrieved from http://www.shrm.org/templatestools/hrqa/pages/whoisacoveredfamilymemberunderfmla.aspx

Tracy, S. J., & Rivera, K. D. (2010). Endorsing equity and applauding stay-at-home moms: How male voices on work-life reveal aversive sexism and flickers of transformation. *Management Communication Quarterly, 24,* 3–43. doi: 10.1177/0893318909352248

Young, M. B. (1999). Work-family backlash: Begging the question, what's fair? *The Annals of the American Academy of Political and Social Science, 562*(1), 32–46. doi: 10.1177/000271629956200103

Twenty-Four

"The Serial-Killer Application": Email Overload and the Dark Side of Communication Technology in the Workplace

Ashley Barrett

With the creation of the World Wide Web in 1991 (Chandler, 2000), electronic mail has rapidly diffused into every corner of our society, and more importantly, it has quickly developed into the backbone of organizational communication infrastructure. While the perks of email are evident—rapid transmission and flexibility, egalitarian decision making (Sproull & Kiesler, 1991), geographical and administrative decentralization (Hiltz & Turnoff, 1978) and speed and accountability (Gimenez, 2005), among others—past scholarship has consistently found that email has drastically increased the number of messages sent in the workplace (Bishop & Levine, 1999). Additionlly, it has increased people's vulnerability and susceptibility to requests, favors, and work demands. (Ducheneaut & Bellotti, 2001). Referred to as "the serial-killer application" (Ducheneaut & Bellotti, 2001), email has the capacity to multiply the most established organizational stressors—including towering workloads, deadlines, and feelings of being undervalued ("Number Crunching," 2006).

That said, email overload/burnout is not a prominent problem or setback to individuals in all organizations. On the contrary, perceptions of email overload and email-related stress are a consequence that is contingent upon the organizational culture in which one works. Depending on how employees are socialized during organizational entry, the same amount of incoming and outgoing emails could be problematic for some and inconsequential for others. This points to the significance of organizational socialization practices for employee newcomers, especially in regards to how information communication technologies (ICTs) are used in the workplace (Flanagin & Waldeck, 2004). If organizations are conscious of the strategies they adopt and utilize to "break in" their employees (Van Maanen, 1978), they must train their employees on how to appropriately manage and prioritize their email messages upon organizational entry. This training can have powerful effects including mitigated feelings of burnout for new employees. Effective socialization reduces inevitable role-orientation uncertainties for newcomers, and satisfies their impulses for information seeking (Flanagin & Waldeck, 2004; Jablin, 1987).

However, there is still much to learn about how socialization procedures regulate/stimulate the infiltration of ICTs, such as email, into organizational routines. Moreover, research is needed to discover how socialization practices provide employees with the coping mechanisms, technology skills, and cognitive frameworks that are essential for survival during the "Information Age" or "Digital Revolution" (Dreyer, Hirschorn, Thrall, & Mehta, 2006)—which has produced a substantial increase in messages. To date, scholarship lacks an understanding of how socialization-related communication alters the content and practice of advanced organizational technologies in organizations. As a result, I answer Flanagin and Waldeck's (2004) call to increase understanding of "how and why people [in organizations] become socialized to use technology competently and consistently with their colleagues" and how to effectively use the technologies that constitute the base of contemporary organizational communication and the intricacies therein (p. 158–159). Specifically, this study unveils which methods of socialization are used to assuage perceptions of email overload and enhancing employee satisfaction.

After reviewing extant literature on the origin, causes, repercussions, and solutions to email overload and stress, I present a brief review of the socialization tactics and information seeking procedures that newcomers typically use/experience during organizational entry. This material serves as a theoretical framework for this study. Afterwards, qualitative results of this research will be discussed, and I end with a set of remarks that explore the relationships between email-related stress, organizational type, and perceptions of email as a socialized, behavioral skill.

Email Overload: Presenting a Dark Side of Workplace Communication

Scholars might consider email-related stressors to be integrated with work-role stressors and find it hard, if not impossible, to distinguish between the two. However, past studies have demonstrated how the strain produced by email systems is not different from work-related stressors, but a catalyst for additional pressures that often push workers over the edge. For example, one study found that email use increases employees' levels of stress (Murray & Rostis, 2007, p. 249). Labeling employees "technologically tethered workers," this study found that the coping strategies employees used to minimize the harsh consequences of work stress were "threatened by the ubiquitous communication delivered by ICTs" (Murray & Rostis, 2007, p. 249). This in turn has a negative impact on employees' mental health and consequently may lead to negative organizational outcomes. Thus, email-related stress is unique in its potential to jeopardize workers' physical and psychological well-being. More recent research analyzing Gmail, which offers its users a considerable 15 GB of free storage, claims email overload is ever-growing along with storage capacities and the idea of email threaded conversations (Grevet, Choi, Kumar, & Gilbert, 2014). It seems as though symptoms of email overload will continue to manifest within the workplace, targeting the weak and central figures.

While scholars have defined email overload as a problem generated by the growing multiplicities of email's functional uses (Whittaker & Sidner, 1996), Dabbish and Kraut (2006) conceptualize it as "email users' perceptions that their own use of email has gotten out of control" (p. 431). Regardless of how it is defined, many argue that the influx of incoming emails carries with it so many new tasks and communicative complexities that users struggle to keep up with the unprecedented electronically set deadlines imposed by this newer form of organizational communication (Bellotti, Duchenaut, Howard, & Smith, 2003). Scholars have discovered that email is quickly becoming the virtual workplace where employees surrender the greater portion of their workdays. Their office transforms into an email "habitat" where they are compelled to check, wait, refresh, check, wait, refresh (Ducheneaut & Bellotti, 2001, p. 30).

In summary, email has the potential to spawn emotional exhaustion, leading to depersonalization, and ultimately decreased personal accomplishment and organizational commitment. Thus, it resembles the sequential stages that are associated with organizational burnout (Leiter & Maslach, 1988). Along the same lines, in order to refrain from email burnout, newcomers need help when adjusting to a new work setting (Leiter & Maslach, 1988). As an early step in examining email pressures in the organizational context, it is important first to clarify and thoroughly investigate factors associated with perceptions of email overload and email-related stress; hence, the first research question.

RQ1: What are the primary causes of email overload and email-related stress in the workplace?

Although email is often viewed positively and depicted as a useful medium, the stresses of imitating immediacy and instant reply to others can be a harassing feature (Hair, Renaud, & Ramsay, 2006). Therefore, solutions to mitigating email overload should be incorporated into organizational behaviors, yet no sophisticated, consistent, or striking solution has been discovered.

Socialization: A Proactive Solution to Email Overload

Socialization, Information Seeking, and Uncertainty

Van Maanen (1978) describes organizational socialization as "people processing," or how others structure experiences for newcomers within an organization (p. 19). It is important to note that different organizations utilize unique socialization tactics, and subsequently, the methods in which people acquire the social knowledge vital to a particular organizational role vary between organizations. For instance, companies may create strong or weak situations under which new employees must adapt, or they could appoint goals concentrated in roles of conformity or innovation (Bauer et al., 2007).

The first postulation of socialization theory (Van Maanen & Schein, 1979) states that upon organizational entry, newcomers will simultaneously attempt to reduce uncertainty through social interactions with superiors and peers (Saks & Ashforth, 1997). According to Berger's Uncertainty Reduction Theory (1979), employees do this to create structure in their relationships with others that will create a sense of predictability and comfort in unfamiliar contexts. In order to seek these and other key performance proficiency messages, Comer (1991) claims newcomers will acquire information in three ways: (a) *active explicit*: overt information seeking and direct communication, (b) *passive explicit*: indirect questioning and disguising conversations, listening to others' conversations, and testing limits, and (c) *passive*: or acquiring information through *observation or surveillance* (Miller, 1989).

Regardless of what role—passive or active—an organization enacts during the socialization process, and more specifically to the current study, what role it embraces in order to teach employees how to effectively manage and maneuver electronic mailing systems, organizations must arm their new employees with adequate, accurate, and pertinent messages to help employees locate their place within a new, unaccustomed setting (Chen & Kozlowski, 1992; Van Maneen, 1978). Unfortunately, socialization will automatically occur at any point a new face enters through the organizational doors, despite organizational leaders' or members' decision to take part in it (Van Maneen, 1978). Informal socialization measures—in which newcomers are left to negotiate for themselves how they will adapt to email pressures and thus personally select their own agents and coping methods—often induce personal anxiety (Van Maanen, 1978; Van Maanen & Schein, 1979). This type of informal, random, and disjunctive socialization creates a "sink or swim" mentality and leaves newcomers on their own to "walk the tightrope of adjustment" (Ashforth, Sluss, & Harrison, 2007, p. 13–14).

Consequently, a number of past studies have exposed newcomers' dissatisfaction with socialization customs in organizations—or a lack thereof (Comer, 1991; Jablin, 1984; Teboul, 1994). Hence, one of the primary objectives of this study is to determine what socialization tactics, if any, organizations use to teach newcomers how to most effectively take advantage of the most basic ICTs, such as email. Thus, the second research question guiding this study is:

> RQ2: What socialization tactics are organizations using (if any) to educate employees on email norms/use in their workplace?

It is also important to note that newcomers sense organizational socialization efforts on an intensified level when they are undergoing the school-to-work (STW) transition because they have never been responsible to a larger organizational mechanism. Thus, they are likely to experience higher levels of role-related and career uncertainty as they negotiate their work and personal identities (Ng & Feldman, 2007). As a result, genuine newcomers to the business world often need additional support and social acceptance upon organizational entry (Bauer et al., 2007).

Methods

Procedures

This study featured a qualitative research design where interviews were conducted with organizational newcomers who had just undergone the STW transition in a variety of organizations. A qualitative study fixated on email-related stressors in the context of multiple organizations has at least two advantages. First, given the small amount of studies that have been designed to unveil the methods that contemporary organizations are employing to socialize newcomers into distinct ICT use (Flannagin & Waldeck, 2004), a qualitative analysis was able to capture textured accounts of these trends in an unrestricted manner and to explore the rich complexities that take place in the day-to-day lives of office personnel. Thus, I followed the traditional standards of choosing a research methodology that most appropriately reveals the type of information involved in the principal aims of the study (Punch, 1998, p. 244; Silverman, 2005). Second, the communication of advantageous technology use in different organizational settings is essential to productivity, but it may also blur home and work boundaries, thus negating performance in both of these complex social domains. As such, an in situ examination of employees' perceptions of email strain, as a direct result of the organization in which they work, enabled insight into the productivity paradox of ICT use.

I conducted 20 interviews with organizational newcomers who had just transitioned into their first career from the academic environment. The participants worked at a variety of organizations including universities, legal organizations, marketing/event planning/public relations organizations, and sales. The interviews ranged from 32 to 93 minutes (*M*= 53.4), resulting in 13.35 hours of data accumulation. Interviews were conducted over a period of three months, and the ages of the interviewees ranged from 21 to 45 years of age (*M*= 25). Eighteen out of the 20 interviews were semi-structured, audio-recorded and transcribed. (The two remaining were conducted with professionals in other states, and thus were conducted via telephone). Ninety-three percent (759 out of 801 minutes) of the interview text was transcribed. The remaining un-transcribed text consisted of casual conversations that prefaced interview questioning and therefore was not relevant to the researched topic, or audible pauses such as utterances of "um" and "uh."

Data Analysis

After the last interview transcribed, I studied the transcribed text of the interviews and implemented the constant comparative analysis method in order to identify recurrent experiences, paralleling issues, and thematic episodes of discourse and dialogue that consistently arose in participants' comments (Strauss & Corbin, 1998). Next, I constructed a tentative master list of relevant themes and asked a colleague to read and analyze a percentage of the interviews in order to determine if her interpretation of the data and the groupings therein were accurate and/or consistent. My colleague and I discussed the differences and discrepancies that surfaced in the categorization methods, and the interview texts were repeatedly referred to in order to extract direct quotations that exemplified conversational topics salient in the data. Conclusively, two of my original themes were abandoned ("advice for first time professionals" and "advantages and drawbacks of email as a communicative tool") and two others were re-contextualized into sub-themes, as my colleague and I worked to collapse categories.

Results and Interpretations[1]

Factors Contributing to Email-Related Stress

RQ1 sought to reveal the ingredients triggering email-related stress in the workplace and the challenges employees are encountering with their organization's email systems. The first realization that stemmed from the interviewing process is that pressures and anxiety created by email overload is a multidimensional concept that is contingent upon individual personalities, face sensitivities, workplace preferences, and technological dependencies.

Individual personalities and email stress. As the interviewing process progressed, it became clear that some personality types are universally insensitive to workplace stressors, including email. Conflicting with individuals who emulate Type A personalities that are driven by strict deadlines and multi-tasking feats, and who are affectionately awarded the title of "stress junkies," Type B personality employees are unanimously relaxed, easy going, and generally lack the proclivity to be compelled by an overriding sense of urgency on any front. Take for example, Mr. Roberts, a representative at a V.A. office in central Texas. Throughout our interview,

Mr. Roberts made several comments that accumulatively hinted at his devotion to his status as a "Type B" personality—the climax of which came with the utterance of the most fundamental and unambiguous statement of "No. No. I am not a stressful person." To further illustrate my point, when asked if he is an organizer, he responded "I organize whenever I have a lot of stuff"; when asked if he felt overwhelmed with his email when he returned from vacation, Mr. Roberts defiantly claimed, "Honestly, here's the thing—I don't really let things stress me out that much" and followed it up by commenting, "I waste more than a minute of my own time, but I rather waste my time than let an email waste my time." He claimed that if emails are petty, he simply ignores them. Finally, he summed up his lack of frustration with email, declaring, "uh, no. I am pretty passive."

Whereas replying to thirty incoming emails was hardly viewed as troublesome or alaborious task by Mr. Roberts, another participant cringed at the idea of working through ten electronic messages, regardless of content or the time required to appropriately respond to the messages. Even as messages that require no reply, such as Spam, flood the inboxes of stress-inclined employees, their heart-rate accelerates at the sight of another email that steals megabytes and infiltrates their personal space. Barb, an executive director at the Texas Capitol, claims that even though these messages don't necessitate portioning off time to read through, they still provoke a "pain" of irritation. This stems from the time it takes to sift "through and erase them to get to the real items I need…The biggest stress factor is really just wasting time with things that are not pertinent." However, Mrs. Green, a lecturer at a University, is numb to this affliction, claiming, "I think if someone is stressed out because of email, they need to re-evaluate their job and their life."

Indeed, this finding of the diverse working demeanors of Type A and Type B personalities is aligned with past empirical research that has suggested that personality variables are a predictor of almost all the job stressors and strains that have received scholarly attention (Spector & O'Connell, 1994).

Face sensitivities. The majority of participants in this study consistently described their conceptualizations of email overload as a daily hassle to uphold their reputations as responsible, competent, and ever-present subordinates. For example, Jenny, a 22-year-old legislative staff at the Texas State Capitol, describes her obsession with displaying her work ethic through email:

> My email account is open on my computer all day, and I get nervous when I can't check it for a few hours. I check my emails when I wake up in the morning and right before I go to bed at night…The problem is that emailing is so quick that you feel obligated to respond immediately, even to less-important questions. It's a blessing and a curse.

Laci, a sales representative,, described a similar situation: "I am constantly checking email every chance I get…I was not much for email before this job, now it's not a choice. Email is the way that almost all of the people I work with contact me. I have to be available." Laura, a senior communication coordinator, expressed how the overwhelming need to adhere to her emailing schedule is interfering with the completion of other, perhaps more important, work tasks:

> It seems that most people use email as their primary form of business communication, so I feel that if I ignore the email to work on a project or attend a meeting, I am taking a risk that I won't get an important message in a reasonable amount of time. This is especially difficult for me as the success of my position is partially measured by my "responsiveness" to the internal customer.

With the onslaught of information communication technologies and their expansive capabilities, younger generations of professionals are now more than ever falling prey to the exasperating need to be "always on," and always available (Moore, 2000). This is beautifully illustrated in Laura's metaphorical depiction of her company-provided BlackBerry as being "more like a leash than a helpful inbox-management tool." Unfortunately, prestige is increasingly being tied to dismantling the age-old tradition of exalting the home as a family-engaging, work-free zone. Organizational trends expect that employees who are accessible should also be available, and those who respond the quickest will experience the shortest ride up the organizational hierarchy.

Face, or the set of behavioral standards and social status that an individual must maintain in order to function effectively in society, is lost when employees flounder the fundamental social requirements of the position they occupy (Ho, 1976). Therefore, appearing unavailable in any social domain runs the lofty risk of losing face. Moreover, reciprocity is inherent in good-standing face behavior, and as such, the burning desire to reply as instantaneously as possible is not likely to fade from organizational offices any time in the near future (Ho, 1976).

Workplace preferences. As the investigation progressed, I discovered an appealing finding: generally speaking, participants who received and/or delegated work tasks through the medium they preferred for inter-organizational communication, experienced less email-related stress and were more likely to report that they were not a victim of email overload. For exemplification purposes, Laura, a senior communications coordinator who reportedly receives upwards of 60 legitimate emails per day and spends roughly 2 to 3 hours of a 9-hour day harnessed to her computer chair, often burrowed through her inbox. Compared to many of the other newcomers interviewed, this locates Laura on the higher end of the email-receiving spectrum. While Laura agrees that email is "an extremely useful tool," 70% of her work tasks are assigned to her through email. She consequently feels overwhelmed, at least partially because her preference is to be given working assignments in synchronous formats. In her own words, "I prefer receiving work tasks over the phone or face-to-face because it is really the only way to ensure that you have my full attention. Many times, I will glance at an email, and it will get buried, accidentally deleted, or just forgotten." Not only does this hinder her ability to stay "responsive," she also claims that receiving or delegating work tasks through email increases the likelihood that she will sacrifice valuable work time engaging in small exchanges that crowd her inbox, such as (1) Did you get the document? (2)'Oh yes- I got it yesterday. (3) Did you have any changes? (4) Let me look through it and get back to you. (5) Great, thanks. (6) No problem. "That's six emails about basically nothing."

On the other hand, although Katie (MLB ticketing coordinator) receives anywhere from 150 to 200 emails per day that involve "meeting requests" and delegating "numerous work tasks to sales reps." The stress she experiences due to the sheer quantity of emails she receives is minimal because she prefers working through email as it allows her to "communicate very quickly" in a system that allows exchanges to be "saved for later use."

Technological dependencies. Finally, the last overarching theme contributing to email-related stress stemmed from participants' recurring descriptions of their desperate addiction to ICTs. In other words, for the majority of the subjects interviewed, email-related stress was induced not by an excess of email activities, but rather by the inability to access their email at all times throughout the day. In a sense, stress was a by-product of email underload—rather than overload. However, the participants were plainly cognizant of the fact that their interactions with their ICTs were completely unacceptable. Comments such as Jenny's remark, "I always check my email. I wouldn't be able to enjoy time off if I couldn't," accountant Ashley's confession, "Not having email would be the biggest stress factor about email. It's such a crutch," and Maria's cry for help, "I look at it [email on her iPhone] like eight times an hour. It is ridiculous. I have tried [to stop] before....I stuck it in a drawer and I was like 'I'm not gonna look at it,' and an hour later, I was like 'I have to look!' And I hate that. I hate feeling like I always have to look at it and this is the way my life is going," suggest that ICTs are impeding young organizational newcomers' production perhaps as much as they are facilitating it. However, another interpretation of the preceding commentaries could be that organizational employees are hopelessly fastened to their email accounts, whether they access them on their PCs, lap tops, or smart phones, because they suffer from an overwhelming sense of stress and/or anxiety when they merely think about the build up of unseen messages in their inbox. For instance, accountant Ashley briefly discussed her repulsion to returning to her "waiting email" after weekday breaks, explaining, "I am so used to checking emails immediately and clearing them out, the thought of opening my email and having 150 unread messages would really stress me out."

As more and more employees are made aware of this paradox that characterizes the double-edged sword of communication technologies, they are realizing that their voluntary enslavement to their computers and smart phones could act as a "crutch" to their future development. In order to help to alleviate the harsh consequences imposed on newcomers, I next explore employers' tactics to socialize organizational newcomers into using ICTs, in particular email, resourcefully.

Socializing Organizational Newcomers into Efficient Email and Technology Usage Patterns

> I just feel that you are incompetent if you can't use word processing programs these days.
>
> Mr. Roberts, V.A.

> I don't know what kind of [email] training employees could go through to alleviate their stress. From my experience, you can't have cookie-cutter responses ready to deliver.
>
> Mrs. Green, full-time lecturer

These are just a couple of the comments that I received when asking organizational newcomers if email socialization could assuage the "debilitating uncertainty" associated with organizational entry—especially for first-time employees (Hart & Miller, 2005). It appeared as though simply asking this question was somewhat condescending in nature. Who needed help learning the logistics of a communication system as elementary as email? These statements alone breathed life into Van Maanen's (1978) words "since many of the strategies for breaking in employees are taken for granted...they are rarely discussed or considered to be matters of choice in the circles in which managerial decisions are made" (p. 14). Noting the defensive reaction, I backpeddled and indirectly probed into how they learned to use technologies, such as email, to comply with the performance expectations in their workplace. This approach was much more lucrative than the first. Ultimately I discovered that the bulk of suitable technology-use information circulating organizations is learned through trial and error and observational methods.

Trial and error methods / Trial-by-fire experiences. The majority of this study's participants learned email norms, and coped with email excess, through adopting what Hart & Miller (2005) refer to as "trial-by-fire" experiences (p. 303). Unfortunately, for a few of the young employees, this subjective and ambiguous approach to learning behavioral and social norms left them fending for themselves. At times, this created a struggle in establishing their workplace professional identities. Laura was all too familiar with this renowned narrative of "sink or swim": when asked how she learned the norms of technology use in her organization, she paused, and then shrugged her shoulders in disagreement. "Trial and error!" she said, but at first it was hard to detect her words because they were quickly followed by a sigh that appeared to be a sign of her vexation with the unstructured system. "Ultimately, I think most of us here devised our own methods for coping with email, but I took cues from other things that I did not like about the ways other handled their email." Laura elaborated by detailing her frustration with the method, saying, "Now people are making major decisions and conducting day-to-day business through email, so it's important to learn the best ways to manage that influx of information." Katie expressed the staggering sense of bewilderment she encountered when commencing her first professional role: "With email, I devised my own method that was loosely based on the method of the individual who held my position before me. However, no one instructed me on how to properly use this method. I basically had to teach myself based on hints that I collected. I was a scavenger for the first month at my office."

Laura's and Katie's dissatisfaction with the trial and error socialization technique is not surprising to many of the scholars who focus on studying methods of effective socialization. Hart and Miller (2005) propose that trial-by-fire experiences actually cramp a new employee's socialization into the company by increasing role ambiguity.

Significance of Findings

In conclusion, the results of this study validate Dabbish and Kraut's (2006) model of email overload in that it further exemplifies the philosophy that email-related stress is not simply a function of email quantity. Instead, other job-related and communication-related factors such as type of worker, the importance of email for work, the number of face-to-face meetings workers attend, routine email tactics, and email-initiated tasks all contribute to soaring levels of email strain.

However, the significance of this study does not lie in its discovery that organizational type and position influence how much email employees receive and how much temporal resources and effort is sacrificed filtering through their emails. The qualitative measures executed in this study revealed that these factors are accumulatively a robust indicator of whether or not employees perceive email socialization to be a valuable program in their organization, and moreover they prescribe the underlying reasons grounding why workers think it would be a constructive and practical tool for effective assimilation into an organization's culture of technology. Depending on their professional position, personality type, face values, technological dependencies, workplace preferences, and the nature of their organization, some individuals are indeed experiencing email overload in the workplace. And for this segment of the workforce, the general sentiment is that current socialization practices are deficient and their means of seeking information about efficient technology use is colloquial, constrained, and often eschewed as a symbol of incompetence.

Take Laura, for example, the senior communications coordinator at a national aircraft producing company. She spends up to three hours a day sorting and tackling her email messages because she is "one of the owners of the company's corporate mailboxes." Therefore, as compared to many of the other individuals interviewed in this study, not only is her company more substantial in terms of the size scale (and therefore she must be available to more coworkers), she also has to entertain email traffic from the general populous to the organization, in addition to responding to her own personal and work email. This corporate email can range from outright attacks on customer service or missed deadlines to more trivial complaints about parking situations and cafeteria food. Therefore, Laura is a prime supporter of email socialization in organizations because the specific workplace situation she currently encounters would greatly benefit by incorporating it.

On the opposite side of the spectrum, individuals like Tiff and Mrs. Green, who work at jobs where they experience minimal email-related stress, cannot fathom the need for email socialization. Moreover, Barb, who repeatedly spoke about the violations and offenses that often correspond with email messages such as inappropriate messages and the unprofessional, informal modes of electronic communication she is repeatedly subject to at work, sees a need for email socialization even though she is not overloaded with the communication technology. In her opinion, even though professional workers are incapable of controlling or manipulating the "number of items in someone's inbox through training," email socialization can still be considerably beneficial. She claims all new employees should receive training that centrally stresses "the ramifications for poor writing and rash missives" and teaches email etiquette such as writing a "proper email, spell checking, and when people should just pick up the phone."

Given these findings, future scholars should continue to unveil the impact that tactical socialization can have on easing employee's interactions with "the serial-killer application." The stresses employees are encountering in managing unprecedented computer-mediated demands and grasping ever-evolving workplace norms are a primary area of study in organizational communication. They create contemporary organizational customs and challenges that must continue to be analyzed and answered.

Notes

1 An earlier version of this chapter's results was presented at a 2012 Southern States Communication Association top paper panel in the Mass Communication Division.

References

Ashforth, B. E., Sluss, D. M., & Harrison, S. H. (2007). Socialization in organizational contexts. In C. L. Cooper & I. T. Robinson (Eds.), *International review of industrial and organizational psychology*, Vol. 22, (pp. 1–70). West Sussex, England: John Wiley & Sons.

Bellotti, V., Duchenaut, N., Howard, M., & Smith, I. (2003). Taking email to task: the design and evaluation of a task management centered email tool. *Proceedings of the CHI 2003 Conference on Human Factors in Computing Systems.* New York, NY: ACM Press.

Bauer, T. N., Bodner, T., Erdogan, B, Truxillo, D. M., & Tucker, J. S. (2007). Newcomer adjustment during organizational socialization: A meta-analytic review of antecedents, outcomes, and methods. *Journal of Applied Psychology,* 92(3), 707–721. doi: 10.1037/0021-9010.92.3.707

Berger, C. R. (1979). Beyond initial interaction: Uncertainty, understanding, and the development of interpersonal relationships. In H. Giles & R. N. St.Clair (Eds.), *Language and social psychology* (pp. 122–144). Baltimore, MD: University Park Press.

Bishop, L., & Levine, D. (1999). Computer-mediated communication as employee voice: A case study. *Industrial and Labor Relations, 52,* 213–233.
Chandler, D. (October, 2000). Who "created" the Internet? It's a tangled web. *Boston Globe.* Retrieved from http://www.seattlepi.com/business/nett20.shtml.
Chen, O., & Kozlowski, S. W. J. (1992). Organizational socialization as a learning process: The role of information acquisition. *Personnel Psychology, 45*(4), 849–874.
Comer, D. (1991). Organizational newcomer's acquisition of information from peers. *Management Communication Quarterly, 5,* 64–89. doi: 10.1177/0893318991005001004
Dabbish, L. A., & Kraut, R. E. (2006). Email overload at work: An analysis of factors associated with email strain. *Proc. SIGCHI Conference,* 431–440.
Dreyer, K. J., Hirschorn, D. S., Thrall, J. H., & Mehta, A. (2006). *PACS: A guide to the digital revolution.* New York, NY: Springer.
Ducheneaut, N., & Bellotti, V. (2001). Email as a habitat: An exploration of embedded personal information management system. *Interactions, 8*(5), 30–38. doi:10.1145/382899.383305
Flanagin, A. J., & Waldeck, J. H. (2004). Technology use and organizational newcomer socialization. *Journal of Business Communication, 41*(2), 137–165.
Gimenez, J. C. (2005). Unpacking business emails: Message embeddedness in international business email communication. In M. Gotti & P. Gillaerts (Eds.), *Genre variation in business letters. Linguistic insights: Studies in language and communication* (pp. 235–255). Bern: Peter Lang.
Grevet, C., Choic, D., Kumar, D., & Gilbert, E. (2014). Overload is overloaded: Email in the age of gmail. In *Proceedings of the SIGCHI Conference on Human Factors in Computing Systems,* (pp. 793–802)
Hair, M., Renaud, K. V., & Ramsay, J. (2006). The influence of self-esteem and locus of control on perceived email-related stress. *Computers in Human Behavior, 23*(6), 2791–2803. doi:10.1016/j.chb.2006.05.005
Hart, Z. P., & Miller, V. D. (2005). Context and message content during organizational socialization: A research note. *Human Communication Research, 31*(2), 295–309. doi: 10.1111/j.1468-2958.2005.tb00873.x
Hiltz, S. R., & Turoff, M. (1978). *The network nation.* Reading, MA: Addison-Wesley.
Ho, D. Y. (1976). On the concept of face. *The American Journal of Sociology, 81*(4), 867–884. Retrieved from http://www.jstor.org/stable/2777600 .
Jablin, F. M. (1984). Assimilating new members into organizations. In R. Bostrom (Ed.) *Communication Yearbook, Vol. 8.* (pp. 594–626). Beverly Hills, CA: Sage.
Jablin, F. M. (1987). "Organizational entry, assimilation, and exit." In F. M. Jablin & L. L. Putnam (Eds.) *The new handbook of organizational communication* (pp. 679–740). Newbury Park, CA: Sage.
Leiter, M. P., & Maslach, C. (1988). The impact of interpersonal environment on burnout and organizational commitment. *Journal of Organizational Behavior, 9*(4), 297–308. doi: 10.1002/job.4030090402
Miller, V. (1989). A quasi-experimental study of newcomers' information-seeking behaviors during organizational entry. 37th *International Communication Association Convention,* Organizational Communication Division. San Francisco, CA. May 26–29, 1989.
Moore, J. (2000). One road to turnover: An examination of work exhaustion in technology professionals. *MIS Quarterly, 24,* 141–168. doi: 10.2307/3250982
Murray, W. C., & Rostis, A. (2007). "Who's running the machine?" A theoretical exploration of work stress and burnout of technological tethered workers. *Journal of Individual Employment Rights, 12*(3), 249–263. doi: 10.2190/IE.12.3.f
Ng, T., W. H., & Feldman, D. C. (2007). The school-to-work transition: A role identity perspective. *Journal of Vocational Behavior, 71,* 114–134. "Number Crunching." (October, 2006). *Business Communicator, 7*(5), 6.
Punch, K. (1998). *Introduction to social research: Quantitative and qualitative approaches.* London: Sage.
Saks, A. M., & Ashforth, B. E. (1997). Organizational socialization: Making sense of past and present as a prologue for the future. *Journal of Vocational Behavior, 51,* 234–279. doi:10.1006/jvbe.1997.1614
Silverman, D. (2005). *Doing qualitative research* (2nd ed.) Thousand Oaks, CA: Sage.
Spector, P. E., & O'Connell, B. J. (1994). The contribution of personality traits, negative affectivity, locus of control and Type A to the subsequent reports of job stressors and job strains. *Journal of Occupational and Organizational Psychology, 67*(1), 1–12. Retrieved from http://www.bps.org.uk/publications/journals/journaltitles/joop.cfm
Sproull, L., & Kiesler, S. (1991). Connections: New ways of working in the networked organization. Cambridge, MA: MIT Press.
Strauss, A., & Corbin, J. (1998). *Basics of qualitative research: Techniques and procedures for developing grounded theory.* Thousand Oaks, CA: Sage.
Teboul, J. B. (1994). Facing and coping with uncertainty during organizational encounter. *Management Communi cation Quarterly, 8,* 1990–224. doi: 10.1177/0893318994008002003

Van Maanen, J. (Ed.). (1978). *Organizational careers: Some new perspectives.* New York, NY: John Wiley & Sons.

Van Maanen, J., & Schein, E. H. (1979). Toward a theory of organizational socialization. *Research in Organizational Behavior, 1*, 209–264. Retrieved from http://sdpace.mit.edu/bitstReam/handle/1721.1/1934/SWP-0960-03581864.pdf?sequence=1

Whittaker, S., & Sidner, C. (1996). Email overload: Exploring personal information management of email. *Proceedings of CHI*, 276–283.

TWENTY-FIVE

Facing the Green-Eyed Monster: Identifying Triggers of Facebook Jealousy

Trisha K. Hoffman and Jocelyn M. DeGroot

Perceived as one of the most notoriously "dark" emotional experiences, romantic jealousy has long been a topic of fascination and research. As Spitzberg and Cupach (2007) noted, dark side research highlights and explores "deviance, betrayal, transgression, and violation" (p. 5). Certainly, romantic jealousy touches on many of these behaviors and offenses and often leads to personal and relational turmoil. Romantic jealousy is fueled by the presence of a real or perceived rival that threatens the strength of the existing relationship as well as personal self-esteem (Buunk & Bringle, 1987). Although jealousy has been studied extensively in the realm of face-to-face interaction (Afifi & Reichert, 1996; Guerrero & Afifi, 1999; Guerrero, Trost, & Yoshimura, 2005), recent research has begun to explore the role of online communication in relation to romantic jealousy. Facebook use, in particular, has been linked with increased levels of romantic jealousy (Muise, Christofides, & Desmarais, 2009), surveillance behaviors (Utz & Buekeboom, 2011), and relational dissatisfaction (Elphinston & Noller, 2011).

Muise et al.'s findings (2009) revealed a link between the amount of time individuals spent on social media and the level of jealousy-related feelings they experienced as a result of information available on Facebook, termed *Facebook jealousy*. Although a specific definition is unclear, Facebook jealousy has been described as the experience of jealousy that arises from Facebook use (Muise et al., 2009; Utz & Beukeboom, 2011). Marshall, Bejanyan, Castro, and Lee (2012) also explored this research area to focus on how attachment styles contribute to Facebook jealousy, and Utz and Beukeboom (2011) identified the impact of self-esteem on reactions to Facebook content. Most recently, Muise, Christofides, & Desmarais, (2014) found that Facebook could potentially allow heightened surveillance and monitoring of romantic partners, which might expose individuals to jealousy-provoking material. Furthermore, academic and anecdotal evidence point to the potential relational fall-out resulting from Facebook usage. While research is beginning to investigate jealousy that stems from social media, what researchers have yet to identify are the specific triggers of Facebook jealousy.

As social networking sites (SNS), like Facebook, continue to gain popularity, it becomes increasingly necessary for scholars to understand how these types of sites influence the health of interpersonal romantic relationships. This chapter investigates and discusses the various triggers of Facebook jealousy as described by people who self-identified as having experienced Facebook jealousy related to their romantic relationship. Identifying these triggers helps shed light on the principle features of Facebook that harm romantic relationships and/or lead to relational dissolution and mistrust. Knowledge of these triggers could help relational partners address and communicate about potentially harmful Facebook material before problems develop. Additionally, understanding common triggers of Facebook jealousy, and how individuals respond to these triggers, could help develop greater awareness of Facebook usage and improved online communication competence. This chapter also explores implications and recommendations related to Facebook jealousy triggers.

Defining Jealousy

Jealousy has been defined in numerous ways. White (1981a) defined jealousy as:

> A complex of thoughts, feelings, and actions which follows threats to self-esteem and/or threats to the existence or the quality of the relationship, when those threats are generated by the perception of a real or potential attraction between one's partner and a (perhaps imaginary) rival. (p. 24)

Buunk and Bringle (1987) further elaborated, defining jealousy as "an aversive emotional reaction evoked by a relationship involving one's current or former partner and a third person" (p. 124). Scholars have also differentiated between reactive jealousy and preventative jealousy (Rydell & Bringle, 2007). Reactive jealousy occurs as a result of a relational transgression (e.g., infidelity). Preventative jealousy occurs as the result of a perceived, potential threat to the relationship (e.g., talking to a prospective rival). For example, this jealousy could cause one to attempt to prevent a partner's contact with a potential threat.

Beyond these definitions, White and Mullen (1989) argued that jealousy involves both an interpersonal and intrapersonal component. Jealousy is fueled by the real or perceived presence of a third person known as the "rival." The intrapersonal component of jealousy occurs as the jealous individual engages in primary and secondary appraisals of the rival. Primary appraisal is the recognition of a threat to the relationship, and secondary appraisal involves developing strategies for dealing with the threat. According to White and Mullen (1989), primary appraisal could take three different forms: (1) assessing the potential for a rival relationship to exist; (2) judging whether or not a rival relationship does or does not exist; and (3) determining the amount of damage the real or potential rival relationship could inflict. After the primary appraisal has occurred, individuals will then make a secondary appraisal based on their perception of a real or potential rival relationship. Secondary appraisals may occur as assessing the cause behind the partner's involvement/interest in a rival, comparing oneself to the rival, evaluating the options if the partner ends the primary relationship, or evaluating the potential damage and/or loss that will be experienced as a result of relationship termination. Depending upon the type of secondary appraisal that occurred, individuals then attempt to cope with jealousy by acting in ways that could end the relationship, strengthen the relationship, and/or provoke the partner.

Factors Influencing Romantic Jealousy

Researchers have attempted to determine underlying factors that influence jealousy (Afifi & Reichert, 1996; Ben-Ze'ev, 2004; Buunk & Bringle, 1987; Guerrero & Andersen, 1998; Knobloch & Solomon, 1999; White, 1981b). Some of the relational factors that could lead to the experience and expression of jealousy include interdependence, trust, uncertainty, and self-esteem.

Relational interdependence. In romantic relationships, the actions and emotions of one individual in the relationship can influence the actions and emotions of his or her partner (Brehm, 1992). Buunk and Bringle (1987) explained that the ability to evoke jealousy is often reliant on the level of interdependency within a relationship. The researchers explained that interdependency tends to be higher in "committed love relationships" when compared with other types of interpersonal relationships, therefore creating an ideal situation for jealousy to be evoked (p. 126). Many individuals in a romantic relationship are often interdependent of one another to meet certain needs (i.e., support, reassurance of self worth, nurturing, social interaction, intimacy; Brehm, 1992) that can only be achieved through being in an interpersonal relationship. When one partner in the relationship fails to meet any or all of these needs, there may be a chance for jealousy to develop. Regardless of whether or not there is an actual act of infidelity or an actual rival, the perception of a potential romantic rival could leave many of the five relational needs left unmet.

Relational trust. Based on the previous information, one could conclude that dependence on another individual requires a certain amount of trust. According to Rempel, Holmes, and Zanna (1985), trust is defined by "feelings of confidence and security in the caring responses of the partner and the strength of the relationship" (p. 97). The authors also explained that trust in romantic relationships depends on previous experiences in the relationship, the reliability of the partner to also positively contribute to the relationship, and the inherent risk of depending on another.

Guerrero and Andersen's (1998) review of literature revealed that individuals in a less committed relationship tend to be more jealous. They also found married individuals experience less jealousy than those who are dating. This could mean that a more committed relationship involves a higher level of trust in the relationship and between partners. Rempel and colleagues (1985) also cited predictability as an important component of trust. The researchers explained that the predictability of a partner's future actions is determined by an understanding of previous behaviors in the relationship, including communication. Predictability is influenced by consistency and stability of such behaviors. Based on this understanding, an individual may become suspicious of a romantic partner if he or she communicates or behaves in a way that is atypical, causing feelings of jealousy.

Relational uncertainty. The predictability component of trust is closely related to relational uncertainty. Uncertainty about a romantic partner can occur when an individual is not able to "predict the other person's attitudes and behaviors within interaction" (Knobloch & Solomon, 1999, p. 262). Afifi and Reichert (1996) found that an individual's inability to predict his or her partner's behaviors increased the experience of jealousy. As uncertainty increases, emotional intensity increases as well (Ben-Ze'ev, 2004). Therefore, the more uncertain an individual is about the strength of his or her relationship, or about his or her partner's commitment to the relationship, the more intensely one may experience feelings of jealousy.

Self-esteem. Issues of uncertainty and trust could further heighten a sense of threat to an individual's self-esteem. White (1981a) directly identified a threat to self-esteem as a significant component of jealousy in his definition of the jealousy complex. Ben-Ze'ev (2004) believed that romantic relationships could threaten self-esteem because they allow for another's evaluation of the self. White (1981b) argued that someone with high dependence on his or her partner's evaluation of self would experience "a direct threat to self-esteem" if the partner was attracted to a potential rival (p. 131). The partner's attraction to another would be interpreted as a negative evaluation of the self, which could lead to jealousy (White, 1981b). Additionally, when confronted with a rival, an individual is likely to compare himself or herself to the rival (Guerrero & Afifi, 1999), which could ultimately have a negative effect on an individual's self-esteem if he or she does not measure up to the rival (Guerrero & Andersen, 1998).

Certainly, these relational and personality factors might act as antecedents to romantic jealousy in a variety of contexts. Less research, however, has examined the specific triggers of jealousy. More specifically, we know little of the types of online communication that trigger romantic jealousy in media such as Facebook.

Facebook Jealousy

In their seminal article on Facebook jealousy, Muise and colleagues (2009) described Facebook as an all-access pass to romantic partners' online communication. For example, individuals can see what their partners are doing at any point of the day, find out whom their partners are friends with, and observe their partners' exchange of messages with other individuals. This increased amount of information could potentially expose a romantic partner to jealousy-inducing information (Muise et al., 2009). Furthermore, the ability to read, save, re-read, and view an individual's Facebook information at any time could also induce, or re-induce, jealousy. For instance, if an individual witnesses a message from a potential rival on his or her partner's Facebook wall, he or she might re-experience feelings of jealousy each time the message is viewed.

Other Facebook activity that triggered jealousy included seeing a romantic partner add his or her previous romantic, or sexual, partner as a friend, or adding an unknown individual (Muise et al., 2009). These types of friendships could lead to issues of trust, self-esteem, and uncertainty about the relationship, primary partner, and/or rival, which could ultimately induce feelings of jealousy.

As evident, several behaviors and types of communication that occur on Facebook could potentially induce jealousy. However, there is still much to be learned about individuals' Facebook use and how certain types of use can impact interpersonal relationships and the development of jealousy. Furthermore, little research has been conducted to understand specific triggers of jealousy on Facebook, which could help partners avoid behaviors that impede relational strength and satisfaction. Therefore, this study aims to identify triggers of Facebook jealousy as well as reveal how people describe these triggers. As a result, we developed the following research questions to guide the current study:

RQ1: What Facebook behaviors or content trigger romantic jealousy?

RQ2: How do participants describe triggers of romantic jealousy that results from Facebook use?

Methods

As part of a larger project on communicative responses to Facebook jealousy, participants described a time that they felt jealous in their romantic relationship after viewing some type of information regarding their romantic partner on Facebook. We examined these responses by engaging in thematic analysis.

Participants and Recruitment

A total of 145 people participated in this IRB-approved study, including 106 (73.1%) females and 39 (26.9%) males. Participants ranged in age from 18 to 40 years old, with an average age of 22.56 years. One hundred nineteen participants identified as Caucasian/White (82.1%), 18 as African American/Black (12.4%), 3 Hispanic (2.1%), 2 Asian/Pacific Islander (1.4%), and 3 bi-racial or other/not listed (2.1%).

We employed nonrandom, purposive sampling to reach eligible participants by recruiting undergraduate students enrolled in communication courses at a moderately sized Midwestern university as well as through Facebook posts. We required eligible participants to: (1) be 18 years of age; (2) be in a romantic relationship (i.e., dating, engaged, or married); (3) have an active Facebook profile; and, (4) have a romantic partner with an active Facebook profile. A *romantic partner* was defined as, "someone participants had dated for one week or more" (Bevan & Tidgewell, 2009, p. 310) and *active Facebook use* was defined as "logging onto Facebook at least once per week to perform activities such as check wall posts, view pictures, send messages, or update statuses."

Procedure

Using Qualtrics, online survey software, participants were asked to recall and describe a time they had felt jealous in their romantic relationship as a result of something seen on Facebook. We defined jealousy for the participants and provided examples of Facebook-related items (e.g., wall posts or pictures) that might be viewed and cause jealousy. Participants typed their responses into the online survey, which were then downloaded for analysis.

Both authors read through the responses to the open-ended question. Using open coding, we independently coded the responses to our open-ended question by writing descriptive notes next to each entry. We then engaged in the constant-comparative method (Glaser & Strauss, 1967) to identify initial themes in the data. We compared our individual lists of themes for triangulation purposes. The themes were consistent across both sets of themes, and any discrepancies led us to examine the coding together and collectively determine the best coding theme. These initial themes included Artifacts, Active Poaching, Passive Poaching, Vague Posts on Partner's Wall, Relationship Status Issues, Response to Poachers, and New Relics.

Next, we used axial coding (as described by Strauss & Corbin, 1990) to identify connections between the initially determined categories. We determined that triggers were either rival-based or partner-based. That is, triggers of Facebook jealousy were due to a perceived rival making comments or posting pictures (i.e., "Rival-Based"), or the triggers were due to the participant's current partner posting items on Facebook (i.e., "Partner-Based"). The triggers are not necessarily mutually exclusive, as some revealed evidence of two types of triggers. For example, one participant indicated that a rival was sending her partner messages, indicating Rival-Based Active Poaching. The partner responded to these messages in a friendly manner, signifying Partner-Based Response to a Poacher.

Results

Analysis revealed various triggers that we categorized as either rival-based or partner-based. That is, the rival or the partner posted something on Facebook that triggered Facebook jealousy. "Rival" is defined as any person perceived to be a threat to the participant's romantic relationship. "Partner" refers to the participant's romantic significant other. Below, we describe the overarching categories of Facebook jealousy triggers as well as the subcategories.

Rival-Based

Rival-based triggers included artifacts, evidence of active and passive poaching, and vague innuendos. These behaviors or content topics focused on the rival's role in causing the Facebook jealousy.

Rival artifacts. Artifacts refer to remnants of the rival's past relationship with the partner. This includes comments, pictures, and statuses that refer to the rival and partner's relationship that remain on the rival's Facebook profile page. One participant described this trigger: "When my boyfriend and I started dating I was jealous of all the pictures of him and his ex on her Facebook page, they had been broken up for six months and his ex was and is a friend of mine but I still felt threatened." Another person described her experience with an artifact:

> I found some old photographs from Christmas last year of him with his former girlfriend during his family's Christmas. I began to compare myself to her physically and then looked for what was going on in the photograph to determine whether or not that was an activity that I had engaged in with his family during Christmas of this year.

Pictures were the most common rival-based artifact that initiated instances of Facebook jealousy.

Active poaching. Active poaching includes definitive poaching behaviors. That is, the rival is definitely trying to attract the partner by flirting, using nicknames, or bluntly stating that he or she would like to date. Examples include: "His female friend (who is flirtatious with most men and has been unwelcoming of me since the beginning) posted a Christmas greeting to his wall and complained that he never answered his phone," and, "A female that I knew he was in a past sexual relationship wrote on his wall and left a message saying 'hey babe' and a heart symbol following it." Instances of active poaching were often uncovered when the participant logged in to his or her partner's Facebook account and read these private messages between the partner and a rival.

Passive poaching. Passive poaching behaviors are those perceived as threatening to the relationship, but they do not appear to be blatantly seductive. These include the rival posting new pictures with the partner, commenting and liking items on the partner's profile, sending a friend request to the partner, or seeing attractive rivals on the partner's newsfeed. One person described this trigger: "There was a girl that commented to like 5 of his statuses and it was his ex romantic partner." A participant explained her experience with passive poaching: "When we began our relationship 4 years ago, he went to visit some friends at [college]. I noticed that he was tagged in some pictures on Facebook by his ex girlfriend!" Another person said she gets jealous "whenever a person of the opposite sex comments on a picture in a way that is more than just friendly."

Vague innuendos. Posts in this category were not considered poaching behaviors. Rather, they were posts that insinuated that something sexual happened, but this could not be determined. Participants referred to this type of communication as *coded* or *open-ended.* One participant explained,

> A cute girl wrote something on my husband's wall—something like "had a good time with you." I got upset and asked him about it. It turns out he gave her (and the rest of the people in the student organization) a ride somewhere, so it ended up being nothing.

Another person wrote, "I am not a fan when there are open ended wall posts on my girlfriends wall such as 'bet you had fun last night…' and knowing that I was not with her last night."

Triggers in this category also included photos of the partner with a rival in which it was difficult for our participant to determine the nature of the relationship between the partner and the rival. For example, "I saw my boyfriend in a picture with a blonde girl I didn't know on Facebook. I thought I knew all of his friends that were girls so when I didn't recognize the girl, I started to ask questions."

Partner-Based

The other overarching category of triggers was the result of the partner's Facebook behaviors. Partner-based triggers included artifacts, relationship status issues, a poor response to poachers, and new relics. These items stemmed from the actions of the participant's partner.

Artifacts. Partner-based artifacts refer to remnants from past relationships, including photos, messages, or comments left on Facebook from the partner's previous relationship. One woman explained what triggered her jealousy:

> I experienced jealousy after being on my boyfriend's Facebook and seeing messages from his ex-girlfriends from when they were dating. Even though the messages were exchanged prior to when my boyfriend and I dated, or even met, I still felt threatened because he had kept the messages.

Photos from past relationships were also a major trigger of jealousy. One woman wrote, "When we first started dating, I saw old pictures of my partner with his arm around different women." Another participant agreed that photos triggered his jealousy: "A couple months of us being back together, I noticed his pictures still on her page. I got jealous and asked her to delete them. It caused a large fight between the two of us because I was jealous."

Relationship status issues. Facebook allows couples to publicly declare their relationship to each other by linking Partner A to Partner B. In order to link the two people, one party must request the relationship, and the other person must accept the request. Participants indicated that failing to change one's relationship status was a definite jealousy trigger. One woman described her feelings toward this behavior, "My partner still has his relationship status as single on Facebook, and it worries me that his ex girlfriends may take that to heart and attempt to talk to him." Another person agreed: "Her relationship status was listed as single after we had been dating for about a month. When I confronted her she said she didn't feel it was needed to let people know she was in a relationship."

Response (or lack thereof) to poachers. Some participants took issue with how their partners responded, or failed to respond, to would-be poachers. Accepting or initiating a poacher's friend request was one such trigger. One person explained, "The first time my boyfriend added one of his female friends on Facebook I was a little jealous because I did not know who she was and it was early on in our relationship." Additionally, sending messages to, chatting with, or commenting on a rival's comments and pictures caused jealousy; however, failing to respond to advances made by poachers was also cause for concern. A woman described her situation: "An ex-romantic partner of his was leaving comments on his wall that could be conceived of as 'flirty.' He never (openly) responded to her, which made me even more suspicious."

New relics. Examples of new relics included newly posted photos of the partner and his or her ex-partner or rival. For example, one woman described, "He made his profile picture one of him and his ex-girlfriend with his arm around her with their cheeks smashed together smiling ear-to-ear." New posts to the rival's wall are also included in this category. A participant explained, "A previous sexual partner posted on his wall about a night they had been out together (as friends) and it upset me." Another wrote, "My partner has a business relationship with his ex-wife. When we first started dating I felt jealous when I would see pictures of them from business trips."

Discussion

Qualitative data revealed that a variety of Facebook content, including seemingly benign content such as old photos of ex-partners, does indeed trigger jealousy. Given the variety of jealousy triggers, one could argue that the ease of access to information increases the likelihood of experiencing a threat to a romantic relationship (e.g., a rival). This might encourage individuals to spend more time on Facebook searching for information, which could then increase the likelihood of exposure to a jealousy trigger. Muise et al. (2009) explained this process as a "feedback loop" (p. 443). Continuous access to jealousy-inducing information could cause individuals to experience feelings of jealousy more regularly than if he or she is exposed to the same stimuli in a face-to-face setting. These findings suggest that online communication might have the capacity to fuel feelings of jealousy more than face-to-face communication because of the amount of information available to romantic partners at all times of the day.

Jealousy might also be triggered more by uncertainty and ambiguity, rather than the actual content (Muise et al., 2009). In both partner-based and rival-based triggers, participants primarily reported feeling unsure of whether an ex-partner had the potential to become a rival or threaten the strength of the relationship. Contact with past romantic partners on Facebook could pose a significant concern about the health of a relationship and call in to question partners' or rivals' motives for maintaining contact with an ex-romantic partner. Indeed, a past romantic partner may pose an even greater threat than an unknown individual given their romantic history and the potential for old feelings and attractions to reignite.

It would seem that friendship with ex-partners is not an uncommon practice on Facebook. Muise et al. (2009) found nearly 80% of the participants in their study were Facebook friends with past romantic partners. In the current study, we also found nearly 84% of the participants in our study were Facebook friends with past romantic partners. Given these findings, Facebook users whose partners are friends with their ex-lovers might engage more in preventative jealousy (Rydell & Bringle, 2007) in an effort to discourage old flames from rekindling through Facebook communication. One participant admitted, "His ex romantic partner [posted on his wall]. This caused me to call him and tell him what I felt about it. I was in a jealous rage." Another person admitted logging into her fiancé's account and declining a friend request from his ex-girlfriend.

Information shared between ex-partners can also be highly ambiguous when exchanged through Facebook. Facebook content often lacks context, which can enhance a sense of uncertainty about the motives behind partners' interactions with potential rivals, thus increasing feelings of jealousy and negative emotions (Afifi & Reichert, 1996; Ben-Ze'ev, 2004; Knobloch & Solomon, 1999). The ambiguity of communication through Facebook could also make it difficult to predict the nature of the communication that is occurring, which is a key component to reducing feelings of jealousy and enhancing relational trust (Rempel et al., 1985). Therefore, an inability to predict the types of behaviors a romantic partner, or potential rival, will engage in on Facebook could lead to jealousy.

Beyond triggers of Facebook jealousy, our study revealed various reactions to jealousy, including surveillance/monitoring of partner profiles, and confronting the partner or the rival about Facebook content. Indeed, Facebook provides unique opportunities for individuals to express jealousy. The features of Facebook allow for discreet methods of monitoring partners' communication because of the public nature of the content. Jealous individuals might also use Facebook as a tool to seek out information on their romantic partners and/or potential rivals (Muise et al., 2014). In the current study, participants reported closely watching their partner's profiles for new information, especially adding new friends. Many participants reported reacting to a jealousy trigger by monitoring their partners' profiles or even logging into their partners' accounts.

Past jealousy research has identified a number of responses to jealousy. Scholars argue that when uncertainty about the strength of the relationship, or the commitment of his or her partner to the relationship, is present, jealous individuals may resort to more indirect ways of reducing uncertainty (Afifi & Reichert, 1996; Guerrero & Afifi, 1999). Surveillance is a common reaction to a perceived threat or concern about the health or status of the relationship in an online context (Elphinston & Noller, 2011; Marshall et al., 2012; Muise et al., 2009, 2014). In this way, surveillance might serve as a form of both preventative and reactive jealousy. Participants react to potentially threatening content by monitoring or interacting with the partner and/or rival, which serves to reduce the potential for future transgressions to occur (Rydell & Bringle, 2007).

Beyond monitoring, Facebook content might motivate jealous individuals to engage in integrative communication, or directly expressing jealousy to the partner (Guerrero et al., 1995). In the current study, it appeared that integrative communication was more common than what previous jealousy research supports. In their responses, some participants acknowledged confronting their partner to let them know of their jealousy and discomfort with the partner's behavior (or lack of action taken). These are major face-risk behaviors, especially when there is uncertainty about the health of the relationship or how the partner will respond to accusations. Perhaps, however, Facebook makes it easier to have conversations regarding jealousy triggers because the partner sees undeniable proof that some type of interaction has occurred (versus offline contexts where the trigger is not as explicit and observable). Additionally, integrative communication is often the least likely reaction to jealousy because of the heightened emotional responses (e.g., anger, fear, anxiety) that often accompany jealousy (Guerrero et al., 1995; Guerrero et al., 2005; White & Mullen, 1989). Jealousy caused by Facebook, however, might allow a jealous individual time to gain perspective and calm down before engaging in a conversation with the partner. In this way, partner-based triggers can be easier to control, as the participant could directly request that his or her significant other remove the offending posts or respond to poaching messages in a particular way. Concurrently, rival-based triggers can be more maddening due to the utter lack of control by the jealous person. The participant might not be able to directly communicate with the rival and experience greater levels of jealousy as a result.

Conclusion

Overall, the results of this study demonstrate online communication has the power to negatively affect individuals and potentially threaten the strength of romantic relationships. It is important to consider the role of Facebook communication in our everyday lives given the pervasiveness of social networking sites as spaces for connection, relational development, and relational maintenance. As communication technologies continue to gain popularity and normalcy, communication scholars should remain vigilant of the relative gains and losses these media offer to the health of interpersonal relationships.

Jealousy is often examined in relation to personality factors, or trait jealousy. In the current study, we did not test individuals for these personality factors or measure variables such as self-esteem or levels of relational trust, interdependency, or uncertainty. Given the connection of these concepts to the experience of jealousy, we are unable to ascertain the extent participants' jealousy is triggered by Facebook content, relational factors, or personality traits. Along these lines, relationship type and duration might also contribute to individuals' perceptions of relational threats and would therefore influence the types of Facebook content that would trigger jealousy. As one participant aptly stated, "I think maybe relationship status and duration of relationship might make a big difference too. Seventeen and dating for three months is quite different than thirty-five and married for ten years." Another explained, "Since we're married, it's a little different than if we were merely dating."

Future research, then, should explore the role of relationship type and duration in the experience and expression of romantic jealousy. Additionally, some participants described confronting their spouse to inform them of their jealousy. Given that direct communication about romantic jealousy can preserve the health of a relationship, future research should examine these responses to Facebook jealousy behaviors and the factors that encourage this type of response. Future research should also examine the levels of jealousy associated with each trigger and outcomes of these triggers. For example, do pictures of one's partner and his or her ex-partner result in increased levels of jealousy as compared to wall comments from an ex-partner? Then, does that particular trigger lead to arguments or even the end of the relationship?

Finally, based on the findings in this and other related studies, we offer several recommendations for Facebook users. Due to the link between Facebook use and jealousy (Muise et al., 2009), people who identify as being jealous in relationships should limit the amount of time spent on Facebook. Secondly, users should talk openly with their romantic partners about their Facebook use to reduce ambiguity and contextualize content. Sharing feelings of jealousy in conflict situations might improve relational satisfaction and communication competence (Guerrero et al., 1995). Lastly, users should consider how Facebook friendships and residual artifacts with a past romantic/sexual partner will affect their current partners. Open communication about potential jealousy triggers should be discussed as problems arise to prevent incorrect assumptions and develop appropriate relational boundaries.

References

Afifi, W., & Reichert, T. (1996). Understanding the role of uncertainty in jealousy experience and expression. *Communication Reports, 9*(2), 93–103.

Ben-Ze'ev, A. (2004). *Love online: Emotions on the internet.* New York, NY: Cambridge University Press.

Bevan, J. L., & Tidgewell, K. D. (2009). Relational uncertainty as a consequence of partner jealousy expressions. *Communication studies, 60*(3), 305–323.

Brehm, S. S. (1992). *Intimate relationships* (2nd ed.). New York, NY: McGraw-Hill.

Buunk, B., & Bringle, R. G. (1987). Jealousy in love relationships. In D. Perlman & S. Duck (Eds.), *Intimate relationships: Development, dynamics, and deterioration* (pp. 123–147). Newbury Park, CA: Sage.

Elphinston, R. A., & Noller, P. (2011). Time to face it! Facebook intrusion and the implications for romantic jealousy and relationship satisfaction. *Cyberpsychology, Behavior, and Social Networking, 14,* 631–635. doi:10.1089/cyber.2010.0318

Glaser, B., & Strauss, A. (1967). *The discovery of grounded theory.* Chicago, IL: Aldine.

Guerrero, L. K., & Afifi, W. A. (1999). Toward a goal-oriented approach for understanding communicative responses to jealousy. *Western Journal of Communication, 63*(2), 216–248.

Guerrero, L. K., & Andersen, P. A. (1998). Jealousy experience and expression in romantic relationships. In P. A. Andersen & L. K. Guerrero (Eds.), *Handbook of communication and emotion: Research, theory, applications, and concepts* (pp. 155–188). San Francisco, CA: Academic Press.

Guerrero, L. K., Trost, M. R., & Yoshimura, S. M. (2005). Romantic jealousy: Emotions and communicative responses. *Personal Relationships, 12,* 233–252.

Knobloch, L. K., & Solomon, D. H. (1999). Measuring the sources and content of relational uncertainty. *Communication Studies, 50,* 261–278.

Marshall, T. C., Bejanyan, K., Di Castro, G., & Lee, R. A. (2012). Attachment styles as predictors of Facebook-related jealousy and surveillance in romantic relationships. *Personal Relationships, 20,* 1–22. doi:10.1111/j.1475-6811.2011.01393.x

Muise, A., Christofides, E., & Desmarais, S. (2014). "Creeping" or just information seeking? Gender differences in partner monitoring in response to jealousy on Facebook. *Personal Relationships, 21,* 35–50. doi: 10.1111/pere.12014

Muise, A., Christofides, E., & Desmarais, S. (2009). More information than you ever wanted: Does Facebook bring out the green-eyed monster of jealousy? *CyberPsychology & Behavior, 12*(4), 441–444.

Rempel, J. K, Holmes, J. G., & Zanna, M. P. (1985). Trust in close relationships. *Journal of Personality and Social Psychology, 49*(1), 95–112.

Rydell, R. J., & Bringle, R. G. (2007). Differentiating reactive and suspicious jealousy. *Social Behavior and Personality, 35*(8), 1099–1114.

Spitzberg, B. H., & Cupach, W. R. (2007). *The dark side of communication,* 2nd ed. Mahway, NJ: Lawrence Erlbaum.

Strauss, A., & Corbin, J. (1990). *Basics of qualitative research: Grounded theory procedures* and techniques. Newbury Park, CA: Sage.

Utz, S., & Beukeboom, C. J. (2011). The role of social network sites in romantic relationships: Effects on jealousy and relationship happiness. *Journal of Computer-Mediated Communication, 16,* 511–527. doi:10.1111/j.1083-6101.2011.01552.x

White, G. L. (1981a). Jealousy and partner's perceived motives for attraction to a rival. *Social Psychology Quarterly, 44*(1), 24–30.

White, G. L. (1981b). Some correlates of romantic jealousy. *Journal of Personality, 49*(2), 129. doi:10.1111/1467-6494.ep7383249.

White, G. L., & Mullen, P. E. (1989). *Jealousy: Theory, research, and clinical strategies.* New York, NY: The Guilford Press.

Epilogue: (Re)Casting the Dark Side of Communication

Shawn D. Long

The dark side of communication exists in everyday life. We engage, interact, resist, highlight, downplay, thrive, survive, and are consistently baffled by the dark side of human communication interaction, regardless of contexts. Dark communication and behavior are not anomalies, but an integral part of the human condition. We are innately self-centered, self-focused, survivalist, and I-oriented. Thus, we must be taught, at an early age, to share, to be selfless, to listen, to wait our turn, and all of the other niceties that go along with existing in a "civilized" society. Socially manufactured or not, this is how American society has been shaped and created. The premise that there are dark sides of human communication existing within and across a variety of contexts is not a new idea. As the citations throughout this volume indicate, there are rich antecedents to this book that have ignited this conversation for some time, and our hope is that more attention to this intriguing area of study will follow. As presented in this book, the dark side of human communication presents vexing problems and issues for individuals and organizations across a variety of spectrums. For good or for bad, the human condition is complex, and the chapters in this volume reflect the range of non-positive communication interactions that arise when humans engage with each other, organizations, and technology.

This book focuses on five distinct, yet related, contexts: interpersonal, organizational, health, computer-mediated, and a combination of contexts (i.e., blended). Yet, we admit that we have barely scratched the surface in attempting to understand all the dark or non-positive communication aspects in society. This is especially salient pertaining to the explosion of new technologies across the globe. The chapters reflect a range of communication interactions that are less than admirable, but completely human. Jealously, (cyber)bullying, hurtful events, bereavement, birth struggles, insults, retaliation, gossip, and email overload are all things many of us have experienced at some point in our lives.

This book is an interdisciplinary and intra-disciplinary exercise. We were impressed with the authors' ability to make sense of dark side communicative acts through research studies, case studies, theoretical and conceptual chapters, and a combination of them all from a number of perspectives. Although Cupach and Spitzberg (1994)[1] were one of the first scholars to introduce us to the Dark Side of Interpersonal Communication, and our volume follows twenty years later, there still remains fertile and untilled ground in the area of Dark Side scholarship.

Side note: As a critical scholar who attends to language, diversity and culture, I had some difficulty with the use of "dark" side as a normalized term for negative or deviant behavior to frame this book. Negative connotations associated with the word "dark" historically have had and continue to have significant and vicious implications for people of color in America and worldwide. The use of the term "dark" has been reified as a latent negative denotation, which many view as troubling and problematic. The binary of "light" equals purity and "dark" equates to deviance has been used to create and manipulate human divisions for generations. The language of color and the value placed on color need more and deeper exploration. However, the reconciliatory aspect is that there is a corpus of literature, under the auspice of the dark side, that we cannot neglect in our efforts to more deeply explore non-compliant and non-positive communication. Dr. Gilchrist-Petty and I had long and interesting conversations about the use of the term "dark side" in our title and the message that it will send to our readers that we equate darkness with negativity; and we are not blindly ignoring the long history of this communication issue and its manipulation. However, in the end, we see this book as advancing the conversation about non-positive and non-normative communication to a readership interested in this familiar and intriguing swatch of scholarship. We fully appreciate and understand that language and words matter and this could potentially serve as a pivot point for us and others to consider our language usage in the future. Our collective hope is to use our voice for good, whatever that means, and not as a divisive tool.

We have learned, in this project, that ill-intended communication may have long-term consequences on individuals and organizations. The contributors outlined these issues, but more importantly, provided a pathway toward resolution in many of the chapters in this book. The human tinder box lit by the dark side of communication is important to continue to uncover.

Directions for Future Scholarship

The collection of original research, theoretical, case study, and conceptual chapters in this book explores the multifaceted scholarship in the dark side of human communication. The twenty-four chapters in this volume are diverse, vast, intriguing, and timely. However, this book only captures a small fraction of the ways in which we can better understand negatively affected communication. We capture five primary contexts in contemporary ways, but there are certainly more contexts and deeper engagement with the ones advanced in this book that should be investigated in the very near future. Studying the human and technological complexities of social media should serve as a single-focused book in the future. The shadow and deviant use of informatics in health care can be another volume. The dark side of organizational engagement, including but not limited to: outsourcing, corporate off shoring, human trafficking, economic manipulation, organizational romance, insider trading, and contemporary red-lining, are all contemporary topics that should be investigated and shared through academic publishing. Certainly, issues of privacy, surveillance, virtual work, diversity, and new organizational forms are all exciting research areas that need to be studied from the dark side paradigm.

As we debate whether to continue using "the dark side" as the primary nomenclature and an umbrella term for this line of scholarship, for the reasons I addressed in the side note above, what is non-debatable is that the dark side is an intellectual movement that needs continuous nurturing with nuanced ideas. We assert that the dark side of communication is a scholarly paradigm that has epistemological weight and meta-theoretical depth. There are unique methods, as evidenced in this book, which contribute to the better understanding and explanation of dark side communication. There are tenets, language, jargon, theories, methods, and assumptions that drive the dark side paradigm and the present time is a great opportunity to strategically expand this important canon.

In closing, our hope is that this book provides readers with an opportunity to engage in important discussions regarding non-positive communication. These communicative acts are not unnatural, but are central to who we are as humans. If we push dark communication to the fringes of our scholarship, we are, in essence, denying a central part of who we are as individuals operating in a diverse and highly populated society. To understand what it is to be human is to embrace that humans also have dark side tendencies. These tendencies may manifest in dark side communication—the premise of this book. We certainly should strive to minimize dark side behavior and communication, but we are a better society when our scholarship reflects the human condition in its totality.

Notes

1 Cupach, W. R., & Spitzberg, B. H. (1994). *The dark side of interpersonal communication.* Hillsdale, N.J: Erlbaum.

Biographies

Editor Biographies:

Eletra S. Gilchrist-Petty (Ph.D., University of Memphis) is an Associate Professor of Communication Arts at the University of Alabama, Huntsville. She earned her B.A. and M.A. degrees in Communication Studies from The University of Alabama and her Ph.D. in Communication Studies from the University of Memphis. Gilchrist-Petty is the Spring 2016 Interim Communication Arts Department Chair at the University of Alabama, Huntsville. Her programs of research focus on instructional communication, interpersonal communication, and African American communication from both the quantitative and qualitative perspectives. Gilchrist-Petty's premier edited text is *Experiences of Single African-American Women Professors: With This Ph.D., I Thee Wed* and was published in 2011 by Lexington Books. She has also published in many leading scholarly outlets, including *Review of Communication, Journal of Intercultural Communication Research, Communication Reports, Journal of the National Academic Advising Association (NACADA), NIDA Journal of Language and Communication, Communication Teacher, Southern Communication Journal,* and *The Northwest Journal of Communication.* Gilchrist-Petty has authored a number of book chapters and presented scholarly research at many academic conferences spanning the international, state, regional, and local levels. Additionally, Gilchrist-Petty has held several offices with the National Communication Association (NCA), including Chair of the African American Communication and Culture Division (AACCD), Legislative Assembly, Nominating Committee, and Affirmative Action and Intercaucus Committee. Gilchrist-Petty is the 2012 winner of the Top Research Journal article presented by the AACCD of NCA, and she received the 2012–2013 Martin Luther King Jr. Service Award presented by the University of Alabama in Huntsville's Minority Graduate Student Association. She also serves as a motivational speaker and organizational trainer/consultant. Gilchrist-Petty was previously employed as an Assistant Professor at Middle Tennessee State University and teaches a wealth of communication classes, including Interpersonal Communication, Dark

Side of Interpersonal Communication, Research Methods, Public Speaking, Media Writing, Small Group Communication, Persuasion, Senior Seminar, African American Communication, and Culture and Communication.

Shawn D. Long (Ph.D., University of Kentucky; M.P.A., Tennessee State University) is Interim Associate Dean for Academic Affairs in the College of Liberal Arts and Sciences at the University of North Carolina at Charlotte. He is the former Chair of the Department of Communication Studies and Graduate Director of Communication Studies at the University of North Carolina at Charlotte. He is currently Professor of Communication Studies and Professor of Organizational Science at the University of North Carolina at Charlotte. An organizational scientist with numerous peer-reviewed publications and several professional awards, Long's teaching and research span organizational communication, organizational science, virtual work, diversity communication, virtual team assimilation and socialization, and health communication. Long studies the utility and development of communication practices and processes in virtual work. He has consulted several organizations on communication, technology, culture, diversity, and structure. He has written, presented and published several peer-reviewed papers around issues of organizational technology, diversity, virtual work in organizations, health communication, and organizational culture. He has appeared as a featured guest on several media outlets including National Public Radio and the Canadian Broadcast Corporation. His most recent research appears in *Communication Monographs, Qualitative Research in Organizations and Management, Human Resource Development International, Journal of National Medical Association, The Electronic Journal of Communication, Clinical Transplantation, Health Communication, Journal of Health Psychology, Journal of Health Communication, Communication Teacher, Health Communication, Information and Science Technology, The Encyclopedia of Organizational/Industrial Psychology, Case Studies for Organizational Communication: Understanding Communication Processes, and Virtual and Collaborative Teams.* He has published two books, *Communication, Relationships and Practices in Virtual Work* (2010) and *Virtual Work and Human Interaction Research* (2012). He serves on a number of journal editorial boards. Prior to arriving at UNC-Charlotte, Long was a Southern Regional Educational Board Doctoral Scholar (SREB) and a Lyman T. Johnson Doctoral Scholar at the University of Kentucky. He has been recognized with several professional awards including the 2012 Outstanding Service Award, National Communication Association, African American Communication and Culture Division/Black Caucus, 2011 Southern States Communication Association Outreach Award, 2009 Organizational Science Outstanding Service Award, 2010 Southern States Communication Association Minority and Retention Award, Chancellor's Award for Outstanding Teaching at the University of Kentucky, The Multicultural Summer Fellowship at the University of Nebraska-Lincoln, Outstanding Teaching Assistant in the College of Communication and Information Studies at the University of Kentucky, Outstanding Graduate Teaching Assistant recognized by the International Communication Association, and Who's Who Among American Teachers. Long received his undergraduate and M.P.A. degrees from Tennessee State University and his Ph.D. from the University of Kentucky.

Contributor Biographies:

Courtney Anderegg (M.A., The Ohio State University) is a Ph.D. student in the School of Communication at The Ohio State University. She is interested in examining mediated portrayals of romantic relationships and their impact on interpersonal relationships as well as the role of communication technology in relational maintenance.

Chris Anderson (M.A. University of Montana) is a Ph.D. student in the Department of Communication at the University of Oklahoma. His research interests include envy and revenge in interpersonal contexts and the use of computer-mediated communication in conflict.

Ashley Barrett (M.A., Baylor University) is a doctoral candidate (ABD) in Organizational Communication and Technology at the University of Texas, Austin. She received her M.A. and B.A. degrees from Baylor University in Communication Studies and Political Science respectively. Barrett's research centers on how new technologies are being incorporated into and utilized in contemporary workplaces in times of change and/or crisis. She is also interested in how these new technologies can create and/or manipulate conceptions of meaningful work. Her work can be found in academic journals such as *Human Communication Research* and the *Journal of Business Communication.*

Erin Basinger (M.A., University of Georgia) is a doctoral candidate in the Department of Communication at the University of Illinois, Urbana-Champaign. She studies interpersonal and family communication, focusing specifically on how people cope with stressors alongside their family members and other relational partners.

Megan Bassick (M.A., University of Oklahoma) is a Ph.D. student in the Department of Communication at the University of Oklahoma. Her research interests include long-distance relationships, conflict in intimate relationships, computer-mediated communication, the dark side of interpersonal relationships, and siblings.

Eryn Bostwick (M.A., University of Oklahoma) is a Ph.D. student in the Department of Communication at the University of Oklahoma. Her research interests include family secrets, interpersonal conflict, aggression, and teen parents.

Kenon A. Brown (Ph.D., University of Alabama; M.S. & B.S, The University of Tennessee) is an Assistant Professor of Public Relations at The University of Alabama. Brown's research primarily focuses on athlete image repair and audience trends and reactions to mediated sports communication. His work has been featured in *Journalism and Mass Communication Quarterly, Journal of Public Relations Research, Mass Communication and Society,* and *Communication and Sport,* among other journals. As a fellow for The Plank Center for Leadership in Public Relations, Brown also conducts research that focuses on building an initiative to recruit and retain minority public relations practitioners. Brown also serves as an independent consultant for various organizations in addition to his academic work.

Rockell Brown-Burton (Ph.D., Wayne State University) is an Associate Professor in the Film Department and Director of Graduate Studies in the School of Communication at Texas Southern University. Brown-Burton earned her Ph.D. in Media Studies from Wayne State University, an M.A. in Mass Communication from Howard University, and a B.A. in Mass Communication at Xavier University of Louisiana. Her research interests concern race and representation, mediated representations of women, communication and culture, and health communication. She teaches undergraduate and graduate courses in mass communication and health communication. She received the School of Communication Teacher of the Year Award in 2009 at Texas Southern and co-authored the 2011 book, *Race and News: Critical Perspectives.* Brown-Burton is a member of the National Communication Association (NCA), the Broadcast Education Association (BEA), and The Association for Education in Journalism and Mass Communication (AEJMC).

Heather J. Carmack (Ph.D., Ohio University; M.A., Ohio University; B.A., Truman State University) is an assistant professor in the School of Communication Studies at James Madison University. Her research interests include communication about patient safety and communication about death and dying.

Marcus J. Coleman (Ph.D., University of Georgia; M.A., University of Kentucky; B.A., University of Southern Mississippi) is an Assistant Professor of Communication and Interdisciplinary Studies at the University of Southern Mississippi. Most recently, he served as the Senior Research Analyst for the Washington, D.C., Department of Behavioral Health and a Research Fellow in the Civic Engagement and Governance Institute at the Joint Center for Political and Economic Studies. His research interests include voting identification legislation, patriotism, civic engagement, and political ideology.

Elaine L. Davies (Ph.D., University of Missouri) is an Assistant Professor of Communication at Southwestern Oklahoma State University, where she specializes in relational communication, the dark side of communication, and gender communication. In addition, she teaches a variety of classes including public speaking, interpersonal communication, intercultural communication, gender communication, group communication, and leadership. Her research examines the intersection of romantic relationships, destructive communication, and the use of technological devices within these contexts. She has published in *Human Communication*, *The Journal of Business Communication*, and *Communication Quarterly.*

Jocelyn M. DeGroot (Ph.D., Ohio University; M.S. & B.S., South Dakota State University) is an Associate Professor in the Department of Applied Communication Studies at Southern Illinois University Edwardsville. Her research interests include computer-mediated communication and communicative issues of death and dying.

Jenny Dixon (Ph.D., University of Missouri) is an Assistant Professor in the Communication Arts Department and the Coordinator of the Gender and Sexuality Studies Program at Marymount Manhattan College. Dixon is featured in *Journal of Applied Communication Research*, *Research on Aging*, and *Communication Quarterly*. Her research focuses on the navigation of public and private identities communicated in workplace settings. Specifically, her current research focuses on work and family balance for employees of non-traditional family structures (LGBTQ, polyamorous, non-married, etc.).

Jessica J. Eckstein (Ph.D., University of Illinois, Urbana-Champaign) is an Associate Professor of Communication at Western Connecticut State University specializing in relational communication. Her research focuses on intimate partner violence from unexplored or understudied perspectives (e.g., love, male victims, stigma). Emphasizing the importance of applying scholarship to people's lives, Eckstein not only publishes in journals related to violence, communication, and relationships, but also maintains ongoing collaborations with national and community agencies.

Jesse Fox (Ph.D., Stanford University) is an Assistant Professor in the School of Communication at The Ohio State University. She received her M.A. from University of Arizona and her doctorate from Stanford University. Fox is interested in the social implications of communication technologies, including their role in the initiation, maintenance, and dissolution of romantic relationships.

Annette Madlock Gatison (Ph.D., Howard University) is an Associate Professor at Southern Connecticut State University. Gatison's work The Pink and The Black Project® focuses on health communication. Gatison serves as a women's health advocate and has participated in continuing education programs that focus on issues such as cancer biology, genetics, and political advocacy. She is the author of the forthcoming book, *Embracing the Pink Identity: Breast Cancer Culture, Faithtalk, and the Myth of the Strong Black Woman.* Other select articles are included the following SAGE References: *Encyclopedia of Cancer and Society, 2nd Edition* and *Multimedia Encyclopedia of Women in Today's World.*

Dianne Gravley (M.A., University of North Texas) is a doctoral candidate at Regent University, Strategic Communication. She earned her M.A. at the University of North Texas and her B.A. at Texas A&M University. She is an instructor of communication studies at North Central Texas College in Flower Mound, Texas, where she teaches business communication and public speaking. She conducts research in the organizational and interpersonal communication field, with a focus on organizational communication within public schools. Her research has appeared in *Management Communication Quarterly.*

Melvin Gupton (M.A. & B.A., Wayne State University) is a doctoral student and research assistant in the Department of Communication at Wayne State University in Detroit. He is currently completing his doctoral dissertation on mass shooting crises, how they evolve, and their impact on issues management and public policy deliberations. Gupton serves as an adjunct instructor at both Wayne State University and Walsh College in Michigan, where he teaches courses in business communication, professional presentations, technical writing, and research. His research interests include public relations studies, issues and crisis management, intercultural communication, communication response strategies, media framing, and organizational renewal and recovery following a crisis. He is also interested in examining the role and relative influence of organizational and community stakeholders during and following focusing events such as crises.

Leandra H. Hernandez (Ph.D., Texas A&M University) is an associate faculty member in the Department of Arts & Humanities at National University. Her research interests fall into two main categories: health communication with a focus on women's reproductive health, and media representations of gender. She enjoys researching reproductive politics; Hispanic/Latina health experiences; patient-provider communication; women's pregnancy experiences; media and news coverage of reproductive politics; and media representations of masculinities and femininities, particularly in reality television shows. She has two publications about gender performances in *Toddlers & Tiaras* and *Duck Dynasty* in *Reality Television: Oddities of Culture* (2014) edited by Alison Slade, Amber J. Narro, & Burt Buchanan. She also has a chapter about *Toddlers & Tiaras* fan culture and mediated interactivity forthcoming in *Television,*

Social Media, & Fan Culture (2015) edited by Alison Slade, Amber J. Narro, & Diedre Givens-Carroll. She enjoys teaching courses about health communication, gender and the media, and feminist theories.

Trisha K. Hoffman (M.A. & B.S., Southern Illinois University, Edwardsville) is a doctoral candidate in the Hugh Downs School of Human Communication at Arizona State University.

Her research interests include health communication and relational well-being as well as applied qualitative methods.

Felecia Jordan Jackson (Ed.D., West Virginia University) is an Associate Professor in the School of Communication at Florida State University. She received her doctorate and M.A. degrees from West Virginia University in Education and Communication Studies, respectively, and her B.S. from Georgia Southern College in Communication Arts/Broadcasting. Her areas of teaching and research include interpersonal, interracial/intercultural, and instructional communication.

Amy J. Johnson (Ph.D., Michigan State University) is a Full Professor in the Department of Communication at the University of Oklahoma. Johnson's research focuses on long-distance relationships and computer-mediated communication, friendships, stepfamilies, and interpersonal arguments. She has published in such venues as *Communication Monographs*, *Communication Yearbook*, *Journal of Computer-Mediated Communication*, and *Journal of Social and Personal Relationships*.

Falon Kartch (Ph.D., University of Wisconsin-Milwaukee) is an Assistant Professor of Communication at California State University, Fresno. Her research focuses on the experiences of diverse family types including stepfamilies, LGBT families, and nonresidential parenting, as well as the dark side of close relationships.

Devlon Jackson (Ph.D., Howard University) is a Health Communications Intern/Cancer Research & Training Award Fellow in the Health Communication and Informatics Research Branch (HCIRB) within the Behavioral Research Program (BRP) of Division of Cancer Control and Population Sciences (DCCPS). She earned her Ph.D. in Communications and Culture from Howard University, her M.A. of Public Health degree from Florida International University, and her B.A. in Organizational Communications from the University of Central Florida. Jackson is interested in assessing communication facilitators and barriers (specifically, patient-provider communication, social media, and health information technology) in the medical shared decision-making process within and outside of the clinical care setting and determining existing health disparities and social determinants. Her doctoral dissertation assessed the current state of shared decision making among Black men and their health care providers regarding prostate-specific antigen (PSA) testing. NOTE: This research was conducted prior to Jackson's term as a Health Communications Intern/Cancer Research & Training Award Fellow for the National Institutes of Health's National Cancer Institute (NIH-NCI) and was not a part of her duties, responsibilities, or training at NIH-NCI.

Kimberly L. Kulovitz (Ph.D., University of Wisconsin-Milwaukee) is an Associate Professor at Carthage College and received her Ph.D. in communication at the University of Wisconsin-Milwaukee. Her research and teaching interests include cyberbullying, close relationships, video games, and how these phenomena connect behaviorally. She has published in *Misbehavior*

Online in Higher Education: Cutting-Edge Technologies in Higher Education, *Star Trek Fan Culture*, and *The Journal of Popular Culture*.

Corey Jay Liberman (Ph.D., Rutgers University) is an Assistant Professor in the Department of Communication Arts at Marymount Manhattan College. His research spans the interpersonal communication, group communication, and organizational communication worlds, and he recently coauthored a textbook dealing with organizational communication (*Organizational Communication: Strategies for Success*, 2013) and edited a case study book dealing with persuasion (*Casing Persuasive Communication*, 2014). He is currently working on his next two book projects, both of which deal with risk and crisis communication. Liberman is most interested in the social practices of dissent within organizations, specifically the antecedents, processes, and effects associated with effective employee dissent communication.

Edward A. Mabry (Ph.D., Bowling Green State University) is Emeritus Associate Professor of Communication of the Department of Communication at the University of Wisconsin-Milwaukee. His research and teaching encompass the contexts and effects of mediated communication. He has served on the editorial boards of *Communication Monographs*, *Human Communication Research*, *Journal of Applied Communication Research*, and the *Journal of Computer Mediated Communication*. He has published in *Misbehavior Online in Higher Education: Cutting-Edge Technologies in Higher Education*, *Interpersonal Relations and Social Patterns in Communication Technologies: Discourse Norms, Language Structures and Cultural Variables*, *Facilitating Group Communication in Context: Innovations and Applications with Natural Groups: Vol. 2.*, *Facilitating Group Task and Team Communication*, and *Journal of Communication*.

Amin Makkawy (M.A., UNC Charlotte; B.A., University of Nebraska) is a Ph.D. candidate in the organizational science program at the University of North Carolina. Charlotte. He received his M.A. in communication studies and his B.A. in psychology and sociology with high distinction and honors. Makkawy conducts research on diversity and voice in organizational contexts. His latest research has focused on the experience of employees with visual impairments in the virtual workplace.

Stacie Wilson Mumpower (M.A., University of Texas at El Paso) is a Ph.D. student in the Department of Communication at the University of Oklahoma. Her research interests include leadership communication, high-reliability organizations, organizational learning, competing discourses and normative influences in the military, and long-distance relationships.

Creshema Murray (Ph.D., M.S., & B.S., University of Alabama) is an Assistant Professor of Corporate Communication at the University of Houston–Downtown. Murray teaches courses in leadership, organizational training & development, and destructive organizational communication. As a faculty fellow for the Center for Critical Race Studies, Murray's first area of research focuses on the lived experiences of women of color in workplace organizations. The second area focuses on the manner in which organizations foster destructive workplace practices with employees through organizational hazing and bullying. In addition to her role at the University of Houston–Downtown, Murray engages in overseeing the creation and implementation of a multitude of strategically focused communication campaigns.

Mark P. Orbe (Ph.D., Ohio University; M.A., University of Connecticut; B.A., Ohio University) is an internationally known educator, author, and consultant/trainer in the area of communication and diversity. In addition to his award-winning teaching, research, and service accomplishments, Orbe is CEO of Dumela Communications, an international communication consulting company that promotes communication competence in an increasingly diverse world. At present, Orbe is Professor of Communication & Diversity in the School of Communication at Western Michigan University, where he holds a joint appointment in the Gender and Women's Studies Program.

Donyale R. Griffin Padgett (Ph.D., Howard University, M.A. & B.A., Wayne State University) currently serves as Associate Professor of Diversity, Culture, and Communication in the Department of Communication at Wayne State University in Detroit, Michigan. Padgett holds a doctorate degree from Howard University, in Washington, D.C., in rhetoric and intercultural communication. She also holds a master's degree in organizational communication/public relations and a bachelor's degree in journalism—both from Wayne State University. Her research examines how crises are negotiated between institutional leaders and other audiences, particularly around issues of race and culture. She is specifically interested in the public dialogue that stems from crises and how meaning is shaped in the aftermath of these events.

Brian K. Richardson (Ph.D., University of Texas at Austin) is Associate Professor of Communication Studies at the University of North Texas. He earned his Ph.D. at the University of Texas at Austin, his M.A. at Louisiana Tech University, and his B.S. at Lamar University. He conducts research in areas of whistle-blowing, peer reporting of unethical behavior, and disaster/crisis communication. Richardson's research has appeared in *Management Communication Quarterly*, *Human Communication Research*, *Western Journal of Communication*, *Communication & Sport*, *Communication Studies*, *International Journal of Mass Emergencies and Disasters*, *Journal of Communication & Religion*, and other outlets.

Idrissa N. Snider (M.A., University of Alabama at Birmingham; B.A., Georgia State University) is a doctoral student in the Department of Communication at Wayne State University in Detroit, where she studies rhetorical criticism, identity, and media effects. Snider earned her bachelor's degree in Broadcasting & Journalism from Georgia State University and holds a master's of arts & sciences from the University of Alabama at Birmingham in Communications Management. Before returning to academia, Snider enjoyed a professional career in broadcasting and television for nearly a decade.

Calvin Spivey (B.S., College of Charleston) is a graduate student in the Organizational Science program at the University of North Carolina at Charlotte. He received his B.S. in Psychology.

D. L. Stephenson (Ph.D., University of Massachusetts, Amherst) is an Associate Professor in the Department of Communication and Media Arts at Western Connecticut State University where she teaches courses in communication ethics, critical studies in rhetoric, communication theory, persuasion and propaganda in media, and interpersonal communication. She spent 14 years working as a newspaper reporter in western Massachusetts. After earning an M.S. degree in Rhetoric and Technical Communication at Michigan Technological University, she earned a doctorate in Communication from the University of Massachusetts, Amherst.

Katie Margavio Striley (Ph.D., Ohio University) is an Assistant Professor of Interpersonal Communication at the University of North Carolina, Chapel Hill. Her research examines the discursive construction of exclusive realities, the myriad ways language enables and constrains our sense of the possible, and the transformative potential of communication to make better social worlds. Her primary research interests include exclusive and inclusive communication and the communicative construction of systems of exclusion. She explores the creation, maintenance, and termination of exclusive communication, such as stigma, ostracism, bullying, and other forms of social rejection, as well as inclusive communication like dialogue, deliberation, and other forms of egalitarian communication. Her work has been published in journals such as *Health Communication, Journal of International and Intercultural Communication*, and *Communication Studies.*

Susan L. Theiss (MA, University of Arkansas) serves as University Ombuds person, Oregon State University. She previously served as a volunteer Ombudsperson at the University of Arizona for eight years while employed as the Department Administrator for Family and Community Medicine. She co-chaired Arizona's Ombuds program before becoming the University of Arkansas's first Ombudsperson, where she served for over nine years. She is a licensed mediator and serves as a mentor for the International Ombuds Association.

Jason Thompson (Ph.D., University of Nebraska) is an Associate Professor in the Department of Speech Communication Arts and Sciences at Brooklyn College–City University of New York. He earned both his M.A. and Ph.D. in Communication Studies from the University of Nebraska in Lincoln, NE. He earned his B.A. in Communication Studies at Rowan University in Glassboro, NJ. His line of research has mainly focused on how people in personal and family relationships communicatively provide and obtain support both during and after the experience of various challenging life events. He teaches undergraduate courses in interpersonal communication, business/professional communication, introduction to communication, public speaking, as well as communication theory. His research has appeared in various journals such as *Communication Studies, Journal of Issues in Intercollegiate Athletics, National Academic Advising Association Journal (NACDA), Communication Education*, and *Ohio Communication Journal.* Thompson has been a member of the National Communication Association (NCA) and Eastern Communication Association (ECA).

Kathleen S. Valde (Ph.D., University of Iowa) is an Associate Professor of Communication at Northern Illinois University. Her research focuses on organizational communication, interpersonal relationships within the organization, and communication ethics.

Michael Vicaro (Ph.D., University of Pittsburgh, M.A. University of Colorado–Boulder, B.A. Rutgers College) is an Assistant Professor of Communication at Penn State University–Greater Allegheny campus. He currently serves as director of the Civic and Community Engagement program and teaches courses in Rhetorical Theory, Media Criticism, and Communicative Ethics. His work has appeared in the journals *Rhetoric & Public Affairs, Argumentation and Advocacy,* and *Mediatropes,* among others.

Lynne M. Webb (Ph.D., University of Oregon) is Professor of Communication Arts, Florida International University. She held previous tenured appointments at the Universities of Florida, Memphis, and most recently Arkansas, where she was named a J. William Fulbright Master Researcher. Her research examines interpersonal communication in a variety of forms, venues, and relationships. Webb has co-edited three scholarly readers and published 75+ essays including multiple theories, research reports, methodological pieces, and pedagogical essays. Her work has appeared in numerous prestigious journals at the national and international levels including the *Journal of Applied Communication*, *Health Communication*, *Computers in Human Behavior*, and the *International Journal of Social Research and Methodology* as well as in prestigious edited volumes including the *Advancing Research Methods with New Technologies* (IGI Global, 2013), *Virtual Work and Human Interaction Research: Qualitative and Quantitative Approaches* (IGI Global, 2012), and *Producing Theory in a Digital World* (Peter Lang Publishing, 2012). Webb is a past president of the Southern States Communication Association and has received a Presidential Citation for her Service to the National Communication Association.

Erin Wehrman (M.A., Missouri State University) is a doctoral student in the Department of Communication at the University of Illinois, Urbana-Champaign. Her research interests focus on interpersonal relationships and families during periods of transition and crisis (e.g., bereavement and military service).

Haley Woznyj (B.A., Temple University) is a doctoral student in the Organizational Science department at the University of North Carolina at Charlotte. She holds a bachelor of arts in Psychology. Her general research interests center around diversity concerns in organizations and positive job attitudes.

Index

A

B

C

D

J

L

M

N

O

S

T

U

V

W

LIFESPAN COMMUNICATION

Children, Families, and Aging

Thomas J. Socha, *General Editor*

From first words to final conversations, communication plays an integral and significant role in all aspects of human development and everyday living. The Lifespan Communication: Children, Families, and Aging series seeks to publish authored and edited scholarly volumes that focus on relational and group communication as they develop over the lifespan (infancy through later life). The series will include volumes on the communication development of children and adolescents, family communication, peer-group communication (among age cohorts), intergenerational communication, and later-life communication, as well as longitudinal studies of lifespan communication development, communication during lifespan transitions, and lifespan communication research methods. The series includes college textbooks as well as books for use in upper-level undergraduate and graduate courses.

Thomas J. Socha, Series Editor | *tsocha@odu.edu*
Mary Savigar, Acquisitions Editor | *mary.savigar@plang.com*

To order other books in this series, please contact our Customer Service Department at:

(800) 770-LANG (within the U.S.)
(212) 647-7706 (outside the U.S.)
(212) 647-7707 FAX

Or browse online by series at www.peterlang.com